A Theory of Forest Dynamics

Herman H. Shugart

A Theory of Forest Dynamics

The Ecological Implications of Forest Succession Models

With 81 Figures

Springer-Verlag
New York Berlin Heidelberg Tokyo

Herman H. Shugart
Environmental Sciences Division
Oak Ridge National Laboratory
Oak Ridge, Tennessee 37831
U.S.A.

Library of Congress Cataloging in Publication Data
Shugart, H. H.
A theory of forest dynamics.
Bibliography : p.
1. Forest ecology—Mathematical models. 2. Plant succession—Mathematical models. 3. Forest ecology—Data processing. 4. Plant succession—Data processing.
I. Title.
QK938.F6S47 1984 574.5′2642′0724 84-5301

Typeset by Bi-Comp, Incorporated, York, Pennsylvania.
Printed and bound by R. R. Donnelley & Sons, Harrisonburg, Virginia.
Printed in the United States of America.

9 8 7 6 5 4 2 1

ISBN 0-387-96000-7 Springer-Verlag New York Berlin Heidelberg Tokyo
ISBN 3-540-96000-7 Springer-Verlag New York Berlin Heidelberg Tokyo

To
Ramona Jeanne Kozel Shugart
and our daughters,
Erika Christine Shugart
and
Stephanie Laurel Shugart
for their support for so many years,
and particularly during the writing of this book

Preface

This book is a consideration of the dynamics of forested systems at the time and spatial scales that I feel are implied by our present-day use of the term "succession." The investigation will be conducted by exercising a set of ecological models called "gap models," which have been in a state of development and improvement for the past 15 years. It is the intent of this book to use these models as tools for exploring theories of ecological succession.

Ecological succession is one of the most obvious and demonstrable features of natural systems when viewed from outside the field of ecology. Succession is used by teachers as a theory that introduces young people to the interactive and dynamic nature of ecosystems. Succession theory and examples of succession are proclaimed from legions of nature trail guidebooks and placards. It is a pleasant classroom exercise to discuss how ecological systems change as the product of internal mechanisms that can be demonstrated by observation. The deductive explanation of how a particular place came to have a given assemblage of tree species has a pleasing "Sherlock Holmesian" touch that can be challenging to puzzle through.

With all the appeal that the notion of succession has for the nonecologist, the concept probably is even more appealing to the ecologist. Most ecologists have a fairly well-developed idea both of what succession is and to what sorts of phenomena succession should be applied. It is in the rigorous application of succession as a full-fledged scientific theory that difficulties seem to occur. What exactly is the theory of succession? Where does it apply? How does it originate? How general is it? These are the questions to which disagreements about succession tend to turn. These are questions that are beyond the particular details of the monologue of the camp nature counselor, "I wonder why there's a grove of pine trees in the middle of this hardwood forest? Look, here's an old brick. There must have been a house or a cabin or something here once. I wonder how long ago that was?"

The present representation of ecological systems uses the digital computer to organize the elaborate detail that comes from incorporating the natural history of several tree species in models of the dynamics of a forest's size-structured set of tree populations. My interest in these models stems from work in the early 1970s in conjunction with the Eastern Deciduous Forest Biome Project of the United States International Biological Program. At that time, J. M. Hett, T. R. Crow, and I were interested in constructing what we envisioned as a spatially nested set of forest succession models. While some progress was made in this direction, it was not until about 1975 (while working initially with the winter quarter systems ecology class at the University of Tennessee, and later with D. C. West at Oak Ridge National Laboratory) that the JABOWA model (developed by Botkin, et al [1972a,b]) was modified to produce a model of an Appalachian deciduous forest called the FORET model. Since that time, and with the help of many collaborators, the FORET model has been modified and used in a variety of forest ecosystems. The original intent in this proliferation of models was to cross-compare forests by using the models as a device for synthesis. The process of developing the models for the various systems typically consisted of starting with the FORET model, modifying the model to be appropriate to the forest, testing the model against independent data (if possible), and using the model in some practical application. This book is a synthesis of what has been learned about ecological succession from this process.

Work with these models was a collaborative effort, and I would like to acknowledge several people for their help in this endeavor. D. C. West, D. L. DeAngelis, W. R. Emanuel, W. M. Post, and A. M. Solomon were sources of information and encouragement throughout the development, testing, analysis, and application of the FORET model, and they worked on a series of papers stemming from this work. Collaboration and help of a similar nature came from I. R. Noble on the BRIND model, M. S. Hopkins, A. T. Mortlock, and I. P. Burgess on the KIAMBRAM model, D. L. Mielke and L. K. Mann on the FORAR model, J. E. Swierzbinski on the FORMIS model, and T. M. Doyle on the FORICO model. M. L. Tharp worked to improve the computer program for the FORET model and its descendents. Her aid in developing the code that appears as the Appendix was invaluable. I have particularly enjoyed discussions with several graduate students on many of the topics reported in this book. Some of these students are mentioned above and others include T. M. Smith, D. A. Weinstein, S. W. Seagle, J. G. Saldarriaga, D. E. Weller, D. L. Urban, M. D. MacKenzie, and D. M. Moorhead. This research has been supported in the greatest part by grants from the National Science Foundation's Ecosystem Studies Program under Interagency Agreement Nos. BSR-76-00761, BSR-77-26722, BSR-77-25781, and BSR-80-21024 with the U.S. Department of Energy (and its antecedent agencies), under contract W-7405-eng-26 with the Union Carbide Corporation. I also received support from the U.S. Atomic Energy Commission, the U.S. Energy Research and Development Administration, the U.S. Department of Energy, the Australian Commonwealth Scientific and Industrial Research Organization, the Australian National University, and the University of Tennessee.

R. L. Burgess, A. M. Solomon, W. M. Post, J. R. Krummel, S. B. McLaughlin, G. M. Lovett, T. J. Blasing, D. L. DeAngelis, J. W. Ranney, D. S. Johnson, S. W. Seagle, M. I. Dyer, R. J. Luxmoore, R. I. Van Hook, D. E. Reichle, and S. I. Auerbach all were kind enough to read and comment on the chapters in this book. Following my changes in response to the advice of these colleagues, G. R. Carter, K. E. Gibson, L. J. Jennings, L. W. Littleton, and D. D. Rhew retyped and reassembled manuscripts that were cut, pasted, and scribbled on more times than any of us probably care to recall. N. T. Millemann's help in technical editing was indispensable. Finally, I am particularly indebted to R. V. O'Neill, W. K. Lauenroth, J. Pastor, and L. B. Brubaker, who critically read and reviewed the entire book and whose comments were a very great help to me. While I was fortunate to receive so much aid and good advice from so many people, any errors that remain in this work most certainly are my own.

Contents

Prologue

Heart-halt and spirit lame,
    City-opprest,
Unto this wood I came
    as to a nest;
Dreaming that sylvan peace
offered the harrowed ease—
Nature a soft release
    from men's unrest.

But, having entered in,
    Great growths and small
Show them to be men akin—
    Combatants all!
Sycamore shoulders oak,
Bines the slim sapling yoke,
Ivy-spun altars choke
    elms stout and tall.*

It often is difficult to extract from the apparent tranquility of a forest a sense of the actual dynamism of trees struggling over centuries to gain landscape for their species. To the human eye, a forest is a slowly changing ecosystem that superficially looks alike from one year to the next. Yet, this seeming quiet is, in fact, a balance between the tremendous progenerative potential of trees and an equally tremendous mortality rate. A single tree can produce tens of millions and (in some species) hundreds of millions of potential offspring over its lifetime. This reproductive po-

* Excerpted from *The Complete Poems of Thomas Hardy* edited by J. Gibson. New York, Macmillan, 1978.

tential by far outstrips the reproductive rates of the vast majority of animals, and it makes the multiplicative powers of organisms such as rabbits and lemmings pale by comparison. When one couples a tree's reproductive potential with the fact that trees can grow to be the most massive organisms on earth, then a notion of the ecological potential of trees begins to emerge. For example, if the progeny of just one white oak tree (*Quercus alba*) all survived and grew to the size of a typical canopy tree, the mass of these trees after two generations would exceed the mass of all living matter that is currently on earth.

One does not need to belabor the potential reproductive powers of trees, because it requires only a bit of mental algebra and the recollection of how large trees are, and of how many acorns can lie on the ground beneath an oak in the fall, to be convinced of this capability. These forces of increase must be logically counterbalanced by the forces of decrease, death, and predation. When one adds to this the greater longevity of trees relative to that of humans, then an understanding of forest dynamics can be seen as being very hard to actually perceive in full detail at a single point in time. For example, if (in a population) the mean fitness of an individual tends to be one, then of the 200 million seeds of a canopy oak tree's lifetime productivity, only one individual within this cohort could be expected to become a canopy tree. It is possible that an energetic plant ecologist might somehow tabulate this number of seeds and, perhaps, train children and grandchildren to help accumulate the data as one of the seeds grows to the canopy stage 100 years later. At present, however, such a study has yet to materialize in the ecological literature.

Thus, the intrinsic nature of the trees that comprise a forest make it all but impossible to collect complete data sets on the dynamics of natural forests. What one typically has is a collection of bits of observations from different points in time on segments of forest dynamics. Attempts are then made to connect these bits of information with an inference to form the complete picture. It is, in part, due to the fact that so much of forest ecology is inferred from partial data sets that so many interminable debates rage in the field. Each scientist who strives to extend our understanding of forest ecosystems is working with a seemingly diabolically carved jigsaw puzzle with many of the pieces missing. Only after a lifetime of observation can a few additional pieces of the puzzle be obtained. Unfortunately, there often is no assurance that these new pieces will fit the old. One can collect data on the results of forest dynamics instead of on the dynamics themselves. Our knowledge of the long-term dynamics of forests is based largely on scientific inference.

Because of this reliance on inferences, mathematical models of forest dynamics offer a valuable and manipulatable formalization of what we believe to be the important mechanisms involved in forest succession. Toward this end, in the early 1960s, several forest biologists began to design mathematical models of changes in forest composition. Foresters realized that certain changes in current forest practice (e.g., change in trees due to genetic improvement and the use of fertilizers to increase forest yields) might cause the stand yield tables (that had been laboriously developed over the past several decades) to be less useful. Some foresters began to develop models of forest growth and yield that could be calibrated on the empirical stand tables and that also could be used to incorporate some of the changes taking place in the forest industry. At the same time, ecologists had become dissatisfied with the static notion of forest typology, and they developed extensive investigations of the dynamics aspects of ecosystems that perhaps were epitomized by the International Biological Programme. This increased interest in ecosystem dynamics naturally led to the development of forest models. By the mid-1970s, there were two fairly distinct approaches to the modelling of forest dynamics: forest models and tree models.

Forest Models*

These models consider the forest as the focal point of the simulation model. Attributes of forest stands (e.g., biomass, numbers of trees, indices of diversity, or timber volume) are the state variables that are used to develop these models. One interesting feature of these models is that they contain the tacit assumption that the important determining factors for predicting stand dynamics are factors that emerge at the forest-stand level of organization.

Tree Models**

These models attempt to simulate the dynamics of a forest by computing changes in each individual tree in a forest stand. The degree of complexity of these models ranges from a simple tabulation of the probabilities of an individual tree of one kind being replaced by an individual of another kind to extremely detailed models that include the solid geometry of canopy shapes of individual trees to determine competition among individuals.

It is tempting to look at these two contrasting modeling approaches as being mathematizations of Clement's (1928) (or Odum's [1969]) holistic view of ecosystem dynamics versus Gleason's (1939) individualistic view of succession. It is preferable to think of each modelling approach as a

* Reviewed in Shugart and West (1980), a category equivalent to the "stand" model of Munro (1974).
** See Shugart and West (1980) and Munro (1974) for reviews.

view of different facets of the same reality. Part of this preference is based on an involvement that use both sorts of models in studies of succession at different scales (e.g., Shugart et al [1973], Shugart and West [1977]) without finding a sharp working discontinuity between the two modelling approaches. Furthermore, as will be discussed in Chapter 7, one sort of model (the tree model) can be used for special cases to derive the other (the forest model).

In this volume, mathematical models of forest dynamics will be inspected as tools for the understanding of the dynamics of forests, which reflect more generally on the phenomenon called "succession." Involvement in the detailed definitional battle that has centered on the meaning of the term "succession" will be avoided as much as possible. However, concepts of succession as they seem to be used currently by ecologists will be discussed. These conceptualizations are not being codified, but they (hopefully) will be exact. Because this book deals with mathematical models of forest system dynamics, this exactness takes the form of equations and computer programs.

Chapter 1

Forest Succession

Succession, to most ecologists, involves the change in natural systems and the understanding of the causes and direction of such change. There is considerable confusion about the precise meaning and mechanisms that are associated with succession. This confusion originates, at least in part, from the lack of common agreement as to what scale of nature one should apply the succession concept. At present, as was the case in the past, there is a wide spectrum of individual ecologists' opinions as to what ecological succession actually consists of. So much has been written about succession in the modern ecological literature that it is difficult either to develop a unifying synthesis of the theory that is associated with the word or to think of new things to say that (when said), unite the myriad of meanings attendant to the succession concept. Unfortunately, the confusion and obfuscation surrounding the succession theory do not necessarily indicate that the topic is unimportant. Many of our present-day laws and policies on the management of natural systems originated in the ideas and influence of Clements (1916, 1928). It is important that ecologists resolve what they know about succession and identify what needs to be discovered.

This volume will use a set of computer models of forest succession (called gap models) to explore the long-term dynamic responses of forests. The investigation will center on the use of models to develop theory and to resolve some of the problems in our understanding of succession. The diverse array of individual-based forest-dynamics models will be reviewed in Chapter 2 to relate gap models to other forest models of both greater and lesser degrees of abstraction. The assumptions that are used in gap models will be discussed in some detail in Chapter 3 and the testing of gap models will be elaborated in Chapter 4. These three chapters are designed to provide the reader with the biological and mathematical details of gap models.

Chapters 5, 6, and 7 will investigate the theoretical implications of gap models at increasingly larger spatial scales. The model investigations in Chapter 5 will be restricted to spatial scales of the order of 0.10 ha. Many of the phenomena that are discussed at this spatial scale are related to concepts that McIntosh (1980, 1981) has called "the individualistic school of succession" after the classic work of Gleason (1917, 1927). In Chapter 6, the spatial scale of interest will be 100- to 1,000-fold larger; the theories that will be discussed will tend to pertain to questions that are normally considered to be problems in ecosystem dynamics. Chapter 7 will treat a theory of categorizing the dynamics of landscape systems.

Chapter 8 will investigate the relationship between landscape dynamics and the pattern of animal responses on landscapes that are composed of dynamic vegetation mosaics. The applied aspect of the chapter involves the dynamic management of animal habitats and the more theoretical aspect is concerned with the theory of island biogeography (MacArthur and Wilson [1967]). Chapter 9 develops three gap model applications in some detail and is intended to identify some of the potential practical uses of gap models. The final chapter will summarize the previous nine. It remains the task of the present chapter to discuss some of the current ideas in succession theory as a preamble to the investigation using models that follow.

There is a recognizable (though hazy) dichotomy in how ecologists consider the nature of succession (McIntosh 1980, 1981). One position in this dichotomy emphasizes the importance of individual attributes of the organisms in forming the pattern of succession. Gleason (1917, 1927) is an early formulator of the constructs that are related to this view. The other position emphasizes the generality of successional dynamics in a variety of ecosystems (regardless of the detailed nature of the organisms comprising the ecosystem). The features of these two positions and their relation to gap models will be discussed in the remainder of this chapter.

Succession from Individual Attributes

In the past decade or so, a number of papers have attempted to reformulate successional theory. These papers (McCormick [1968], Drury and Nisbet [1973], Horn [1976], Pickett [1976], Connell and Slatyer [1977]) all are critical of Clements' (1916, 1928, 1936) theory of succession, as well as the supposed successors to Clements' holistic theories—particularly the works of Margalef (1968) and Odum (1969). It probably is not particularly useful to dwell on the arguments regarding Clements' theory other than to note that the disproof of an older theory provides space for a newer theory. This need for scientific *Lebensraum* is reflected by the papers dealing with a revised succession theory that often contain many examples counter to a putative Clementsian theory of succession.

Fundamental elements of the individual-attributes-derived theory that will be discussed in more detail below are:

(1) An emphasis of the importance of competition as an underlying mechanism in species composition dynamics. This recognition frequently is coupled with a vigorous denial of the importance of facilitation (the idea that a successional species alters the environment so that it is less able to survive than another species, which then replace it).

(2) A recognition of the importance of understanding the population dynamics of the important species in a community. Reasons for this recognition include the ease with which modern evolutionary theory can be attached to a population-based succession theory, and the logic of using population mechanisms (e.g., birth, death, competition, predation, and so on) to derive a theory of community dynamics.

(3) A denial of the concept of a climax community and the recognition of the nonequilibrium nature of the vegetation that comprises most modern landscapes.

(4) Formulation of mathematical models that can be manipulated to explore the long-term theoretical implications of interactions of the dominant organisms (e.g., trees in the case of forests), and that can be used to test theory against data.

Competition Versus Facilitation

Egler (1954) presented two opposite views of old-field succession. One (to which he did not particularly subscribe) was "relay floristics," in which one floristic group relays predominance to another until a stable stage was reached. Whittaker (1970) described this process: "one dominant species modified the soil and the microclimate in ways that made possible the entry of a second species, which became dominant and modified the environment in ways that suppressed the first and made possible the entry of a third dominant, which in turn altered its environment." Egler (1954) also published a second hypothesis—"initial floristic composition"—in which abandoned land, having received an initial load of propagules, develops its vegetative cover from this initialization without additional inputs of seeds or sprouts. According to this hypothesis, the observed stages of secondary succession simply are a consequence of the different rates of growth of the plants involved.

Egler's (1954) theory was restricted to the successional sequences that were associated with abandoned cultivated fields, which is an important subset of the successional theory. There are several cases of initial floristic composition succession noted by Drury and Nisbet (1973), which include Niering and Egler's (1955) studies of plant succession along power line rights-of-way in Connecticut, Hack and Goodlett's (1960) research on forest succession following flooding in Virginia, and Marquis' (1967) investigation of the regeneration of northern hardwoods following clear-cutting. Drury and Nisbet also cited Clements (1916, 1928) as an early reference with "secondary areas such as burns, fallow fields, drained areas, etc., contain a large number of germules often representing several successive stages. In some cases it seems that the seeds and fruits for the dominants of all stages, including the climax, are present at the time of the initiation."

The importance of these considerations is in the light they shed on the phenomenon called "reaction" by Clements (1916) and "facilitation" by others (e.g., Connell and Slatyer [1977]) which serves as a theoretical mechanism to drive Clements' successional seres. If one restricts the discussion to secondary successions (Clements [1928] provides an abundant compilation of archival as well as more recent examples of "reaction" in primary succession), it is clear that in some ecosystems strong competitive relations are of equal or greater importance than facilitation. Drury and Nisbit (1973) listed several systems, including early old fields (McCormick 1968), *Viburnum* thickets (Niering and Egler [1955]), and forests of early stage trees (Raup and Carlson [1941], Lutz and Cline [1947], and Olson [1958]), all of which appear to seriously retard the development of climax-stage vegetation. One mechanism, allelopathy, has been presented as an opposite phenomenon to facilitation (Brian [1949], Harley [1952], Brown and Roti [1963], Rice [1964], and Whittaker and Feeny [1971]).

The gap models (see Chapter 3), as well as several of the other extant individual tree-based models (see Chapter 2), consider the competition among trees as an explicit part of the model formulation. This competition can be for space, light, or nutrients depending on the model and the simulated conditions under which it is being used. The explicit consideration of both individual organisms and competition makes gap models useful tools for investigating facilitation. Facilitation seems to be most appropriately treated as a phenomenon of smaller spatial scales (approximately 0.10 ha), and it can occur in competition-driven systems (at least in theory). This topic is discussed in Chapter 5 in the context of tree-species' ecological "roles."

The Importance of Population Dynamics

The island biogeographic theory that was developed by MacArthur and Wilson (1967) provided an appealing model that could be used as a theoretical base for a population-biological theory of succession. The MacArthur-Wilson model abstracted the dynamics of the numbers of species on an island as a linear, first-order, and constant-coefficient ordinary differential equation. This model is a simple dynamic model of an input/output relationship. Its theoretical merit was in its elegance of development—not in the sophistication of its mathematics. MacArthur and Wilson (1968) modelled the input of species to an island as a function of the size of the island, of its distance from the species source, and of the attributes of the taxa considered. The loss rates of species from an island was a constant proportion related to the nature of the island and the taxa. Pickett (1976) applied these ideas to succession and noted that disturbance creates patches of successional environments whose size depends on the type and severity of the disturbance. He felt that these patches are analogous to islands with different physical and biotic selective arrays that are related to the time and type of the disturbance. The speed of invasion, the maximum population sizes, and the species richness of various successional patches all might depend on the age of the patch, its size, and its distance from other patches of slightly greater age. The species are sorted by biotic and physical selective arrays, so that a patch supports species with appropriate adaptations.

By viewing a landscape as a mosaic that is randomly (or even regularly) disturbed by a variety of forces, it is possible to incorporate the theoretically rich island biogeographical literature of the late 1960s and early 1970s with the older succession theory (*see also* Chapter 8). Clements (1916, 1928), as well as other early ecologists (Forbes [1880] and Gleason [1917, 1927]) recognized and elaborated the various adaptations of species to different successional stages. Along with the injection of evolutionary theory into succession theory (Pickett 1976), and attendant with the more direct consideration of population processes, the population biology/succession synthesis has provided a fairly explicit consideration of spatial scale in studies of disturbance (Table 1.1 and Chapter 7; *see also* Whittaker and Levin [1977] and White [1979]).

The Nonequilibrium Nature of Vegetation

One of the principal problems with the Clementsian view of vegetation dynamics seems to be the concept of the climax. This is the idea that a succession process culminates in a stable community with homeostatic

Table 1.1. Some sizes, cases, and frequencies of disturbance in North America Landscapes (from Pickett & Thompson 1978)

System	Type of disturbance	Mean size	Maximum size	Frequency	Reference
Beech maple	Windfall	0.010 ha	0.025 ha		Williamson (1975)
Beech maple	Tornado track		400 m[b]		Lindsay (1972)
Northern deciduous	Wind		15533 ha		Irving in Curtis (1959)
Deciduous	Hurricane	1.61-2.4 km[b]			Beebe in Goodlett (1954)
Grassland forest	Tornado track	77 ha	217 ha	8 year	Eshelman & Stanford (1977)
Northern deciduous	Hurricane			1635. 1815. 1938[c]	Spurr (1956)
Deciduous	Flood		60.96 m[b]		Hack & Goodlett (1960)
Deciduous	Landslide	15.24 m[b]	304 m[b]		Hack & Goodlett (1960)
Boreal	Fire		809716 ha		Schmidt (1970)
Pond pine	Fire	11860 ha[a]			Wade & Ward (1973)
Sand hills	Fire	2995 ha[a]			DeCoste et al. (1968)
Boreal	Fire	5922[a]			Sando & Haines (1972)
Montane conifer	Fire	20242 ha[a]			Anderson (1968)
Interior Alaska	Fire (total)	1226 ha		218 year	Barney (1969)
Interior Alaska	Fire (lightning)	3095 ha		83 year	Barney (1969)
Boreal	Fire	1739 ha		33 year	Johnson & Rowe (1975)
Lodgepole Ponderosa	Fire			0.4 year	Franklin et al. (1972)
Boreal	Fire			4.8 152700 ha year	Martinka (1976)
Montane conifer	Fire	99071 ha[a]			Neiland (1958)
Pine spruce fir	Fire	Up to 97° of site showed fire evidence			Patten (1963)
Southern Appalachians	Fire (lightning)	3.4 ha	33 ha	6 400000 ha year	Barden & Woods (1973)
Southern Appalachians	Fire (man-made)	5.4 ha			
Mixed conifer	Fire			0.13 year	Biswell (1967)
Deciduous	Fire	6.6 ha		10 198138 ha year	Haines et al. (1975)
Deciduous	Fire	4.4 ha		10 349635 ha year	Haines et al. (1975)
Deciduous	Fire	10.4 ha		39 504655 ha year	Haines et al. (1975)
Deciduous	Fire	12.6 ha		212 622307 ha year	Haines et al. (1975)
Deciduous fir	Fire	2.0 ha		7 153441 ha year	Haines et al. (1975)
Deciduous	Fire	11.1 ha		140 407165 ha year	Haines et al. (1975)
Deciduous	Fire	8.1 ha		22 482348 ha year	Haines et al. (1975)
Deciduous conifer	Fire	6.4 ha		148 381255 ha year	Haines et al. (1975)
Deciduous conifer	Fire	3.9 ha		16 495303 ha year	Haines et al. (1975)
Deciduous	Fire	12.8 ha		41 198623 ha year	Haines et al. (1975)
Deciduous boreal	Fire	5.0 ha		60 1102024 ha year	Haines et al. (1975)
Deciduous boreal	Fire	1.9 ha		7 319068 ha year	Haines et al. (1975)

[a] Single event.
[b] Width.
[c] Years of occurrence.

properties. Indictments of the climax concept come from several quarters, as discussed in the following paragraphs.

(1) Small landscape units definitely are unstable with respect to their vegetative cover (see Chapter 5). For example, a 1/10–1/20-ha quadrat in a southern Appalachian deciduous forest is disturbed on a regular basis as a dominant tree falls from the canopy every 1 or 2 centuries and the recruitment of new canopy trees is initiated. Disturbances of different spatial scales can predominate in different ecosystems (e.g., fire, hurricanes, or tree falls; *see* Table 1.1 and White [1979]). Also, the size of the spatial unit that is associated with the climax community can be thought of as varying as a function of these different scales of disturbance. Whittaker (1953, 1974) and Whittaker and Levin (1977) add rigor to this observation and point to a landscape-level average of disturbed patches that are analogous to climax vegetation. On a sufficiently large landscape, this average approaches a constant over time. Landscape size is discussed in this context in relation to gap models in Chapter 7.

(2) Climate change has altered the species distributions of trees. In the northeastern United States, for example, the forests have been developing and changing over the past ten of thousands of years. Davis (1981 (*see* Fig. 1.1) collated pollen data from lakes and bogs in various locations in the eastern United States to generate dynamic range maps of important tree taxa. Clearly (*see* Fig. 1.1), several important tree species only recently have been present in the northeastern United States. It is quite probable that the ranges of some trees are still changing in response to recent climate shifts, such as the "Little Ice Age." Similarly, pollen profiles from the British Isles show vegetation of different compositions during interglacial periods of the Pleistocene, depending on which species became established as successive glaciers retreated (Davis [1981]). Thus, it appears that large segments of the Northern Hemisphere's forests in Quaternary time (and of course during the climatically dynamic Pleistocene) are in a nonequilibrium state due to continuing changes in species ranges. Both Bernabo and Webb (1977) and Delcourt and Delcourt (1980) developed maps of the natural vegetation in the eastern United States based on pollen chronologies and evidence from ^{14}C-dated macrofossils. These maps clearly demonstrate that the forests (and other ecosystems), at given points in space have changed dramatically over recent geological time. Chapter 9 provides an example of using a gap model to simulate compositional dynamics under different climatic regimes that are associated with the Pleistocene full glacial condition.

(3) Regardless of the changes in the composition of the vegetational part of ecosystems over recent antiquity, the vegetational composition for the extant ecosystems currently found over most of the earth's surface are (in fact) in a nonequilibrium state due to timber harvest, blights, and changes in natural disturbance regimes. For example, the chestnut blight *Endothia parasitica* effectively removed the chestnut *Castanea dentata*

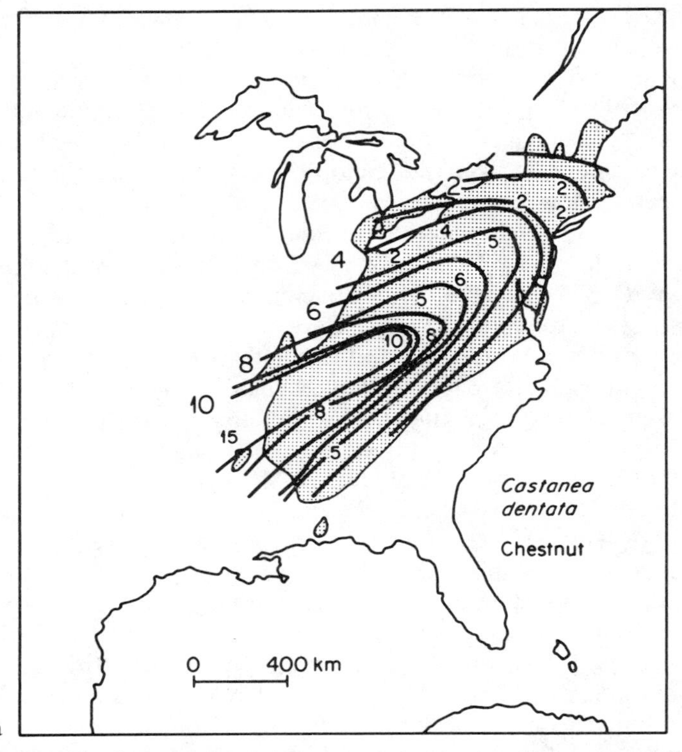

Figure 1.1. Migration maps for three species. The numbers refer to the radiocarbon dates (in thousands of years before present) of the first appearance of the species in a site, as evidenced by an increased pollen abundance or the presence of macrofossils at a site. Isopleths connect points of similar age and represent the leading edge of the expanding population since the Wisconsin glaciation. The dotted area is the modern range of (a) *Castanea dentata*, (b) *Fagus grandifolia*, (c) *Tsuga canadensis*. (From Davis [1981]).

from its former state of importance in the eastern deciduous forest of North America.

These three problems with the climax concept all involve the application of the concept and not its basic theoretical nature. Relative to point (1), the spatial scale at which to apply the climax concept is a difficult problem (McIntosh 1980); however, the spatial scale is not currently resolved for many theoretical conceptualizations of ecological systems. Considering points (2) and (3), the apparent disequilibrium state of the world's ecosystems surely is in opposition to Clements' (in Weaver and Clements [1938]) assertion that landscapes in their natural state are composed mostly of climax systems. The climax condition could be the point toward which a change in disequilibrated systems is directed. Thus, the

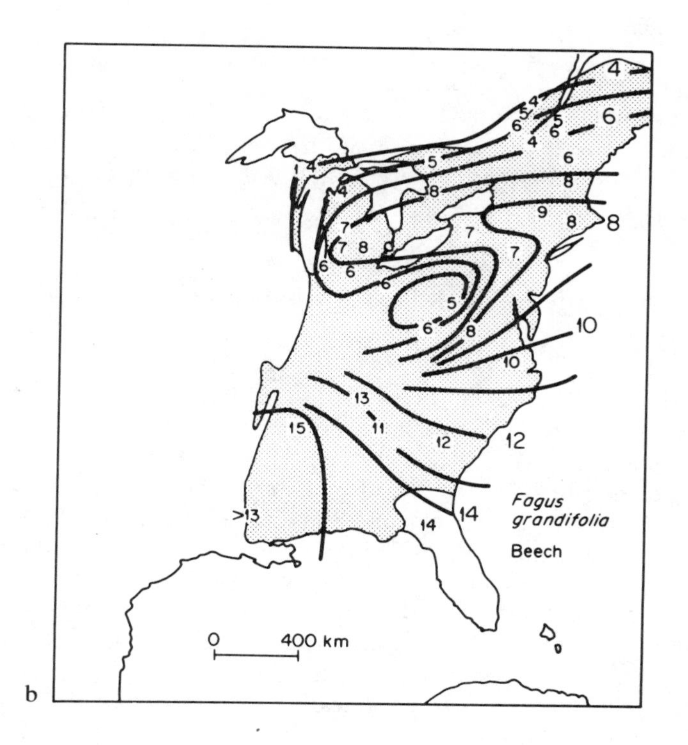

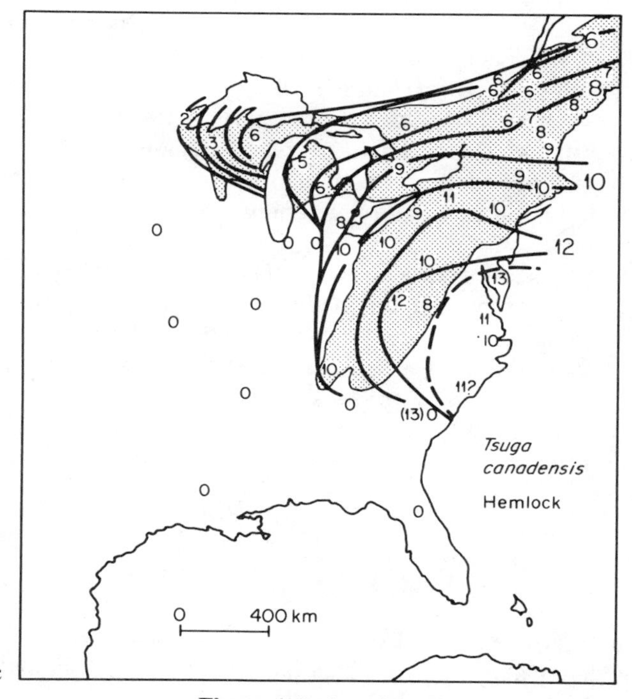

Figure 1.1. (continued)

climax concept still could be a useful theoretical construct. It also is appropriate to reiterate a point made by McIntosh (1981) that the term "succession" does *not* necessarily require the progressive development to a climax or mature state.

Mathematical Models of Individuals as Succession Models

The fourth feature of recent investigations of succession has been the use of models to elaborate the dynamics of plant populations. These models have been successful in their heuristic value, and they are discussed in Chapter 2. The models are used for theory development and often are highly abstracted (e.g., Connell and Slatyer [1977] and Grime [1979]). The models are, as McIntosh (1980, 1981) contends, little more than formalizations of verbal models that were developed by ecologists in earlier times. One notable success has been the Markov model (Horn [1975a,b, 1976]) of a forest in Princeton, New Jersey. This model has been tested on its ability to predict the composition of mature forests by using a matrix of transition probabilities of one species of canopy trees that replace another (*see* Fig. 1.2) *see also* discussions in Chapter 2). Although the model is relatively simple, Horn has used this abstraction to provide considerable insight into the long-term dynamics of forests (Horn [1975a,b, 1976]).

Noble and Slatyer (1978, 1980) similarly developed a purposefully restricted set of species adaptations (which they call vital attributes) and used grammar theory (Haefner [1975]) to produce expected successional sequences for Tasmanian ecosystems. In this case, their result can be tested against the inferred patterns of successional dynamics in Tasmania. The Noble and Slatyer Tasmanian model is used in Chapter 8 as a landscape simulator in a study of animal responses to dynamic landscapes.

In general, a school of succession theory that emphasizes individual organism attributes has begun to develop solid contributions as it has matured over the past decade. An initial tendency to bombast "Clementsian" succession theory has been replaced by a theoretical investigation of the significance of the attributes of important species in determining ecosystem function and dynamics. This individualistic school of succession theory probably is the dominant view of succession at present.

Ecosystem Succession

There is an alternate view of succession that emphasizes the dynamics of the ecological system as a whole as the object of successional theory. In this view, the ecosystem (not a collection of changing populations) is

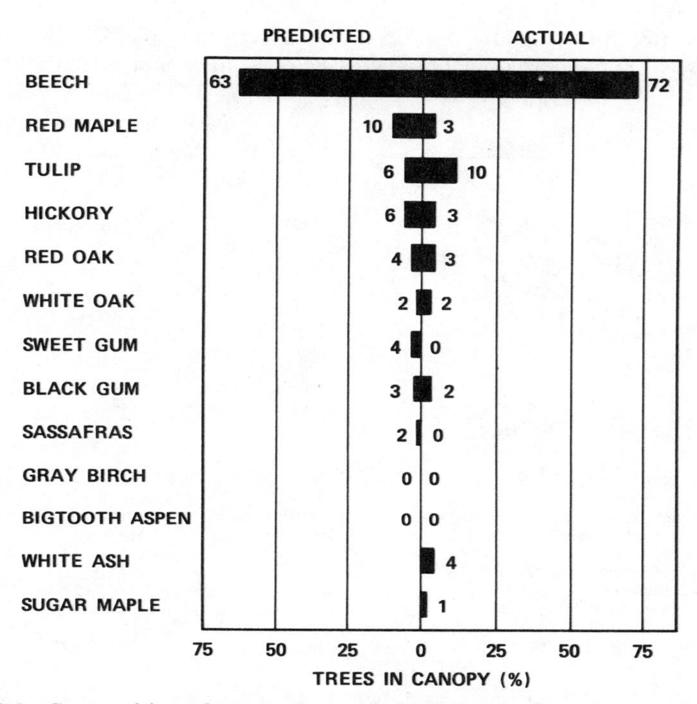

Figure 1.2. Composition of mature forests in New Jersey that were measured and predicted by Horn's Markov model of forest dynamics. (From Forest succession, Horn, H.S. Sci. Am. May 1975, p. 98, all rights reserved.)

viewed as the object of study. According to McIntosh (1981), this interest in holism had its modern origin in ecology with Lindeman's (1942) classic conceptualization of the trophic-dynamic concept and the promise that other ecologists (Hutchinson [1942]; Odum [1960, 1962, 1968], Odum and Pinkerton [1955], Margalef [1963], and Van Dyne [1966]) saw in Lindeman's work. This new conceptualization of ecological systems sought to reduce to complexity to a few variables and strove to identify ecological laws that would relate these variables to each another. This approach has ample precedents in other sciences, particularly in astronomy and physics. It also has a well-developed classic philosophical context. An example set of hypotheses for an ecosystem theory of succession can be found in Odum's (1969) paper (Table 1.2). The ecosystem trends that are expected in this view of succession are concerned with production, standing crops, efficiencies of various sorts, trophic structure, and nutrient cycles.

The formulation of a succession theory for ecosystem ecologists who are concerned with succession has several objectives and features:

(1) The recognition of regularity in ecosystem pattern, and an interest in the development of a theory that allows succession to be viewed as a regular process in a number of different ecological systems. Most ecolo-

Table 1.2. A tabular model of ecological succession: trends to be expected in the development of ecosystems (from Odum 1969)

Ecosystem attributes	Developmental stages	Mature stages
	Community energetics	
1. Gross production/community respiration (P/R ratio)	Greater or less than 1	Approaches 1
2. Gross production/standing crop biomass (P/B ratio)	High	Low
3. Biomass supported/unit energy flow (B/E ratio)	Low	High
4. Net community production (yield)	High	Low
5. Food chains	Linear, predominantly grazing	Weblike, predominantly detritus
	Community structure	
6. Total organic matter	Small	Large
7. Inorganic nutrients	Extrabiotic	Intrabiotic
8. Species diversity--variety component	Low	High
9. Species diversity--equitability component	Low	High
10. Biochemical diversity	Low	High
11. Stratification and spatial heterogeneity (pattern diversity)	Poorly organized	Well organized
	Life history	
12. Niche specialization	Broad	Narrow
13. Size of organism	Small	Large
14. Life cycles	Short, simple	Long, complex
	Nutrient cycling	
15. Mineral cycles	Open	Closed
16. Nutrient exchange rate between organisms and environment	Rapid	Slow
17. Role of detritus in nutrient regeneration	Unimportant	Important
	Selection pressure	
18. Growth form	For rapid growth ("r-selection")	For feedback control ("K-selection")
19. Production	Quantity	Quality
	Overall homeostasis	
20. Internal symbiosis	Undeveloped	Developed
21. Nutrient conservation	Poor	Good
22. Stability (resistance to external perturbations)	Poor	Good
23. Entropy	High	Low
24. Information	Low	High

gists believe that ecosystem science is still in a formative state; they expect quite a number of theories to fall by the wayside as more is learned about ecosystem dynamics. It is important to understand that the compositional dynamics of the vegetation in an ecosystem are not the sole objective of ecosystem theories.

(2) The recognition that processes involving more than plant-environment or plant-plant interactions also are important in ecosystem dynamics. Bormann and Likens (1979a,b) pointed out that the modelling of species interactions (the living fraction of ecosystems) is not equivalent to modeling whole ecosystems. McIntosh (1981) reiterated that a key distinction between species attributes and biogeochemical processes as the basis of succession is not presently resolved.

(3) The incorporation of indices, methods, and approaches from engineering, applied mathematics, cybernetics, and general systems theory in studies of ecological systems. Margalef's (1958) introduction of the information theory-based index of diversity represents an example of this type of incorporation that has produced a rich body of investigations.

(4) The inclusion of human activities as part of the ecosystem. This is analogous to the recognition of the disturbed state of most of the world's present vegetation (discussed earlier) in connection with the individualistic succession theory. The first three points will be elaborated in more detail.

The Regularity of Ecosystem Pattern

There can be striking similarities between the ecological systems at different points in time that have similar geological and climatic features. Historically, this concept permeates the 19th century publications of the great German plant geographers, from Von Humboldt through Drude and Graebner to Warming and Schimper. The similarity of ecosystem patterns in the face of taxonomic dissimilarity is used by ecosystem theorists as strong evidence of a higher order organization. Major (1951) postulated that the vegetation could be taken as a direct function of the soil, climate, animals, and flora in an area. This same type of idea can be found in the rationale of several geographers' algorithms for mapping climate and vegetation (e.g., Koppen [1931], Thornthwaite [1948], Troll [1948], Holdridge [1967], Walter [1971], and Box [1981]). Table 1.3 is a quantification of how successful these approaches can be; in this case, the method of Box (1981) is used to predict the vegetation from the climatic variables at a site. Box's method is able to meet a strong test of predictability in 50% of the test cases and a weaker test in 42% of the remaining cases. Table 1.3 details the strong and weak test criteria and Table 1.4 provides examples of sites that meet these criteria.

This regularity of pattern and relative predictability of gross features of

Table 1.3. Ability of Box's (1981) ecosieve method in predicting vegetation at 74 validation sites. Sites are grouped by continent and by elevation. To meet the strong criterion, the predicted dominant vegetation would have to include the exact combination of dominant life forms at the validation site. To meet the weak criterion, the predicted dominant vegetation would have to be present at the validation site. Examples of sites that meet the strong and weak criteria are found in Table 1.4

Site-class	No. of sites	Sites that meet Strong criterion	Weak criterion	Neither
By continent				
North America	30	18	12	0
South America (including Central America)	13	7	4	2
Europe (excluding USSR)	4	1	3	0
Africa	14	10	4	0
USSR	3	1	1	1
Asia (exclusing USSR)	5	0	2	3
Australia and Oceania	5	0	5	0
By elevation				
Lowland or upland sites	51	26	22	3
Mountainous sites	23	11	9	3
Total	74	37 (50%)	31 (42%)	6 (8%)

ecosystems at a site, when given equivalent environmental conditions, leads ecosystem theorists to postulate that successional responses of the systems will have the same types of general regularity (*see* the discussion on Noble and Slatyer's [1980] model in Chapter 2). The basic idea is that the ecosystem is a self-ordering system that develops toward a stable configuration over time. Embryogenesis—the process by which a single fertilized egg cell becomes a functional multicelled individual—can be taken as a model of this process. One can view the development of an individual as the product of interactions among the multiplying cells of the developing embryo.* In an ecosystem context, there are analogous interactions among ecosystem components that are postulated to be the organizing mechanisms of succession. Ecosystem theory has concerned itself with the proper categorization of ecosystem components and with the

* E.P. Odum, one of the prime architects of ecosystem theory, produced his Master's thesis on the fate of the primordial germ cell in the dogfish shark. This analogy to embryology also is found in: "as viewed here, ecological succession involves the development of ecosystems; it has many parallels in the developmental biology of organisms . . . (Odum [1969]).

quantification of the magnitudes of these components (e.g., "producers," "decomposers") in different systems.

One of my original goals, in developing a series of gap models of widely different forests, was to look for regularities in the dynamics. The rationale was to use a particular gap model as a single sample of the structural and functional pattern of a given forest. The investigation would proceed by cross comparisons of several of these models. The comparability of ecosystem attributes should best expected to be made at larger spatial scales; these comparisons are treated in Chapters 6 and 7. The problem of the proper categorization of functional ecosystem components is a smaller scale problem, which is discussed under the topic of "Roles" in Chapter 5.

Consideration of the Whole Ecosystem

Ecosystem theory may take considerable notice of the vegetation that makes up the ecosystem, but this interest usually is directed toward the system's productivity—not its composition. At the same time, there is an equivalent interest in the decomposition, element cycling, and soil kinetics of the system. While neoindividualist theorists tend to view succession as a consequence of interactions of the life histories of the plants that dominate a system, the ecosystem theorist might view the successional change as being manifested in the balancing of production and respiration or of equilibration of input and output of major nutrients. It is derived from the physical laws of mass balance that if a system has unbalanced input versus output of matter, the amount of matter in that system must change. If one postulates that a dynamic equilibrium is a necessary attribute of any steady-state ecosystem, then one can compare ecosystems relative to this criterion (Fig. 1.3).

Basic processes that are involved with carbon fixation in plants, microbial decomposition rates, and soil mineralization rates all may vary with the species composition of ecosystems; however, they are strongly controlled by abiotic factors (e.g., light, temperature, and moisture). Taken as a whole, the material content and input/output balance of an ecosystem can be studied as a response to such factors. Gorham, et al (1979), in a review of the successional dynamics of chemical budgets in terrestrial succession, present a list of the following ecosystem processes for determining the pattern of biogeochemical cycling in developing ecosystems:

1. Processes affecting input:
    a. Rock and soil weathering
    b. Nitrogen fixation
    c. Particle impaction and gas absorption

Table 1.4. Examples of predicted and actual vegetation at validation sites for Box's (1981) ecosieve prediction of vegetation from climate. Strong and weak criteria as in Table 1.3

Site	Predicted vegetation	Actual vegetation
Sites meeting strong criterion:		
Mazatian (Mexico, Pacific coast at Tropic of Cancer)	Dry raingreen woodland and dense scrub: *Raingreen thorn-scrub *Tropical evergreen sclerophyll trees *Xeric raingreen trees *Broad-raingreen small trees Xeric evergreen tuft-treelets Bush stem-succulents Evergreen arborescents Leaf-succulent evergreen shrubs Short bunch-grasses Raingreen forbs and vines	Lowland dry raingreen thorn forest: *Ipomoea arborescens*, *Acacia*, *Zizyphus*, *Bauhinia*, *Ceiba*, *Prosopis*, *Caesalpinia*, *Jatropha*, *Cordia*, *Mimosa*, and many others, plus *Cassia*, *Croton*, *Fouquiera*, *Cereus*, *Pachycereus*, and many grasses, forbs, and vines.
Kericho (Western Kenya 2042 m)	Tropical montane rainforest *Tropical montane rainforest trees *Tropical linear-leaved trees Tropical evergreen microphyll-trees Palmiform tuft-trees Tropical cloud-forest small-trees Tropical evergreen sclerophyll trees Treelets, tree ferns, shrubs, bamboos, other grasses, forbs, lianas and vines, epiphytes.	Equatorial montane rainforest *Ocotea*, *Podocarpus*, with *Aningeria*, *Casearis*, *Ficus*, *Pygeum*, *Schefflera*, and many others, plus understory trees, *Croton*, *Dracaena*, *Ficalhoa*, and numerous vines, forbs, and epiphytes.

Sites meeting weak criterion:

Site	Predicted vegetation	Observed vegetation
Mahadday-Weyne (Somalia, 70 km N of Mogadisho)	Raingreen thorn-scrub with tall stem-succulents: *Raingreen thorn-scrub and small trees *Arborescent stem-succulents Typical and bush stem-succulents Xeric rosette shrubs and tuft-treelets Leaf-succulent evergreen shrubs Xeric cushion-shrubs Short bunch and desert grasses Evergreen arborescents Raingreen and succulent forbs Raingreen vines Xeric cushion-herbs Ephemeral desert herbs	Dry raingreen thorn-scrub: _Commiphora_, _Acacia_, _Capparidaceae_, _Anacardiaceae_, _Grewia_, plus short _Chrysopogon_, _Aridtida_, _Sporobolus_, _Cenchrus_, etc., including scrub and forbs.
Barinas (Venezuela, east base of Cordillera de Merida	Tropical semi-deciduous forest: *Tropical evergreen sclerophyll trees Palmiform tuft-trees and treelets *Tropical evergreen microphyll-trees *Monsoon broad-raingreen trees *Tropical rainforest trees Raingreen and evergreen understory forbs, and vines.	Tradewind raingreen forest with scattered evergreens: _Spondias_, _Pterocarpus_, _Bombacopsis_, _Sapium_, _Hura_, _Brosimum_, _Terminalia_, _Swietenia_, with understory palms and arborescent shrubs, many grasses, forbs, and epiphytes (_Ficus_, _Phoradendron_, etc.).

*Predicted dominant vegetation.

2. Hydrologic processes affecting outputs:
   a. Loss of dissolved substances
   b. Erosion
   c. Regulation of redox potential
3. Biological processes affecting the balance of inputs and outputs:
   a. Net ecosystem production
   b. Decomposition and element mobilization
   c. Regulation of soil solution chemistry
   d. Production of allelochemics
   e. Variability in use of elements.

By this conceptualization, the change of vegetative composition is considered to be a secondary factor that might influence net production, allelochemical production, nitrogen fixation, or other types of these processes. Vegetative change is not seen as a prime factor controlling dynamic change. For the boreal forest of Alaska, Van Cleve and Viereck (1981) diagramed the processes that control the production of biomass and nutrient cycling (Fig. 1.4) as a web of process interactions. Again, in this case, there is no great emphasis on vegetation change over time, except for the effect of moss on the nutrient conservation and quality (with respect to decomposability) of the ecosystem organic matter.

Gap models, when taken as a whole, include several ecosystem processes in their underlying formulation. Thus, in Chapter 3 (where the model development is presented), one finds functions that simulate, (for example) net production and element cycling. The treatment of such processes is at annual or near annual time scales. The spatial scales are on the order of 0.10 ha. Even if the representations of these processes are restricted in their spatial and temporal range, the spirit of these representations is to include ecosystem processes in a model that includes the dynamic response of individual trees. In gap models, both the demographic changes in the tree community and the growth of individual trees alter abiotic conditions (e.g., light levels, available nutrients, and soon; *see* Chapter 3). As the abiotic conditions are altered, these changes are (in turn) feedback to the trees as changes in growth rates, death rates, and regeneration rates. These sorts of biotic/abiotic feedback cycles are important ecological mechanisms that are thought by many ecosystem ecologists to be the important organizing forces that underlie their view of succession. Gap models do not treat these cycles across a wide range of spatial or temporal scales, but they do present a bridge between demographic models without such feedback (including many of the models discussed in Chapter 2) and ecosystem models that do not explicitly consider demography.

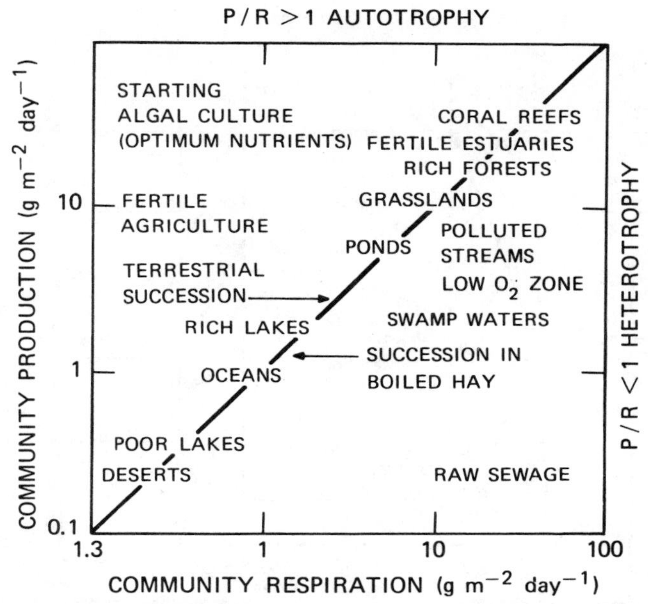

Figure 1.3. Position of various types of communities based on community metabolism. Photosynthetic production (P) exceeds the community respiratory consumption (R) in the upper left side of the diagram. In the lower right side of the diagram, the activities of the respiratory processes exceed photosynthesis. Such communities are importing organic matter or living off previous storage (P/R < 1; heterotrophic type). Over a 1-year average, communities along the diagonal line tend to consume about what they make. Such communities often have an autotrophic regime in the spring and an heterotrophic regime in the winter. In general, succession proceeds toward the diagonal line from either extremely autotrophic conditions or extremely heterotrophic conditions. The diagram emphasizes two kinds of biological fertility: one based on concurrent plant photosynthesis and the other based on accumulations of organic matter from the past. It is possible to have a steady-state with any P and R anywhere on the diagram, if proper amounts of substances are flowing into and out of the community regularly. The term "climax" most often is used for the more self-sufficient communities along the diagonal line. (Redrawn from Odum [1956].)

Systems Theoretical Approaches

In an individualistic approach to successional dynamics, there is a focus on the mathematical models of individuals. A parallel development in an ecosystem approach is in the use of ecosystem models, which are coupled with a general interest in systems approaches to ecological problems. An initial modelling approach used in ecosystem studies was the compart-

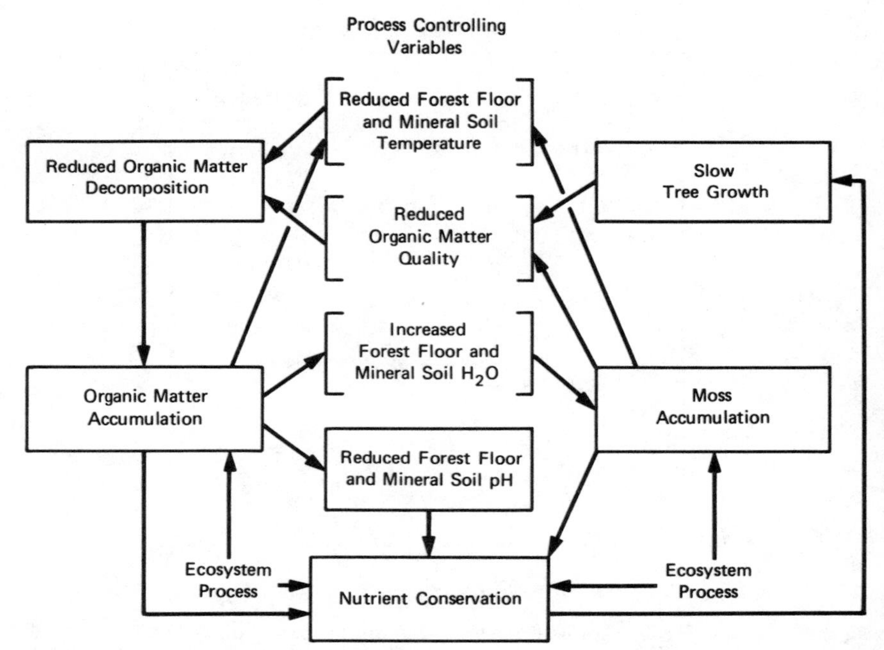

Figure 1.4. Hypothesized processes that control the distribution of biomass and the distribution and cycling of nutrients between ecosystem compartments in an Alaskan boreal forest. (From Van Cleve and Viereck [1981].)

ment model. These models often are diagramed with boxes, which represent amounts of material in the compartments that make up an ecosystem, and arrows between the boxes, which indicate the flux of material. Such a conceptualization of a system naturally leads to the use of differential (or difference) equations as the appropriate mathematics (Fig. 1.5). There is ample precedent for this approach in other biological fields (see Shugart and O'Neill [1979] for examples) and in science in general. The role (or potential role) of gap models as a bridge between ecosystem models and models of individuals was discussed in the preceding section.

Related to an interest in systems approaches is an interest in the use of analytical techniques, particularly from the engineering disciplines, in ecological applications. There also has been considerable work to perfect the use of microcosms (laboratory-sized microecosystems) to develop and test ecosystem theory (e.g., Van Voris, et al [1980]). An important interest has involved forming indices of ecosystem performance that can be used to study attributes of ecosystems in terms of temporal or spatial patterns. For example, Figure 1.6 shows the relation of the heterotroph/autotroph biomass ratio plotted against ecosystem carbon turnover time; it is a good example of the possible regularity of indices of system perfor-

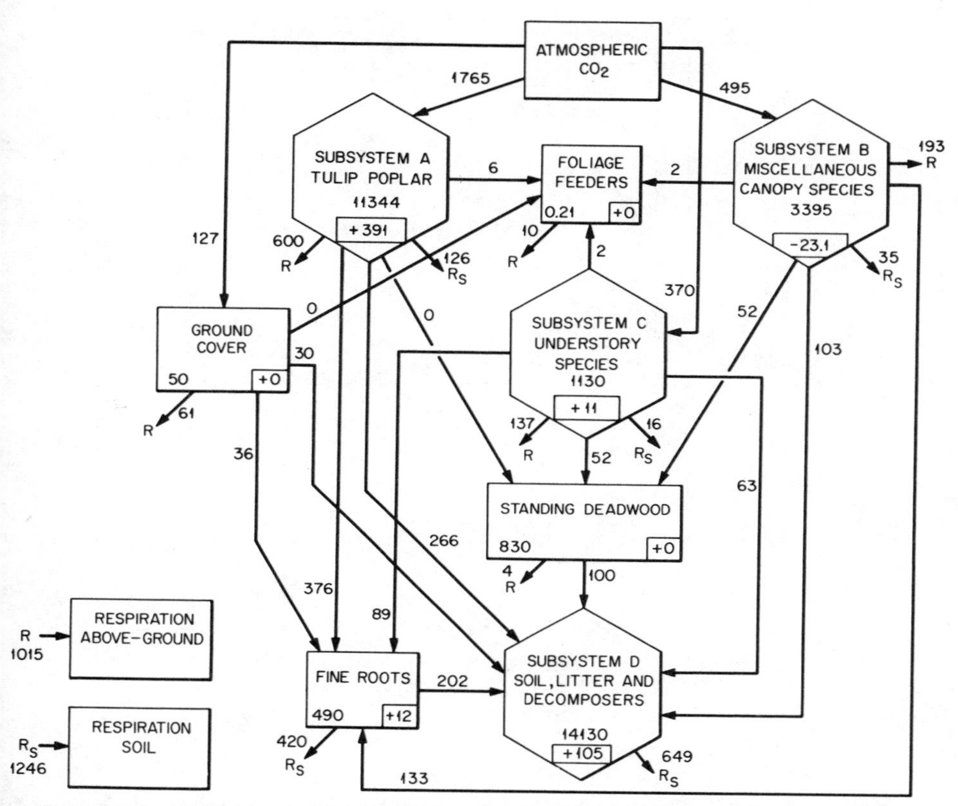

Figure 1.5. Organic matter budget for a *Liriodendron*-dominated mesic forest ecosystem at Oak Ridge, Tennessee. All compartment values are grams dry weight per square meter; all transfer and increment values are grams per square meter per year. Compartment increment values are shown in boxes within the compartment. Values for all compartments except leaves are for the 1965–1966 dormant season. Leaf values are at peak biomass. Transfers are annual totals based on an average of the 1965 and 1966 data. R and R_s refer to above ground and below ground respiration, respectively. (From Sollins, et al [1976].)

mance (in several very different ecosystems). One index that has fueled considerable debate over the past decade has been that for species diversity. Currently, there is little agreement as to what exactly is meant by diversity as an ecological index (Hurlbert [1971]); arguments about diversity in successional time appear to be rumbling into a semantic morass. The relation between functional diversity and biomass dynamics is discussed in Chapter 6.

In general, ecosystem theory is a relatively young and still-forming paradigm that happens to have some well-known ecologists as elder statesmen, spokesmen, and founders. There are no data sets to test the

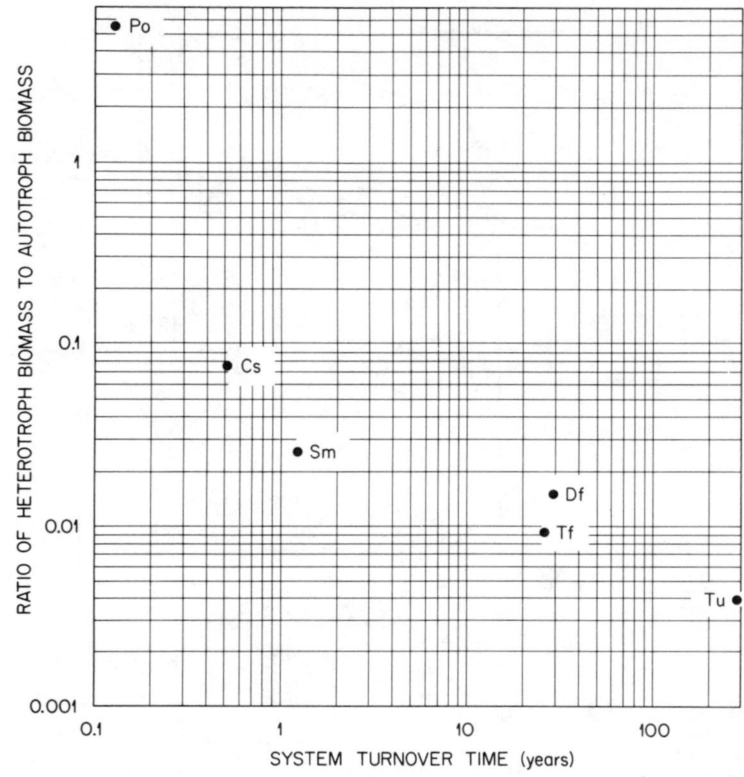

Figure 1.6. Ratio of heterotroph to autotroph-standing crop against ecosystem turnover time for six different ecosystems. Abbreviations for points on figure and data sources: Tu, tundra (Whitfield [1972]); Tf, tropical forest (Odum and Pigeon [1970]); Df, deciduous forest (Reichle et al [1973]; Sm, salt marsh (Teal [1962]); Cs, cold spring (Tilly [1968]); Po, pond (Emanuel and Mulholland [1975]). (From O'Neill, R. V., Ecosystem persistence and heterotrophic regulation. *Ecology* 57:1244–1253, 1976. Copyright © 1976 by The Ecological Society of America. Reprinted by permission.)

trends, which are shown in Odum's (1969) table (Table 1.2 in this volume), collected for any one ecosystem. MacMahon, et al (1981), are collecting such sets of data for successional systems in Utah, but such observations require considerable effort. An ecosystem theory does not proceed logically from Clements' work, although there is a philosophical kinship to these earlier studies. The most important aspect of ecosystem theory is the recognition of holism—the consideration of whole-system phenomena and an explanation of how these phenomena arise from interactions among lower-level phenomena.

Conclusions

In this chapter's introduction, a dichotomy was differentiated between organism-oriented and ecosystem-oriented theories of succession. It almost goes without saying that such artificial categorizations merely are devices for framing a discussion, also, the positions discussed, in many ways, are stereotypes within a more general set of theories and ideas. Most ecologists associate succession with change and the response to change in ecological systems. An ecologist's fundamental idea about succession varies with the space or time scale over which a system is considered and with the actual research interests of that particular scientist. Some might form a dichotomy between the set of ecologists who are particularly interested in whole system behavior and those who are interested in the consequences of an individual organism's adaptive strategies. Others ecologists might prefer a different dichotomy (e.g., holist versus reductionist, botanist versus zoologist, and so on). The point is that, at present, there is a wide array of opinions about the nature and generalizability of successional phenomena. Some of these opinions undoubtedly are incorrect or restricted to special cases; perhaps others will indeed have generality.

The source of at least some of the debate on the nature of ecological succession has its origin in the fact that the ecological system that one ecologist has in mind when a hypothesis is set forth may differ significantly from the system of reference of another ecologist. Clearly, some of the differences in what have been called the neoindividualist school of succession and the ecosystem school originate in the study of both plant communities by the former and ecosystems by the latter. Even if the ecological system under discussion is established, the spatial or temporal frames of reference of the discussants still can be very different. One of the advantages of ecological models is that these sorts of definitional problems usually are explicit in the model formulation. Models of ecological succession may not resolve the current debates on the nature of succession, but they have the potential to be used as tools to explore the consequences of ideas at a given scale with regard to how a given ecological system might work.

The chapters that follow will treat the rationale and application of forest dynamics models. These models initially will be considered generally (Chapter 2). Then, specific models will be examined in greater detail at multiple spatial scales (Chapters 5, 6, and 7). The spirit of these investigations is to use the behavior of these realistic models of forest dynamics to understand more of the successional responses of forested ecosystems.

Chapter 2

Computer Models of Forest Succession

A pronounced trend in ecological studies in the 1960s and 1970s was the progressive mathematization of the field. This can be demonstrated simply by surveying the contents of an ecological journal, such as Ecology and noting the most mathematical paper in each of a sequence of volumes over the past 20 years. One obtains the impression from this exercise that the researchers in ecology were working through a mathematics course sequence; the most mathematical paper in a given year moves progressively through algebra and solid geometry to calculus, to differential equations, to calculus of variations, and to the statistical-mechanical applications that currently are an important theme. Perhaps, because of this accelerated use of mathematics, ecologists sometimes perceive their science as undergoing an evolutionary progression that parallels the historical development of physics. By this perception, physics, astronomy, and chemistry are "old" and "highly evolved" sciences, as evidenced by their mutual interactions with the historical development of mathematics. Biology and ecology, in particular, are "young" sciences that are wobbly on their still-weak mathematical underpinnings. Because modern ecology has placed a strong emphasis on quantitative methodology and systems techniques (*see* Burgess [1981] for an historical review), one sometimes tends to view mathematical sophistication as an index of scientific evolution much in the way that the elaboration of sutures is an index of biological evolution in the ammonoids.

In fact, there is evidence that a systems emphasis on ecology may have sufficient antiquity to reach well into prehistory. Many so-called "primitive" peoples have elaborate whole-system perceptions of ecological phenomena. For example, the Hanunoo people of southern Mindoro in the Philippines are swidden agriculturalists whose slash-and-burn fields contain over 400 cultivars. These agro-ecosystems are reared with complex spacings and timings in a successional sequence that lasts for about 16 years (Conklin [1954]). This managed successional sequence contains many more species than are usually tabulated in a modern study of natural succession.

Regardless of the antiquity of systems orientation in ecology, one beneficial consequence of the recent interest in mathematics (as such) has been a sharpening of the verbal theories that have dominated ecological studies in the past. Examples of this process are abundant in such highly descriptive fields as island biogeography (e.g., MacArthur and Wilson [1967]), morphology (e.g., Kauffman, et al [1978]), leaf geometry (Parkhurst and Loucks [1972], Givnish [1978]), and bird-feeding behavior (Craig, et al [1979]). However, until recently, succession has resisted theoretical consolidation in response to the power of mathematical formulations to clarify meaning, to add internal consistency, and to eliminate redundancy. For forest ecosystems, there now exists a spectrum of models that simulate ecological succession or forest dynamics with a reasonable degree of reliability. These models now can be used as tools to develop and inspect ideas about long-term changes in forests. Before reviewing this diversity of models, it is reasonable to consider the modeling of ecological systems in a more general context.

A Speculation on Modelling

Ecological models are as interesting in what they leave out as in what they include. Success in forming ecological models lies in a correct perception of how to combine scales of phenomena, space, and time. Certain phenomena are dynamic at certain temporal (or spatial) scales, and they either are unmeasurable or relatively constant at others scales. The recognition of appropriate scales and their incorporation into ecological models is far from a clear-cut procedure. For example, a commonly used modelling paradigm for populations would be a formulation of the form as follows:

$$\frac{dN}{dt} = f(N,t)$$

where N = the number of individuals in the population, and t = time. How this formulation can be successful as a basis for a general theory of population dynamics when it (typically) ignores the effects of sex ratios on reproduction or the effects of age structure on mortality may seem to be a paradox. In fact, for certain ecological phenomena, it appears that all manner of seemingly important features can be left out of various models and yet one still can obtain similar general results.

Most models of population reproduction ignore sex ratios. Many models of mortality ignore the number of old individuals in a population, models of the number of species on an island ignore insular speciation, models of evolution contain no explicit consideration of individual genetic traits, and so on. These omissions would appear to an outside observer to be a great detriment to the credibility of the models. Some of these apparent paradoxes are imbedded in differences in perception as to what is important in an ecological system. Many populations have regular sex ratios. In large populations near equilibrium, the number of old individuals in a population is a function of the size of that population; thus, it contributes little additional information to the mortality rate.

Not only can certain phenomena be omitted (or at least not explicitly treated) in good ecological models, but other phenomena that are of obvious importance seem to actually defy inclusion in ecological models. For example, most models of photosynthesis, at the level at which they are typically measured, consider a small area of a leaf surface that works to balance heat, H_2O, and CO_2 by opening and closing the stomatal aperture. With the stomata open, CO_2 diffuses in, H_2O transpires out, and the evaporation of this H_2O helps to balance the heat load of the leaf, which is a necessary consequence of trapping light to drive the photosynthetic biochemistry. If H_2O is in short supply, the leaf can close the stomata, reduce the input of CO_2 to the photosynthesis process, and (if the heat load does not become too high) it then can resist wilting. The important parameter in this consideration is the stomatal aperture. The stomatal aperture changes very rapidly (in seconds or less), and it essentially is unobservable under experimental conditions. However, a parameter called stomatal resistance (which is a function of stomatal aperture) can be measured; as a result, detailed models of leaf photosynthesis can be developed (DeMichele and Sharp [1973]).

It has proved to be exceedingly difficult to use this type of photosynthesis model to predict the amount of growth in a tree that is growing in a forest over 1 year. It is not that the mechanisms of photosynthesis are not important in tree growth; it is that the explicit phenomena involved in photosynthesis at a leaf's surface are lost in the higher space/time scale. A forest is a complex geometric arrangement of many thousands of leaf surfaces that interfere with each other's access to light, H_2O, and CO_2. In the course of 1 year, the growth of a tree may greatly depend on the way that the tree allocates the sugars that have been produced by photosynthesis. The allocation of sugars to the entire tree hardly is evident in a leaf photosynthesis model nor is it needed in such a model. Phenomena and their attendant causes at one scale are not necessarily evident at another scale (Allen and Starr [1982]).

Building ecological models is a scientific art that involves the successful blending of time, space, and phenomenological scales. When a method is found to produce models that predict observations of reality in a dependable way, that method can be inspected for its theoretical content.

Scale in Forest Succession Models

In building a mathematical model of forest succession, two fundamental problems must be overcome. First, the appropriate scales for understanding the mechanisms of forest succession are not well-known. Second, new observations, either to elucidate ecological mechanisms or to test the models, most likely will be collected at time intervals that are short in relation to the scale of a forest ecosystem's actual dynamics. This latter problem will be discussed in Chapter 4, which is concerned with model validation.

The problem with the scale of mechanism can be illustrated by considering what causes the biomass of a young forest that is growing back after some disturbance to increase over time. Table 2.1 summarizes several possible explanations (at different scales) for the cause and the variations in this increase. At a given place and for a given forest, any of these mechanisms can be important over some time scale. Unfortunately, all of these mechanisms cannot be included in a single model of forest growth. If one considers leaf energy balance and the higher order geometry of tree canopies in the same model, the consideration of even a few leaves and their interactions (e.g., using up CO_2, shading one another, losing H_2O and changing the humidity, and so on) is difficult. The number of interactions among the leaves increases as a function of the number of leaves squared. It practically is impossible to consider a large tree by simulating one leaf at a time.

In spite of the potential difficulties in bringing certain mechanisms into forest succession models, many of the models of forest dynamics have similar features. These features are the same in a general sense, but they often differ in the details of formulation. Some models are restricted by their underlying assumptions of cases in which one or more of these featured mechanisms are constant or (at least) predictable. For example, a model that assumes the equal spacing of trees may be restricted in use to applications on plantations. The consistent features that are found across forest dynamics models are:

1. Recruitment: sprouting, seed production, seed dispersal, germination, and growth of seedlings until the young plants are large enough to be thought of as trees.
2. Growth: height and diameter increase of trees.
3. Geometric competition: the spatial interactions of trees related to the actual geometry of the trees' structures. Generally, larger individuals are favored in geometric competition.
4. Resource competition: growth-limiting factors that can limit the development of all the trees in a forest at a given site.
5. Mortality: the death of individual trees.

These features are treated with great detail in some forest dynamics models and are absorbed into the model parameters in others. Model

Table 2.1. Explanations of the growth of young forests at differing scales

Scale	Mechanism	Explanation
Very small	Photosynthesis	Within the leaf, the trees maintain a biochemical factory that converts CO_2 and H_2O to sugar by using light energy to drive this synthesis.
Stomata	Energy balance	The stomata of the leaf must optimize the heat, water, and CO_2 balance of the forest. The resistance of the stomata to the inward diffusion of CO_2 tends (in some cases) to be the determining factor on the rate of photosynthesis.
Leaf	Leaf geometry	The shape and orientation of a leaf can have a pronounced effect on the ability of the plant to take in CO_2 to feed the photosynthesis machine while giving up little H_2O. Plants that do this well have a high water-use efficiency (ratio of CO_2 fixed/H_2O lost).
Leaf layers	Light extinction in the vertical	The orientation and layering of the leaves of a tree can alter the rate at which the canopy captures light. Some arrangements of leaves can be quite inefficient relative to others.
Tree shape	Light extinction in the horizontal	Tall, thin trees are efficient in capturing light at high latitudes (where the sun angles are flat); they also capture light in the morning and evening (when moisture relations may be more favorable). The latter can improve the water-use efficiency at lower latitudes. Other geometries have other advantages.

objectives, as well as the mathematical structure of the particular model, are important in determining the level of treatment of these features. For example, foresters often use the "site class" concept in their models of stand development. The site factor is the height that trees would be expected to reach (typically) at a given age at a given location. Using a site factor in a forest model can be compared to having a single model parameter (the site factor) that, in some sense, contains information about soil water and nutrients at a particular site. In an alternative model for the same forest, the dynamics of soil water or the mechanisms of nutrient uptake might be simulated more explicitly. That one model is detailed while the other are simplified cannot be taken as a criterion for using one model and not the other. For example, fairly detailed treatment of seedling recruitment usually is important in models that simulate the compositional change of natural forests. This is due to the fact that recruitment is necessary to determine which species will increase or decrease over several tree generations. In many models of commercial plantations, however, recruitment is not even considered, because plantations are initiated with a given number of seedlings on some fixed spacing scheme and the forests are harvested before the next generation's recruitment is important.

One convenient way to classify the models of forest dynamics that have been developed uses the models' assumptions, structure, and parameters as criteria for categorization (Table 2.2). Tree models (Munro 1974, Shugart and West 1980, Trimble and Shriner 1981) simulate the

Table 2.2. Forest dynamics models. The schematic diagram illustrates assumptions that are associated with model classification. Phenomena typically included in the various models are listed, with ** indicating strong emphasis, * indicating some emphasis, and a blank indicating that there is little or no emphasis on the particular phenomena. The indicated examples are discussed in text.

SCHEMATIC	MODEL CLASSIFICATION				PHENOMENA					EXAMPLE
	CATEGORY	AGE-STRUCTURE	DIVERSITY	SPACE	REGENERATION	GROWTH	GEOMETRIC COMPETITION	RESOURCE COMPETITION	MORTALITY	
	TREE	EVEN	MONO	SPATIAL		**	**		*	HEGYI, 1974
	TREE	EVEN	MONO	NONSPATIAL		**		**		SULLIVAN and CLUTTER, 1972
	TREE	EVEN	MIXED	NONSPATIAL		*	*		**	SOLOMAN, 1974
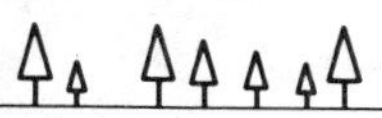	TREE	MIXED	MONO	SPATIAL		**	**		*	MITCHELL, 1975
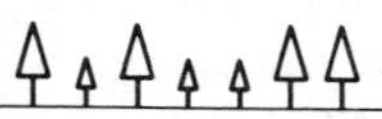	TREE	MIXED	MONO	NONSPATIAL	**	*		*	**	SUZUKI and UNEMURA, 1974a
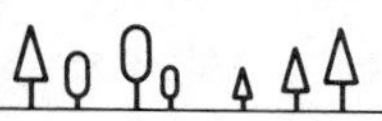	TREE	MIXED	MIXED	SPATIAL	*	**	**	*	**	EK and MONSERUD, 1974
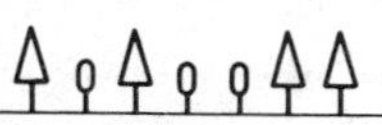	TREE	MIXED	MIXED	NONSPATIAL	**		**		**	HORN, 1975a
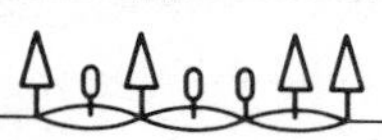	GAP	MIXED	MIXED	NONSPATIAL	**				**	WAGGONER and STEPHENS, 1970
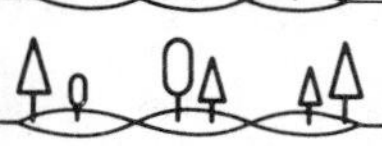	GAP	MIXED	MIXED	SPATIAL	**	**	*	*	*	SHUGART and WEST, 1977

changes in each tree in a forest. Forest models (Munro 1974, Shugart and West 1980) simulate forest changes that summary variables (e.g., total forest biomass and cover as state variables). Some tree models simulate the change of trees that are restricted to a small plot and are called gap models (Shugart and West 1980). Much of this volume will deal with the dynamics of gap models (beginning with the next chapter); forest models will be treated in Chapter 7 in the context of simulating landscape dynamics. The remainder of this chapter and Chapter 3 will discuss the diverse types of tree models that have been used to simulate forest change.

Forests, as simulated by tree models, may be monospecies stands (typically plantations) or mixed stands. The models may treat even-aged forests or forests with a mixed-aged structure. The models may or may not deal with the geometry and location of each tree directly (spatial or non-spatial models). Depending on the forest that is simulated, the models will emphasize recruitment, growth, geometric or resource competition, and mortality to differing degrees. In this chapter, the general outline is discussed and examples of the great diversity of tree models (Table 2.2) are provided. Gap models will be discussed in Chapter 3.

Even-aged, Monospecies, Spatial Tree Models

Highly detailed, spatially explicit models of individual tree growth were developed by Ph.D. students at several universities in the late 1960s and early 1970s (Newnham [1964], Lee [1967], Mitchell [1969], Lin [1970], Bella [1970], Arney [1971], and Hatch [1971]). These works were produced in several different forestry schools (i.e., University of British Columbia, Yale, Duke, and Oregon State University); when viewed in detail, they contain a diversity of modelling approaches. However, at a general level, the approaches are similar, and Hegyi's (1974) extension of Arney's (1971) model can be used as a representative example.

Arney's (1971) basic model for Douglas-fir (*Pseudotsuga menziesii*) stands simulated both the annual height growth and diameter growth of each tree. Competition decreased the maximum expected growth by using a geometric competition index. Hegyi's (1974) modification of this model for jack pine (*Pinus banksiana*) reformulated the competition index (Hegyi [1972]) and considered the growth rate of jack pine in stand conditions to be the appropriate maximum rate of decrement under competition. The model uses several regression equations for height and diameter increments at regular 25.4-cm (10-inch) intervals along the stem as a function of competition, age and size of the tree, and the site index. Mortality is applied at 5-year intervals. Hegyi tested the model against published yield table data (Plonski [1960]) with good success and felt that the model was realistic in stands up to 100 years of age. Figure 2.1 illustrates Hegyi's use of the model to simulate the morphology of 60-year-old trees, and it is indicative of the detail in the output of this model.

Hegyi's model and most of the other even-aged, monospecies, spatial tree models depend heavily on empirical relations that usually are determined by regression. This detailed information generally is available only for commercial species that are managed with at least some intensity. Provided that such data exist and assuming that one is interested in the growth-to-harvest of an established cohort of trees, then (as Hegyi [1974]) notes) it is a rather simple task to build a simulator that will give realistic results.

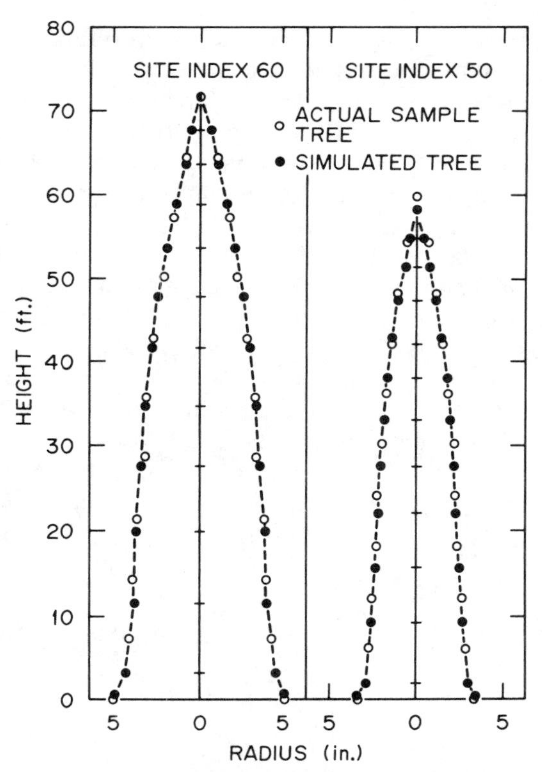

Figure 2.1. Stem profiles of 60-year-old jack pine trees from sites with site indices of 15 m and 12 m (60 ft and 50 ft) compared with simulation results from Hegyi's (1974) model.

Even-aged, Monospecies, Nonspatial Tree Models

The spatial models that were just discussed have their origins in combining empirical relations for individual tree form and growth with geometrically explicit competition indices. The explicit geometric index on an even-spaced plantation is directly related to stand density, so one can disregard the explicit spatial effects by including a density variable. Clutter (1963) took this approach in a plantation model for loblolly pine (*Pinus taeda*), as did Moser and Hall (1969), Dress (1970), and Curtis (1967) for more complex forests (uneven-aged or unevenly spaced).

The modelling problem is to obtain both a statistically reliable estimate of the volume of wood per unit area that is expected at a given time and a simultaneous estimate of the basal area of the stand. These two variables are both functions of the number of trees and individual tree geometry. Much of Sullivan and Clutter's (1972) work deals with this rather difficult statistical problem, but they were able to develop functions that predict the yield of a plantation at different densities as a function of site index and initial densities. These models were found to account for 99% of the variation in mean cubic feet at the first measurement and about 98% of the variation in volume on a subsequent remeasurement of 102 study plots of different ages and stand densities for loblolly pine experimental plots from Georgia, Virginia, and South Carolina.

Methods that relate individual tree growth to forest yield have the important practical advantage of allowing the formulation of a forest dynamics model with parameters that can be statistically obtained from experimental study plots. This is valuable, particularly when the performance of a plantation species is not well-known. For example, the methods are very useful for predicting the plantation response of species that were tested in experimental plots only (e.g., *Pinus radiata* plantations in various parts of the world). The approach is also useful in devising new and untried forestry practices (e.g., Rondeux [1977]).

Even-aged, Mixed-species, Nonspatial Tree Models

The two categories (Table 2.2) of models that have been discussed thus far typically use large data sets (or their synthesis in the form of empirical relations) that only are available for commercially important species and are restricted to forestry applications. Solomon (1974, 1977) was able to use a stand stocking guide (Fig. 2.2) as a basis for forming a similar management model for northern hardwood (mixed-species) forests of the

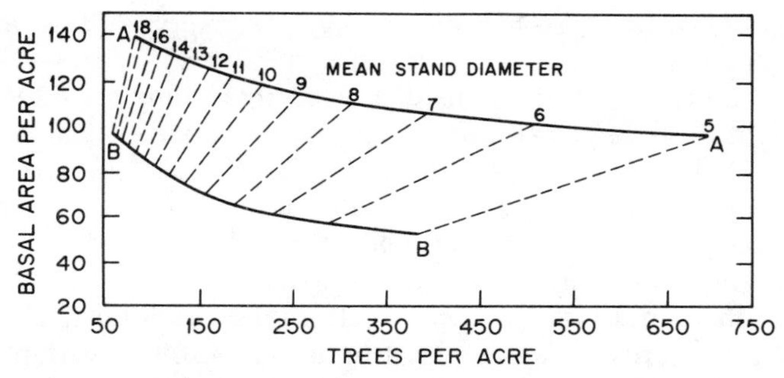

Figure 2.2. Stocking guide for even-aged northern hardwood stands showing basal area per acre, numbers of trees per acre, and mean dbh for trees in the main crown canopy. Line A is for a fully stocked stand; Line B is for a minimally stocked stand. (From Solomon [1974].)

northeastern United States. The stocking guide was restricted to even-aged stands. Basically, the stocking guide is a functional relationship between the size of trees, basal area of trees, and density of trees. The dynamics of the simulator are the changes in the numbers of trees, which is an approach originally explored by Leak (1969a,b, 1970). In this case (Solomon [1974]), the demographic model is for a decrease in the numbers of trees as a function of the differences between the basal area of the stand and the basal areas that are expected for full or minimal stocking (Fig. 2.2). Thus, somewhat similar to Sullivan and Clutter's (1972) approach, demographic processes are coupled with tree growth processes. Solomon's model also includes several options that involve harvesting and thinning. The model is directed toward timber harvests and commercial forestry.

Mixed-aged, Monospecies, Spatial Tree Models

The simulation of the detailed responses of regenerating Douglas-fir (*Pseudotsuga menziesii*) has been a focal problem for forest dynamics models (Newnham [1964], Lin [1974], Curtis [1967], Goulding [1972], Arney [1971], Mitchell [1975a], Reed [1980], and Franklin and Hemstrom [1981]) for the past 2 decades. The great life span of the Douglas-fir proscribes the approach of calibrating experimental plantings of trees to obtain an empirical determination of the growth and yield of the second-growth forest that will replace the forests which are now being cut. This, coupled with the commercial importance of the forests of the Pacific Northwest, has spurred an interest in modeling these forests. Mitchell's

(1975a) model of Douglas-fir provides a good example of how a detailed geometric model can be used to simulate monospecies forests.

This model (Fig. 2.3) contains elaborate algorithms for growing each tree's height, branch growth, and diameter increment. Each tree that is grown by the model interacts with other trees via the geometry of crown overlap. The model also incorporates a variety of factors, which include animal grazing effects, site-quality, genetic variation, and tree spacings. The model draws on empirically calibrated curves for tree geometry to grow trees in shapes that are appropriate to the conditions. The interaction among tree crowns is computed by using a crown overlap competition index that is based on branch pruning by overlapping crowns. The model was designed so that it could be initialized by using low-level stereo photographs of actual forest canopies (Mitchell [1975b]). One test performed on the model was to inspect its ability to simulate the statistical distribution of diameters at year 20 on 16 permanent sample plots that

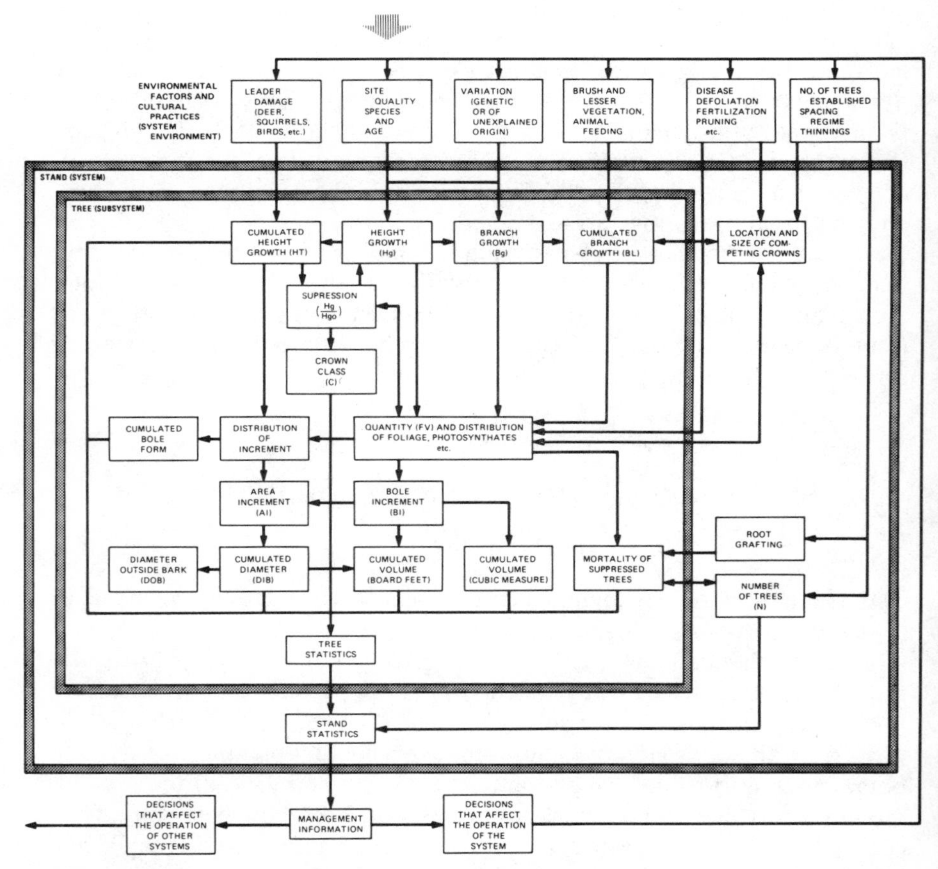

Figure 2.3. Block diagram of a stand of Douglas-fir as modelled by Mitchell (1975a).

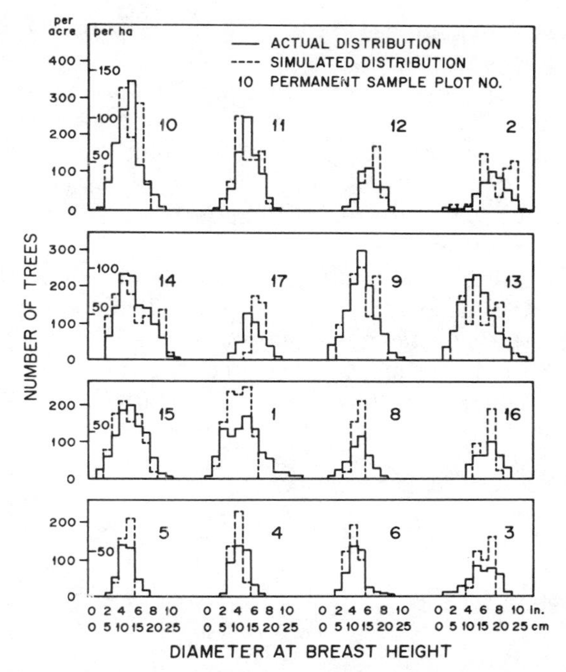

Figure 2.4. Actual and simulated diameter frequency distributions. (From Mitchell [1975a].)

were maintained by the Research Division of the British Columbia Forest Service. These plots were planted in 1941 (2,965 trees/ha) and were thinned in 1951–1952 to a range of densities (990–2,965 stems/ha) with uniform spacings. The plots also varied in their site index. The ability of this model to simulate the statistical distributions of diameters for these 16 different plots is shown in Figure 2.4. The model also was tested for its ability to simulate the details of individual tree height growth, as well as basal area and volume for these sample plots; also, for its ability to predict the responses of the forests to thinnings or fertilization. This model's principal use was to generate stand yield tables for Douglas-fir for specific silvicultural or management regimes.

Mixed-aged, Monospecies, Nonspatial Tree Models

The prediction of the statistical distribution of tree diameters over time (e.g., *see* Fig. 2.4) is an essential goal of any forest model, particularly for purposes of management. One approach to modeling forest structure is to simulate the dynamics of these distributions directly (Preussner [1976], and Suzuki and Umemura [1967a,b, 1974a,b]). The model for changes in the variance about the increasing average tree diameter of the growing trees was an analogy to diffusion. The equations (often partial differential equations) that are used to model diffusion in physics or engineering are modified to apply to forest diameter distribution change. These models are applied to cohorts of trees of the same age in single species stands, partly because this greatly simplifies the formidable mathematical problems in applying this approach.

Suzuki and Umemura (1974a,b), for example, developed equations for the change in the mean (growth of an individual tree), its variance (a diffusion process about the moving mean), and a partial differential equation for the probability of mortality for trees as a function of diameter and time. The behavior of their model and its similarity to a set of plantation data (remeasured over 18 years) is shown in Figure 2.5. The actual appli-

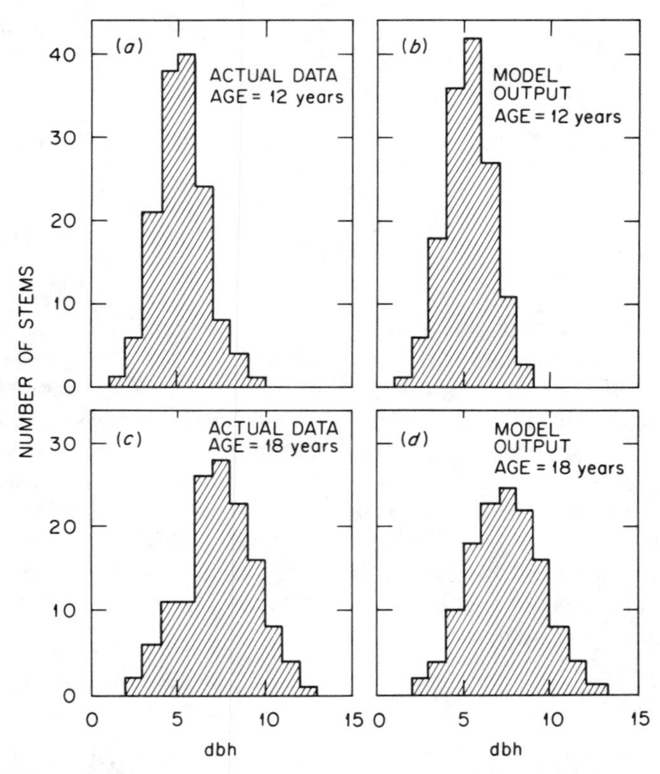

Figure 2.5. Comparison between actual data and model predictions for Hinoki (*Chamaecyparis*) stands in Owase-shi (Mie prefecture), Japan. Model predictions are from Suzuki and Umemura (1974a). (a) Stand diameter distribution (number of stems versus diameter breast height in centimeter midpoints at 1-cm intervals) measured on a 0.02 ha plot at age-12 years; (b) Model prediction for age-12 years; (c) Actual data, age-18 years; (d) Model prediction, age = 18 years. (Data from Table 2, Suzuki and Umemura [1974a].)

cations of this model have been made in even-aged stands. Since Suzuki and Umemura's (1974a) basic equation for tree growth does not compute tree age as such, the model also could be applied to mixed-aged stands of one species. Actual models of mixed-aged, monospecies stands presently are restricted to redwood stands (Bosch [1971], and Namkoong and Roberds [1974]). However, the recently documented discovery that Douglas-fir forests can have an extremely wide variation (approximately 200 years) of ages, in what usually were thought to be virtually even-aged stands (Franklin and Hemstrom [1981]), indicates potential applications for other species.

Mixed-aged, Mixed-species, Spatial Tree Models

Probably, the most elaborate forest dynamic model that uses individual tree growth as an underlying paradigm is the FOREST model (Ek and Monserud [1974a,b]). The model simulates growth and reproduction of mixed-species and even or uneven-aged stands, and it includes natural regeneration and growth as well as a variety of management applications. The model treats tree growth and competition in a manner that is analogous in mechanism and detail, to the spatial monospecies models that already have been discussed. For example, competition among nearby trees is computed by determining the crown overlap with all adjacent trees. One very detailed part of the model is the computation of seed rain (both production and dispersal) from the trees that are simulated on the plot. Table 2.3 outlines the principal subroutines in the model and provides an indication of the level of detail in this model. The model currently is restricted in its application to the northern hardwood forest in or near the state of Wisconsin, but work is under way to modify it for other forests (Vermont, C.W. Newton, personal communication, fide Trimble and Shriner [1981]). The model requires considerable amounts of information about the growth and habits of the trees simulated in a given forest, and it provides output at a level of detail much beyond that collected in most ecological studies (e.g., maps of exact positions and sizes of trees by species).

Mixed-aged, Mixed-species, Nonspatial Tree Models

The Ek and Monserud (1974a,b) FOREST model of mixed-aged, mixed-species stands is an exceedingly elaborate model with considerable detail in the simulation of forest processes. Horn's (1975a,b, 1976) model of the

Table 2.3. Principal subroutines in the FOREST model

MAIN	Determines height, diameter, and crown development of overstory trees.
INPUT	Accepts parameter values for each species, primarily for overstory development.
STANGN	Accepts real tree input data or generates spatial patterns (clustered, random, uniform, and so on) and tree characteristics for each species.
HOWFAR	Determines distance between points on main plot and buffer zone that are needed for evaluation of competition and seed and sprout distribution (eliminates plot edge effects).
COMPE	Evaluates tree competition.
YIELD	Calculates timber product yields based on individual tree dimensions, specific gravity, and bark characteristics.
STAT	Computes parameters of distributions of tree and stand characteristics for summary output.
CUT	Orders trees by size or increment for pruning or harvest treatments and implements these treatments on individual trees by species. Harvest options include row thinning, selection according to specified criteria, spacing rules, cuts to basal area levels, and combinations of the above. The timing and degree of cutting may be set by the user or allowed to vary as dictated by stand development.
OUTPUT	Prepares table, stem map, and graphic output that describe stand development.
REPRO	Accepts input of initial reproduction status, reproduction parameters for each species, and specifications for degree and timing of any changes in reproduction parameters to be implemented during the run.
PSEED	Determines seed and sprout production for each overstory tree as a function of species, size, and threshold age.
SEEDYR	Generates seed year multiplier for each species, (i.e., frequency of good, moderate, and poor seed years).
DSTRIB	Distributes seeds and sprouts (root suckers and basal sprouts) from each overstory tree to subplots within main plot.
GRMIN8	Calculates seed germination as a function of microsite and overstory cover conditions.
GROW	Controls growth and mortality of reproduction until surviving individual reproduction stems reach overstory status—then, MAIN assumes control of stem development.

From Ek and Monserud [1974b].

same kind of forest contrasts sharply in its simplicity. Horn (1975a,b) viewed the species dynamics of a forest as a problem in determining the likelihood of whether a canopy tree of a given species would be replaced by the same or another species in the next generation. Horn (1975a,b,

1976), as well as Acevedo (1978), modelled this view of nature as a first-order Markov process. An important feature of this model is the time-invariant nature of each of the replacement probabilities. If one allows sufficient time, this type of model will converge to a stationary composition (Peden, et al [1973]). This stationary composition can be compared against old forests for a test of the model (Fig. 1.2).

The tendency for mixed-age and mixed-species models to be reasonably abstract is found in several other independent approaches to modeling such forests (e.g., Leak [1970], Forcier [1975], Leak and Graber [1976], and Connell and Slatyer [1977]).

Mixed-aged, Mixed-species, Nonspatial Gap Models

Nonspatial gap models are among the earliest forest succession models developed; they are a logical outgrowth of the now traditional interest in vegetative community classification (Whittaker [1978]). An example case is the model of Waggoner and Stephens (1970) for a northern hardwood forest. While Horn's initial development of a Markov model for individual canopy tree replacement (Horn [1975a,b] was based on the development of a theory to estimate tree-by-tree replacement probabilities, Waggoner and Stephens (1970) used a set of forest-dynamics remeasurement data from the Connecticut Agricultural Experiment Station to directly compute the change of small (0.01 ha) tracts from one class in a classification to another over a 10-year interval (1927–1937). The stands were classified according to the most numerous tree species above 12 cm dbh (diameter at breast height). The performance of this Markov model was reasonable in predicting the composition of the forest types on the landscape at a subsequent remeasurement in 1967 (Fig. 2.6). Waggoner and Stephens' approach is largely empirical, and it indicates the variability that is inherent in forecasting landscape composition based on the observed change in landscape elements over a relatively short period of time. A principal source of deviation in the predicted and observed percentages of the various vegetation types is that oaks (*Quercus* species) generally declined in the 1937–1967 period at a rate much greater (due to drought and defoliation by gypsy moths) than the transition probabilities that were calculated on the 1927–1937 remeasurement period would indicate. The variability in the predictions versus the observed forest transitions also could arise from difficulties in computing the transition probabilities. One problem with Markov models and their parameterization, which uses real forest data, is the fact that the number of parameters to be estimated is $N(N - 1)$ where N is the number of states. In Waggoner and Stephens' case, the 40-

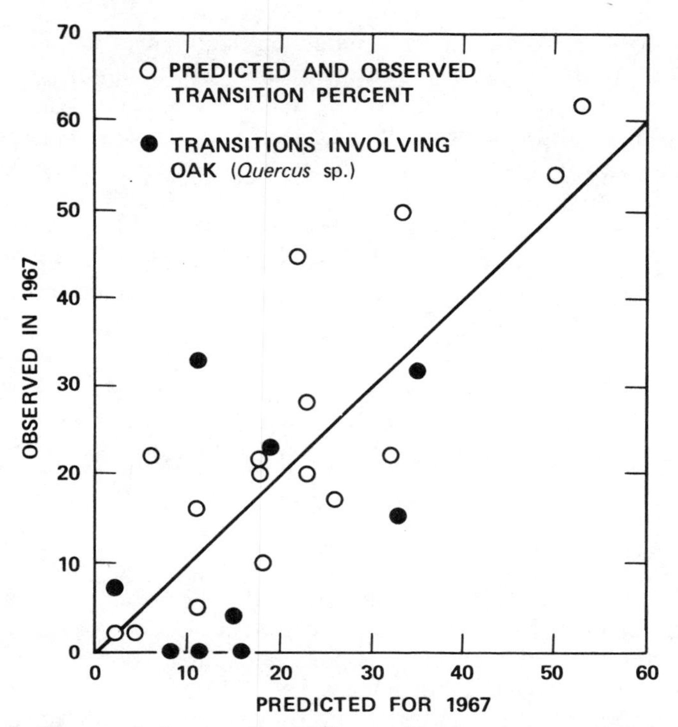

Figure 2.6. Changes in forest composition predicted and observed by using Waggoner and Stephens' Markov model for forest type change on small (1/100 ha) tracts of forested landscape. Each point represents a transition (percentages of forests in one type that should be in another type after 40 years). Transitions involving oak (*Quercus* species) are shaded. (Data from Table 1, Waggoner and Stephens [1970].)

year remeasurement data set was comprised of 327 1/100-ha sample quadrats, or an average of only 16 data points, to estimate each transition probability. If the transition probabilities are functions of environmental variables, or if the process is a higher order Markov process (i.e., knowledge of previous systems states as well as the present system state are necessary to determine the future state), the parameter estimation problems become virtually insurmountable (given the size of most forest remeasurement data sets). In a cautionary concluding statement, Runkle (1981) points out, "As a way of predicting equilibrium community composition a Markov approach should be used with caution. No one way of computing transition probabilities seems to work for all species in all communities," and later, "In many cases the species of tree creating a gap seemed to influence the species composition of its likely successors (saplings within the gap). However, for the most part, these relationships consisted of significant tendencies toward self-replacement. . . ."

What is needed to extend Waggoner and Stephens' very useful conceptualization of a forested landscape as a stochastically changing, perhaps Markovian, mosaic is an efficient way to estimate the nature of the compositional change in the mosaic-element—the gap. Horn (1975a,b) used a theory that the composition of the saplings beneath a tree could be computed to estimate the transition probabilities. This straightforward idea has fueled considerable theorization (Forcier [1975], Fox [1977], Woods [1979], and Woods and Whittaker [1981]), particularly in forests of low species-diversity; however, Runkle's (1981) comments (in part quoted above) would indicate that caution is appropriate in applying this theory to diverse forests. Kessel (1976, 1979a,b, 1981a,b) and Kessel and Potter (1980) took the idea of viewing a landscape as a dynamically changing mosaic and (using an intuitive approach to estimating the transition probabilities) applied landscape models to actual management in the United States, Canada, and Australia.

There presently is a need for a predictive understanding of how the vegetation on a small gap (or "patch," or "mosaic" element, and so on) will change over time. One approach to this problem was put forward by Noble and Slatyer (1978, 1980) and is called the "vital attributes" approach. Noble and Slatyer's theory has historical antecedents in Humboldt (1807), Grisebach (1838), and (particularly) Warming (1909) and Raunkiaer (1934). It is philosophically allied with Gleason's (1927) emphasis on the individual properties as important determinants in succession. Noble and Slatyer (1980) recognize three vital attributes of a species:

1. The method of arrival or persistence of a species at a site during and after a disturbance.
2. The ability to establish and grow to maturity in the developing community.
3. The time needed for an individual of the species to reach critical life stages.

The exact biological mechanisms to attain the particular set of vital attributes that are associated with a species potentially varies among species. It is the end result of these varied mechanisms that actually is manifested as a species vital attributes. Noble and Slatyer (1978, 1980) encode each of these vital attributes with a standard symbol set. For example, a species that has a short-lived seed pool (C) and is intolerant (I) would be denoted CI. This type of species would need to be established (i.e., would need to have individuals at a critical stage of bearing seeds) at the time of an initial disturbance to survive. Noble and Slatyer's method has proven to be useful in predicting major changes in species composition and dominance patterns in regularly disturbed plant communities in Australia (Noble and Slatyer [1978, 1980]) and in North America (Cattelino, et al [1979]). This method essentially is an abstract theory for determining, from species attributes, the same sort of transition probabilities that Waggoner and Stephens (1970) attempted to compute directly from remeasurement data. However, Noble and Slatyer's (1980) method appears to have its best applications in cases in which the successive communities differ in their physiognomy. In this sense, the success of the method in predicting the temporal sequence of community change is directly analogous to a geographer's (*see* Chapter 1) success in predicting a spatial pattern of community change as a climate/physiognomy response (e.g., Raunkiaer [1934], and Box [1981]).

Mixed-aged, Mixed-species, Spatial Gap Models

Because the spatial gap models are discussed throughout the remainder of this volume, their development and underlying assumptions will be treated in detail in Chapter 3. The fundamental approach to developing these models was outlined by Botkin, et al (1972a,b), and it was reiterated by Shugart and West (1977). Basically, gap models simulate the annual change on a small plot by calculating the growth increment of each tree, by tabulating the addition of new saplings to the stand (both from seeds and by sprouting), and by tabulating the death of trees. All of these processes are considered to be stochastic. The philosophy that underlies the model construction of spatial gap models is to strive to represent dynamic phenomena by using general equations that can be parameterized from a knowledge of basic physiology, morphology, or forestry; thus they do not require elaborate data sets for parameter estimation (note the kinship with the Noble and Slatyer [1978, 1980] approach discussed above). The usual approach is to reserve data for independent tests on the models (*see* Chapter 4 for more details) and to use the models for applications, when possible. In Chapter 3, several different gap models are discussed, and some of the diversity of approaches that are taken within the gap model framework also will be indicated.

Conclusions

In this chapter, an attempt has been made to communicate an impression of the diversity of approaches that are available for simulating the development of forests by simultaneously tracking the growth of trees and the demography of forest tree populations. It seems inevitable that a review of such a variety of approaches to such a fundamentally similar problem invites competitive comparisons. One is tempted to ask, "Which model is best?'' This question may be logical, but it is not particularly fruitful at present. Depending on what criterion one uses to judge a ''good'' model, there are good models that are very complex or that are very simple. There are forest models that clearly are beyond the present level of data collection for forests. There are models that are designed to mimic the behavior of forests only within the constraints of particular data sets. The number of parameters in these models may be large or small; they may increase as a linear function of the number of tree species (as in the complex Ek and Monserud [1974a,b] model) or as the square of the number of species (as in the Horn [1975a,b] model). The quality of a model in this context always is judged in relation to the purposes for which the model was designed. For this reason, comparisons across models of very different sorts often are onerous.

In the next two chapters, the development of gap models (Chapter 3) and their testing (Chapter 4) will be discussed. Gap models are constructed with a similar fundamental approach to the other models that were discussed in this chapter; namely, the unification of population processes with the growth processes of individual trees. The models are reasonably complex, but they are less so than some of the other cases discussed in this chapter (Table 2.2).

Chapter 3

Gap Models

The concept of a forest gap or "gap phase" is attributed to Watt (1947) who used the term to refer to a patch in a forest created by the death of a canopy tree. Gaps become localized sites of regeneration and subsequent growth. Watt (1947), Bray (1956), and Curtis (1959), as well as the earlier observations of Jones (1945), indicated that mature forest ecosystems could be seen as the relatively consistent average of the responses of the dynamics of such gaps. Several tree species appear to depend heavily on their ability to exploit locally favorable conditions to survive in mature forests. These species include: 1) Yellow poplar, *Liriodendron tulipifera* (Skeen [1976]); 2) Pin cherry, *Prunus pensylvanica* [Marks 1974]; and 3) Black cherry, *Prunus serotina* (Auclair and Cottam [1971]). Runkle (1981), in an extensive study of gap regeneration of forests in the eastern United States, found that a "regeneration in small gaps was sufficient to perpetuate the current canopy tree species of the stands studied."

If the concept of gap dynamics as a driving phenomenon of the larger compositional pattern in temperate forests has support, then the case is even stronger for tropical systems. Several reviewers (Jones [1950], Richards [1952, 1969], Poore [1964, 1968], Ashton [1969], Baker [1970], Gomez Pompa, et al [1972], and Whitmore [1975]) have noted the importance of gap dynamics in the formulation of a dynamic climax forest concept for tropical forests. Medway (1972 for Malaysian forests), Brunig (1973 for swamp forests in Sarawak), and Leigh (1975 for several tropical forests) actually have determined rates of gap creation. Williams, et al (1969), Webb, et al (1972), and Richards and Williamson (1975) used gaps (both naturally and artificially generated) as objects of intense study for understanding tropical rain forest dynamics.

In addition to the importance of gap phase replacement in forests as a natural phenomenon, the notion of representing a spatially distributed dynamic (e.g., the dynamic changes in a forest ecosystem that are distributed across a landscape) as an aggregate response of homogeneous mosaic patches is a standard simplification for mathematical modeling. If the time and space dynamics of a forest were represented by partial differential equations (Smith [1980] provides an example in his study of gradient responses of forests), one might solve such equations by using a finite element approximation. Such an approximation amounts to solving the spatially continuous processes that are implied by the partial differential equations over regular patches.

Environmental patchiness or graininess is a fundamental concept in modern theoretical ecology. Spatial heterogeneity has been implicated as a major factor in species survival (Andrewartha and Birch [1954], and MacArthur and Wilson [1967]). The size of a theoretical grain or patch in natural environments has been variously defined by boundaries in space (e.g., edges between forests and fields, Curtis [1956]) or by the properties of a given species (e.g., environmental graininess quantified by foraging patterns in grassland birds, Cody [1968]). Levins (1968) combined these two approaches by considering a coarse-grained environment to be one in which an individual may be expected to spend its entire life in one habitat. Shugart and West (1979) found that, for gap models, the assumed size of the modelled gap could have a profound effect on the model dynamics; they noted that these effects could be considered to be a "phenomenological basis for grain-size quantification based on community-level interactions." The importance of spatial scale in developing gap models will be discussed at the end of this chapter after the mechanisms that are used in the models have been presented.

The Growth of Trees

As with most individual tree-based models, gap models require a basic equation to increment the size of each tree on the modelled stand. For all gap models developed to date, there have been two approaches to developing these equations. The initial approach (and the one used in most gap models) was developed by Botkin, et al (1972a,b). The size variable of interest is the diameter of each tree and its annual increase as a function of weather, competition, and other factors. The growth equation is developed by assuming that tree volume is a function of a tree's diameter (D) squared times the trees height (H), and that tree growth is based on the annual volume increment. The tree volume increment:

$$\frac{d[D^2H]}{dt} = r\,La\left(1 - \frac{DH}{D_{max}H_{max}}\right) \tag{3.1}$$

where r is a growth rate parameter, La is a tree's leaf area (m^2/m^2), D is the diameter breast height, H is tree height, and D_{max} and H_{max} are the maxima for diameter and height, respectively. The structure of this equation implies that instantaneous volume increment is a linear function of leaf area (*see* Schulze [1982]), and that there is a cost associated with tree size that decreases tree growth. This approach approximates any number of finer time and space-scale interpretations of net photosynthesis as the difference between the photosynthate synthesis (a complex function of leaf area) and photosynthate demand for nongrowth-related respirations. Kozlowski (1971a) states this proposition well in summarizing the relation of leaf development and cambial growth (particularly the work of Satoo and Senda [1958], and Satoo, et al [1956, 1959]). He states, "Although stem wood production was not linearly related to the amount of leaves per tree, when branchwood and leaf production were added to it and hence dry weight increment considered, growth and the amount of foliage approached a straight line relationship. This suggested that the lower efficiency of foliage of dominant trees in producing stem wood was caused by differences in distribution of dry matter in various parts of the tree and by respiratory losses of nongreen tissues."

The basic growth equation can be simplified by noting that the height (H) is a function of tree diameter (D). One useful formulation for this relationship is (Ker and Smith [1955]):

$$H = 137 + b_2D - b_3D^2 \tag{3.2}$$

where b_2 and b_3 are parameters quantifying tree form and the constant 137 (in cm) is breast height (so that a tree of 0.0 dbh is 137 cm tall). If we assume that a tree has its maximum height when it has its maximum diameter ($dH/dD = 0$ and $H = H_{max}$ when $D = D_{max}$), then it is possible to solve for the two parameters,

$$b_2 = 2\left(\frac{H_{max} - 137}{D_{max}}\right) \tag{3.3}$$

and

$$b_3 = \left(\frac{H_{max} - 137}{D^2_{max}}\right) \tag{3.4}$$

as functions of the fairly readily obtainable observations, D_{max} and H_{max}. If it is further assumed that,

$$La \sim c\, D^2 \tag{3.5}$$

where c is a constant, then Equation 3.1, on substitution and differentiation, becomes:

$$\frac{dD}{dt} = \frac{GD(1 - DH/D_{max}H_{max})}{(274 + 3b_2D - 4b_3D^2)} \tag{3.6}$$

with G (a growth parameter) being used for the product of constants, rc. One advantage of Equation 3.6, along with its straightforward albeit simplistic derivation from a tree's energy balance, is the ease of parameter estimation. Given a knowledge of the maximum size of a tree species, one needs only to estimate a single growth parameter, G. Presumably, H_{max} can be estimated from D_{max}, based on the structural-mechanical constants of tree geometry (MacMahon [1973]); however, such an estimation has not been required in any gap models to date.

Along with the obvious application of fitting the G parameter to actual data (Fig. 3.1), the parameter G also can be determined by two fairly straightforward means. If a tree grows in the manner indicated in Equation 3.6, and if dD/dt = 0 at the maximum age, its optimal growth curve will reach approximately two-thirds of its maximum diameter (D_{max}) in one-half of this its maximum age (AGEMAX). Botkin, et al (1972a,b) solved Equation 3.6 for G to find:

$$G = \frac{4H_{max}}{AGEMAX}\Bigg\{\ln[2(2D_{max} - 1)] + \frac{a}{2}\ln\left(\frac{9/4 + \alpha/2}{4D_{max}^2 + 2\alpha D_{max} - \alpha}\right) + \left(\frac{\alpha + \alpha^2/2}{\sqrt{\alpha^2 + 4\alpha}}\right) \times \ln\left(\frac{(3 + \alpha - \sqrt{\alpha^2 + 4\alpha})(4D_{max} + \alpha + \sqrt{\alpha^2 + 4\alpha})}{(3 + \alpha + \sqrt{\alpha^2 + 4\alpha})(4D + \alpha - \sqrt{\alpha^2 + 4\alpha})}\right)\Bigg\} \qquad (3.7)$$

where $\alpha = 1 - 137/H_{max}$.
This relationship is particularly useful in estimating AGEMAX from G, or vice versa. Another useful approximation is:

$$\delta D_{max} \sim .2G\ D_{max}/H_{max} \qquad (3.8)$$

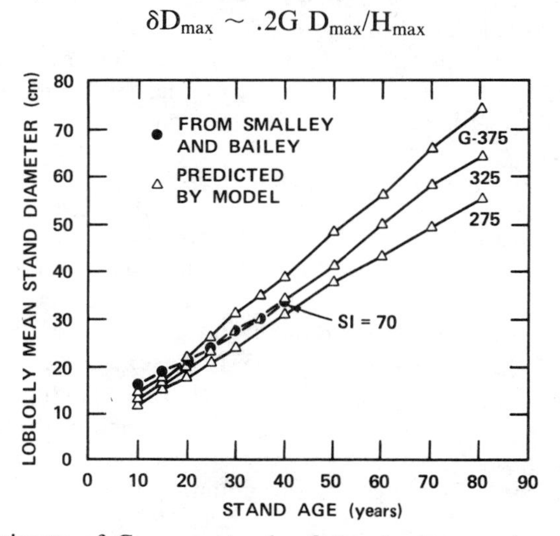

Figure 3.1. Estimate of G parameter by fitting basic growth equation against growth data. Site index (SI) = 70.

where δD_{max} is a maximum diameter increment that can be obtained from tree-ring or remeasurement data (Botkin, et al [1972a,b]).

Equation 3.6 is used in all of the currently published gap models as the optimal growth equation with the exception of Phipps' (1979) SWAMP model. Phipps uses a modified paraboloid model of tree geometry (Phipps [1967]), constrained in such a way that the areal cross section of each ring added by a tree is a constant (after the tree has reached a given size). In fact, one probably could use any number of optimal growth formulations if it is given that they approximated the general pattern of the growth of trees in nature. The critical problem is one of estimating the growth parameters for a large number of tree species with more or less equivalent accuracy. Because gap models are stochastic in both their parameters and structure, they defy closed-form solutions and are solved by Monte Carlo techniques. For this reason, the exact mathematical form of the growth equation (given an appropriate basic shape) does not arise as the consideration it might be, if the equations were solved analytically.

Light Limitation of Tree Growth

Given a basic equation for the growth of an individual tree (Equation 3.6), gap models incorporate the interaction of trees with one another and with their environment by modifying the growth that is predicted by this species-specific optimal growth equation. One important consideration is the shading relation among trees; it is in computing these interactions that the spatial element of gap models arises. Gap models are geometrically explicit in the vertical dimension and use the height of trees to compute shading and the consequent reduction in individual growth. The light that reaches a given tree is calculated by attenuating the incident radiation by the sum of leaf areas for all trees taller than the tree:

$$Q(h) = Q(o)e^{-[k\int_{h}^{\infty} La(h')dh']}, \tag{3.9}$$

where La(h′) is the distribution of leaf area as a function of height, Q(o) is the incident radiation, Q(h) is the radiation at height (h), and k is a constant.

Equation 3.9 is Beer's law, and it often is used to compute shading in forest canopies (Monsi and Saeki [1953], Kasanaga and Monsi [1954], Loomis and Williams [1969], and Anderson [1971]). Having computed the light falling on a given tree (usually as a percentage of the ambient), the gap models use photosynthesis light-response curves to decrease the growth of that tree. The form of the equation used in most of the gap models is:

$$r[Q(h)] = c_1(1 - e)^{c_2[Q(h)-c_3]}, \tag{3.10}$$

where Q(h) is the light available to a tree of height (h), c_1 is a scaling constant, c_2 is a parameter that increases as the rate of photosynthesis in response to increased light, and c_3 is the compensation point (i.e., the light level at which photosynthesis and respiration are balanced).

Equation 3.10 provides good fits to many measured photosynthesis curves when scaled so that optimal light conditions are equal to 1.0 (Fig. 3.2). In the gap models Equation 3.10 is used to diminish the growth of each tree as a function of departures from the optimal, which depends on the available light (computed by Equation 3.9). This application has its justification in the hope that the annual, whole-tree response to shading mimics the smaller time- and space-scale photosynthetic response of leaf surfaces (*see* Farquhar and von Caemmerer [1982]). Kramer and Kozlowski (1960) discuss the idea of shade tolerance and note, "In one accepted usage, that which defines tolerance as the capacity to endure shade, a tree which reaches maximum photosynthesis at relatively low light intensity is tolerant, while one whose rate of photosynthesis continues to increase with each added increment of light up to full sun is considered intolerant," and furthermore, "It is not the intent of the authors to defend a chosen definition of tolerance but merely to indicate that reactions of photosynthetic mechanisms of different species to environment are implicit in the term, however it is used." In a more recent review of physiology and plant succession, Bazzaz (1979) notes that the photosynthetic rate generally declines with the advancing successional status of

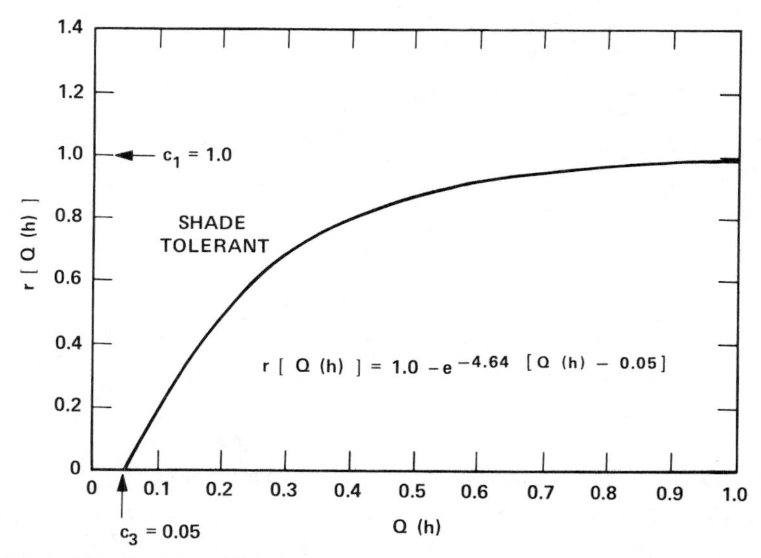

Figure 3.2. The function r[Q(h)] represents the percent of maximum photosynthesis rate plotted against Q(h), which is the percent of incident light available to a tree for a shade-tolerant tree.

the plant; also, there was an expected difference in the shape (e.g., slope, light-compensation point) of the photosynthesis light-response curve (Fig. 3.3a) for plants that were associated with early or late successional situations. Bazzaz further noted that the rate of photosynthesis generally declined with increased shade tolerance.

Horn (1971) inspected the consequences of the geometry of leaf placement on trees in terms of net photosynthesis. While investigating a theoretical model, he found (Fig. 3.3b) that monolayered tree geometries (tree with leaves held in a single layer) differed in the shape of the photosynthesis/light response from multilayered trees. This difference occurred even

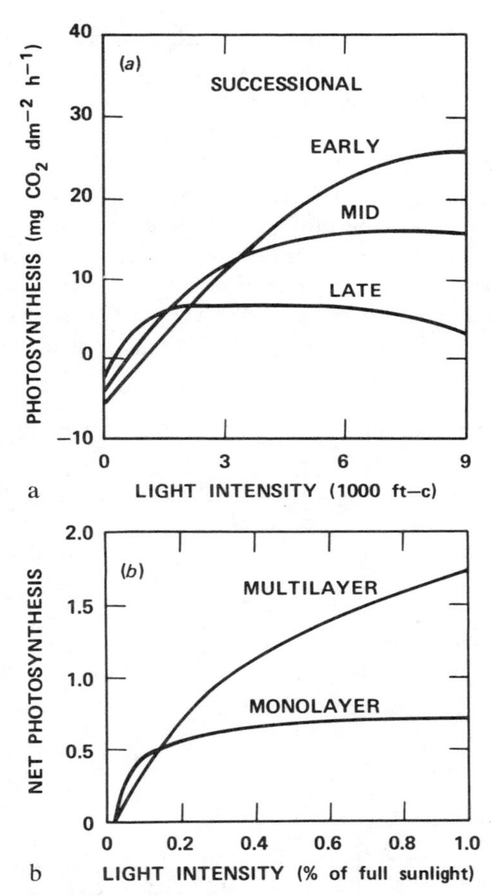

Figure 3.3. (a) Idealized light-saturation curves for early, mid, and late successional plants. (From Bazzaz, F. A.: *The physiological ecology of plant succession in Annual Review of Ecological System*. Annual Reviews Inc., 1979, vol. 10.) (b) The effect of light on net photosynthesis in multilayered and monolayered trees. (From Horn, H. S.: *The Adaptive Geometry of Trees*. Princeton University Press, 1971, Repriented by permission of Princeton Unversity Press.)

when the photosynthetic responses of the leaf surface were identical. Fairly large variations in the parameters of his basic photosynthesis response model did not alter the relative behavior of the multilayer and monolayer curves, as shown in Figure 3.3b.

Thus, it appears that with both physiological and morphological considerations (and probably from a combination of the two), trees can be assigned to different types of light use that are related to the shape of the photosynthesis/light curve. Starting from a basic shape (see as Fig. 3.2), one would expect a shade-tolerant response that was relatively more productive under lower light conditions, but less productive under higher light conditions, than the corresponding shade-intolerant response. The gap models use curves (e.g., Equation 3.10) that incorporate these differences, fit the curve parameters to normalized photosynthesis responses, and then use the forester's concept of tolerance [many of the North American models use Baker (1949)] to assign the appropriate parameterization to the tree species.

Temperature Effects on Tree Growth

In both the Botkin, et al (1972a,b) JABOWA model parameterization and in several subsequent gap models (Shugart and West 1977, Shugart and Noble 1981), the variation in the tree species range with respect to the growing degree-day heat sum (the annual sum of daily departures of temperature above a 40°F base temperature) isopleths used to scale the tree growth response. In most cases (Fig. 3.4), species range boundaries correspond to isotherms of growing degree-days. The growth responses between these extremes is assumed to be parabolic and of the form:

$$T(D_{40}) = \frac{4\left(D_{40} - D\frac{min}{40}\right)\left(D\frac{max}{40} - D_{40}\right)}{\left(D\frac{min}{40} - D\frac{max}{40}\right)^2} \qquad (3.11)$$

where $T(D_{40})$ is the growth reduction due to temperature effects, D_{40} is the 40°F base heat sum for a site, D(min/40) is the minimum degree-day value where the species is known to occur, and D(max/40) is the corresponding maximum.

This parameterization currently is best viewed as an expediency in estimating a generalized temperature growth response across a large number of tree species; Salisbury's (1926) caution certainly is valid: "The mere correspondence of the limits of distribution of a species or community with isothermal lines, or any other climatic limits, is at best presumptive evidence and no proof of any causal connection between them" . Probably, as more data become available, the calibration of Equation 3.8

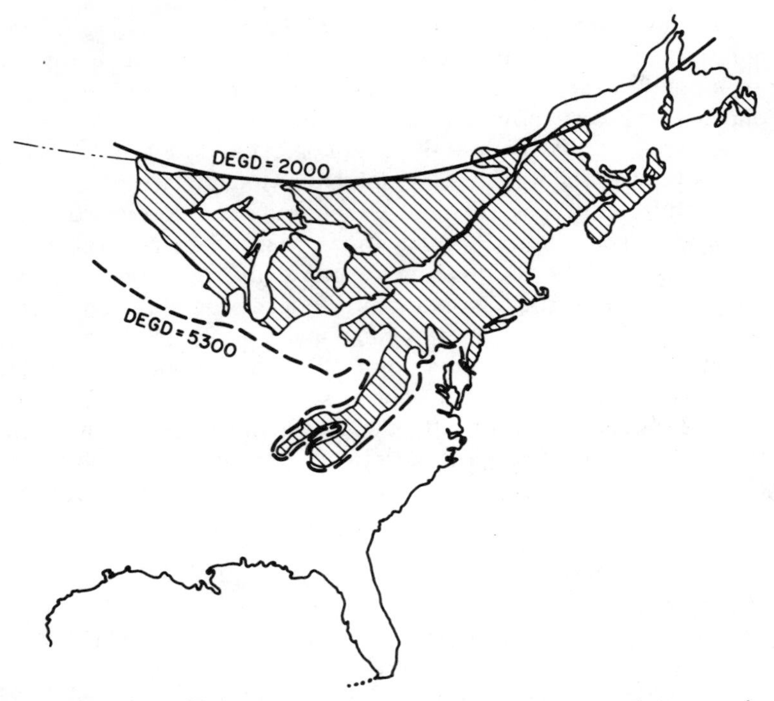

Figure 3.4. The geographic range of *Betula alleghaniensis* and the growing degree-day (variable DEGD) values of 2,000 and 5,300 which closely approximate the north and south boundaries of this species. Range map from Fowells (1965); isotherms used to compute the DEGD lines from the U.S. Department of Commerce publication (1968). (Figure from Botkin, et al [1972b]. © 1972 by International Business Machines Corporation. Reprinted with permission.)

from maximum tree-ring data as a function of climatic data or the direct application of a tree-ring/climate statistical transfer function (Fritts [1976]) could be used to develop the climate/growth response. In this case, one would substitute a statistical predictor of tree ring growth as a function of age, site factors and climate for the growth equation. Phipps (1979) anticipates this latter method in using a ''climatic noise'' modifier for his growth rate equation that is derived from a residual tree-ring variation record to simulate ''natural'' year-to-year climatically induced variations in tree-ring width.

Nutrient Cycling and Growth

All gap models at least give abstract consideration to the growth-limiting effects of shortages of some critical plant nutrient. For example, in the JABOWA model, Botkin et al (1972a,b) used an upper value for the total biomass on the simulated plot to approximate the effects of a hypothetical limiting nutrient. The principle applied was that as the biomass approached this maximum (called SOILQ in the JABOWA formulation), the short-fall in supply of some nutrient would cause a simultaneous reduction of all tree-growth rates as the element became limiting. Most other gap models follow a similar rationale to limit the growth. Aber, et al (1978, 1979) added more explicit consideration of element cycles to the JABOWA model, as did Weinstein, et al (1982) for the FORET model. Both of these additions tie the ability of the gap models to forecast biomass growth (and the consequent demand on soil nutrient reserves) to the traditional ecosystem ecologist's interest in the cycling of important elements in natural systems.

The Weinstein, et al (1982) FORNUT model uses a polynomial function of the form:

$$GMF = a + b[RNA] + c[RNA]^2 \tag{3.12}$$

where a,b,c are constants estimated by regression from field data, RNA is the relative nutrient availability, and GMF is the growth-modifying factor to modify the growth rate of trees under nutrient limitation.

This function initially was used by Mitchell and Chandler (1939) to distinguish three classes of tree response to fertilization. Each of these classes (Fig. 3.5) have different parameter values for the a,b,c parameters of Equation 3.12. They are: 1) Nutrient-tolerant species having a slight response in growth with an increased nutrient availability and little growth reduction with low nutrient levels; 2) Nutrient-intolerant species having marked growth response with increased availability, but low growth rates at low levels of nutrients; and 3) An intermediate between the two other classes. Weinstein, et al (1982) classified some 48 species of trees and shrubs from the southern Appalachians Mountains according to this classification scheme. The curves for each class were applied to limit each tree's growth in each year if any of five elements (i.e., calcium, magnesium, potassium, phosphorus, and nitrogen) were in insufficient supply to meet the amounts needed for tree growth. Following Liebig's (1840) Law, only one element was assumed to be limiting in any year. The actual computation of the growth reductions called for two sorts of considerations.

First, the demand for the five elements was determined. This demand was computed by calculating the biomass and elemental content (five elements) for the fine roots, medium roots, large roots, leaves, branches, and boles of each tree. The demand for nutrients (uptake) is the change in these elemental pools from the growth of new tissue. Another use of these computations is to determine the return of elements to decomposition processes that are due to litterfall, tree death, and so on. The formulations for calculating these elemental pools are based on data from Shanks and Clebsch (1962) and Harris, et al (1973).

Second, the availability of the five elements was determined by calcu-

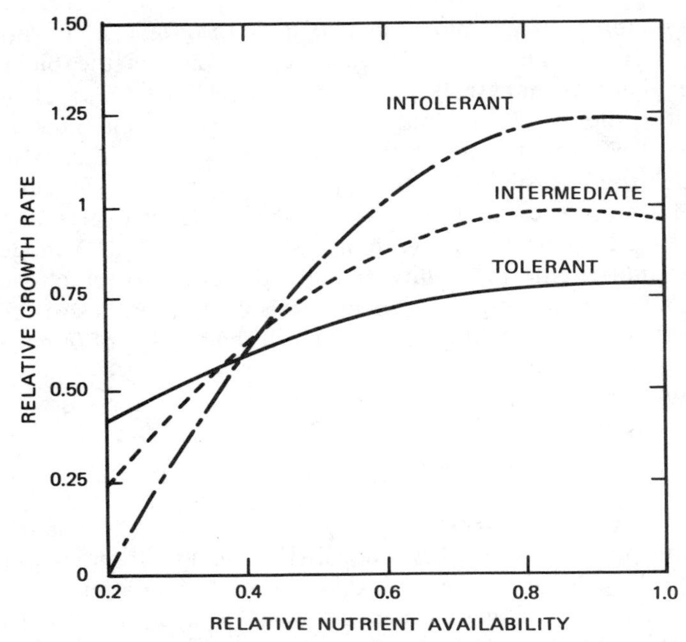

Figure 3.5. Relationships used in the FORNUT model for reducing growth as a function of nutrient availability. (From Weinstein, et al [1982].)

lating the decomposition of six litter types (leaf litter, bole, branch, fine root, midsized root, and large root) by using the Meentenmeyer (1978) equation of decomposition as a function of lignin content and actual evapotranspiration. Decomposition was elevated in wet years (by 35%) and similarly diminished (by 35%) in dry years. The nutrients that were released by primary decomposition are transferred to the secondary decomposition zone for possible nutrient immobilization. The decomposed litter, as well as elemental inputs to the ecosystem due to precipitation, throughfall, and any nutrients released by fire, are subjected to immobilization by the soil microbial organisms. This immobilization is based on the amount of carbon that is available to these organisms. Modelling carbon disappearance as a function of the lignin-nitrogen ratio (Swift, et al [1979]) causes the nutrient concentrations to tend toward an ideal carbon-nitrogen ratio. Nutrients that are not immobilized are released and are available to meet the nutrient uptake demands of trees. The model also accounts for the dynamics of soil carbon, adsorption of used nutrients within the soil, release (or weathering) from the nonexchangeable soil pool, and loss from the system. By this fairly detailed modelling of the soil decomposition processes, it is possible to compute the supply of nutrients that are available to satisfy tree uptake demands.

The RNA index (Equation 3.12) is the nutrient availability expressed as a percentage of a 1-year nutrient demand. When the demand for nutrients is met, then the RNA index is set to a value of 1.0 and the growth is not limited by nutrient shortage. Otherwise, each individual of the various species is subjected to reductions in growth according to the curves shown in Figure 3.5.

The FORTNITE model (Aber and Melillo [1982a]) also determines nutrient availability by simulating decomposition. The model modifies growth according to nutrient availability. FORTNITE simulates the nutrient dynamics of nitrogen in a northern hardwood forest, and it is derived from the JABOWA (Botkin, et al [1972]) model. Decomposition of the annual litter input ''cohort'' is computed on annual time steps. The weight loss of each species' litter within each cohort is determined by using loss rates that are determined from field studies using litterbags. The nitrogen immobilization-mineralization dynamics of each cohort is related to the percent of dry weight remaining at the end of each year by linear functions. The parameters for the rate of litter lost are functions of litter quality (Aber and Melillo [1980, 1982b]). Each cohort of each species' litter is tabulated in a separate simulated forest-floor litter layer as long as it immobilizes N. When N begins to mineralize from the cohort, the material is transferred to a soil humus pool with a mineralization rate of 9.5% per year. Subsequent modifications of the FORTNITE model that are based on more recent field data (Pastor, et al [1984]) tie soil-humus N mineralization to litter carbon-nitrogen and humus phosphorus-nitrogen ratios. Nitrogen availability limits tree growth via growth multipliers (Aber, et al [1979]) that are derived from Mitchell and Chandler (1939).

Moisture Effects on Tree Growth

As was the case with incorporating nutrient effects in a gap model, there are two problems to be solved in incorporating the effects of moisture on tree growth: 1) One must simulate the dynamics of soil moisture at a time and spatial scale that is compatible with the other parts of the gap models; and 2) One must be able to parameterize a model of the effects of soil moisture (or lack thereof) on the growth of each tree in the simulated stand. In treating the first problem, the annual time step of the gap models generally proscribes any number of detailed soil moisture models. In a modification of the FORET model, Post and Mann (unpublished manuscript) used a single layer model of the soil moisture and solved this model at monthly intervals.

The amount of moisture that a soil holds is determined by the depth of the rooting zone (SD) and the properties of that particular soil. The volume of water in the soil that is potentially available for plant growth is the difference between the water volume at field capacity (FC) and the volume when the soil water becomes limiting to growth (DRY). DRY is the volume of water at a soil water tension of −15 atm, since this parameter is reported for many soils and is close to the wilting point for crops. The available water for plants (AW) is defined as:

$$AW = (FC - DRY)\ SD \tag{3.13}$$

The balance between rainfall additions to the soil water and the subtraction that is due to evapotranspiration and runoff is solved on a monthly basis. Rainfall on an unsaturated soil is added to the soil water. If the soil is saturated, then any additional rain is lost to runoff or to subsurface flow. Evapotranspiration is calculated from the average monthly temperature by using the Thornthwaite and Mather (1957) equation.

Given monthly rainfall and temperature data and the soil parameters to characterize the moisture-holding properties of that soil, one can compute the number of days during the growing season in which there is insufficient soil moisture (Fig. 3.6). This computation is used to limit the growth of trees.

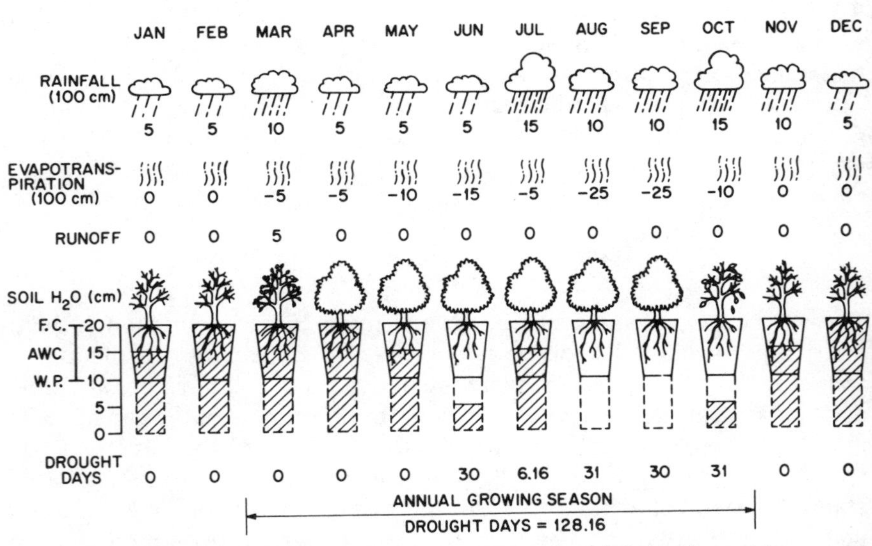

Figure 3.6. Annual drought day calculation. Monthly rainfall indicated in cm and by size of clouds. Evapotranspiration computed by using Thornthwaite (1948) equation based on monthly temperature, rainfall input and latitude. FC is soil field capacity (−0.1 bar). WP is the wilting point (−15 bar). AWC is the available water capacity (based on the difference between FC and WP). Runoff occurs when rainfall is in excess of field capacity (*see* values for March). Drought days are computed by interpolation between points at the middle of each month (hence, the value of 6.16 for July).

Bassett (1964) found a linear reduction in the growth rate of pines ($r^2 = 0.95$) as a function of "no-grow days" or days during the growing season when soil moisture was insufficient. Based on Bassett's observations, as well as other studies (*see* Zahner [1968] for a review), Post and Mann modified the FORET model to reduce the growth of each tree as the function:

$$\text{SMGF} = \begin{cases} 1 - \dfrac{d}{d_{max}}; & d < d_{max} \\ 0; & d > d_{max} \end{cases} \quad (3.14)$$

where SMGF is the soil moisture growth factor, d is the number of no-growth days (days that the soil moisture tension is below the wilting point), and d_{max} is the maximum number of no-growth days that a species is known to tolerate.

To compute d_{max}, Post and Mann developed a map of the United States with no-growth days that were computed from county-level rainfall and temperature data, and also by using a soil description for a soil on the eastern edge of the Great Plains; d_{max} was determined for each species by finding the most extreme d value for a county in the species' range. This use of the species' range to scale the soil moisture response is compatible with the methods used to scale the temperature response of the species (Equation 3.11).

The Ingrowth of Trees

Harper's (1967, 1977) work in plant population biology created a considerable interest in the life history, reproductive strategies, and demography of plant populations—all areas of study which were formerly of interest to zoologists and applied botanists, particularly foresters. Most gap models use mathematically simple formulations that often are little more than computerized verbal descriptions of the establishment of trees. It is appropriate to discuss this simplicity in the face of a developing theory of plant reproductive strategy. It appears that gap models can, in most cases, capture the essential elements of tree reproduction for many species simply by tabulating the environmental conditions at a site during a given year and then matching this tabulation against the conditions that are associated with the species regeneration (e.g., in an account of the tree's silvics—Fowells [1965]). For example, several species of oaks (*Quercus* species) as well as other species, produce seeds that require leaf litter for germination. Also, these seeds are eaten by small mammals and insects; when there is a layer of leaf litter, some of these seeds are hidden and undetected. Thus, the seeds of these species are sheltered to a degree from what Janzen (1969, 1971) termed "seed predation." Other species for a variety of reasons (*see* Kozlowski [1971a,b]), tend to have their most successful germination in the absence of litter and in the presence of mineral soil.

Janzen (1979) reviewed the seed dispersal in *Ficus* species and noted that figs are dispersed as a thin sheet, with occasional peaks of very small seeds over much of the habitat (a diffuse but thorough seed shadow). In relation to one of the logical competitors to seedling trees—the herbaceous forest plants—a broad dispersal pattern would be typical of forest trees. A recent reviewer, Bierzychudek (1982), notes that many deciduous forest herbs lack any special seed dispersal mechanisms, and that the distance at which seeds are dispersed from the parent often is not more than the height of the stem. Bierzychudek also reported that the annual seed production of temperate forest herbs is in the range of 0–1,000 seeds per individual with a typical value of less than 100. Seed production in temperate forest trees is higher by several orders of magnitude (Fowells [1965]). Tree seedlings generally are physiologically less efficient in light use than their nonwoody competitors (Bazzaz [1979]).

These and other factors cause the establishment of trees from seeds to be chance events that are very difficult to develop mechanistically in a computer model. The gap models attempt to capture the important conditions for a given forest ecosystem as factors that filter species from being eligible or ineligible to become established as small trees. Species then are drawn randomly from the list of eligible species (for a given site and year), and a small number (usually 1–8) of individuals of each selected species are planted.

Filters that are used to obtain the eligible species (across all gap models) include:

1. Presence of mineral soil;
2. Presence of leaf litter;
3. Drought or an unusually hot year;
4. Abundance of seed predators;
5. Light level of forest floor;
6. State of forest seed bank;
7. Macroclimatic conditions at site;
8. Fire;
9. Presence of a seed source;
10. Seed phenology;
11. Presence of an elevated site (e.g., tip-up mound, tree trunk);
12. Presence of host for strangler fig;
13. Flooding;
14. Dispersibility; and
15. Episodic disease.

Generally, these filters are applied in a yes-or-no fashion to obtain the eligible species list, but in some models the establishment probability is modified continuously. Also, in most gap models trees are established at the size of small saplings (dbh ~ 2 cm); however, where appropriate, the population dynamics of small trees are followed by using a Leslie matrix approach until they reach an age at which they are large enough to be followed as an individual tree.

Along with planting seedling trees, the gap models also allow trees to sprout. It is taken into account that some species must be of a certain size to successfully sprout, and that some species lose the tendency to sprout with either size or age. Because both the size and age of each individual tree are followed in a gap model, these effects are easily incorporated. In the BRIND model (Shugart and Noble [1981]) of Australian *Eucalyptus* forest dynamics, lignotubers (large woody underground structures) are tabulated; their ability to sprout new trees is included in the model.

The Death of Trees

Due to the longevity and consequent low annual probabilities of mortality, the tree death rate is a difficult parameter to calculate. Gap models assume that tree mortality is a consequence of three processes. First, trees at any age have a probability of mortality that is related to their growth rates. Under optimal conditions, the growth equation (Equation 3.6) produces two-thirds the maximum diameter growth in one half of the maximum age (derived from Equation 3.7). From this relationship or from actual observations of tree longevity (e.g., Folwells [1965]), one can estimate the maximum age (AGEMAX) parameters. The constant basic probability of mortality that is applied each year results in an intrinsic survivorship curve for forest trees in the form of a negative exponential. One method to compute the mortality probability for a species is to assume that a certain percentage of a cohort (under optimal growth conditions) would never reach AGEMAX (*see* Equation 3.7); that is, i.e., the probability that a tree will be dead by the nth year:

$$P_n = 1 - (1 - \varepsilon)^n \tag{3.15}$$

where P_n = probability of mortality by year n, and ε = annual mortality probability.

When n is equal to the maximum age, and if one assumes that 2% of a cohort reach this age, then:

$$\varepsilon \sim 4.0/\mathrm{AGEMAX} \tag{3.16}$$

for a range of values of AGEMAX, including most tree longevities. The comparable probability that is associated with a 1% survival at AGEMAX is:

$$\varepsilon \sim 4.605/\mathrm{AGEMAX} \tag{3.17}$$

The original model of Botkin, et al (1972a,b) is used in Equation 3.16. The Shugart and West (1977) FORET model and the models discussed in this volume use Equation 3.17 to estimate the intrinsic likelihood of mortality.

A second source of mortality is an augmentation of intrinsic mortality when tree growth is below a certain minimum. Generally, the probability of mortality is set at 0.368 for an individual tree, which adds an increment of less than 0.1 cm/year. This results in a suppressed tree that has a 1% chance of surviving 10 years. The rationale for this increased mortality for slow-growing trees is that these trees have reduced photosynthate reserves to ward off disease and repair damage to tree structure; also, they lack vigor. Thus, if growth is greatly limited, the risk of death is increased. Along a similar rationale, the BRIND model increases the mortality of fire-scarred trees (and keeps track of the magnitude and degree of healing of the fire damage) based on the idea that scarred trees become more vulnerable to disease and insect attack. Phipps (1979) used the growth reduction/mortality increase as a sole source of tree mortality.

The third source of tree mortality is episodic. In the various models, trees are killed by harvest (including specific dbh and species-specific harvest prescriptions), by being damaged by felled trees, by wildfire, by hurricanes, by inundation, and by episodic diseases. These effects generally are simulated by tabulating the expected mortality that is associated with the event (by species and dbh when appropriate) and by simulating the occurrence of the event either deterministically (e.g., harvest) or stochastically (e.g., hurricanes). Occasionally, the events are generated by both approaches. For example, prescribed fires added deterministically, with a stochastic background of wildfires.

Size of Modelled Gaps

The important competitive interactions that are considered in gap models are of two types. One is due to the shading of shorter by taller trees and involves computing the canopy leaf areas and the light extinction rates that are associated with different forest canopies. The other type of competition is due to the use of the same potentially limiting resource pool by all of the trees in a forest stand. In general, the light extinction is calculated as if all the leaves on all of the trees on a plot shade the area of that plot; the resource competition is determined as if the limiting resource reserve is the one associated with that plot. This is implied by assuming that the plots are homogeneous with respect to light and nutrient competition. There is a difference in the scale at which these two types of competition can be treated in a gap model; these scales must be compatible over at least part of their range.

By experimentally varying the size of the simulated plot in a gap model, one can produce changes in the stand dynamics that are a function of the crown and resource competition. If the area of the simulated gap is arbitrarily small, then both shading and resource competition effects are made more intense because the negative effects of both types of competition (i.e., associated with the population of trees on the simulated plot) are increased. In the case of competition for light, the dominant canopy tree affects only subordinate trees. As light competition is intensified on small plots, the ability of a canopy-dominant tree to eliminate all subordinate trees also is intensified. This effect is directly attributable to the fact that the leaf area associated with each tree is assumed to shade only the gap in which that tree grows. Thus, with small plots, the leaf areas and the consequent intensity of shading are large. On small plots, and in response to the plot size-related distortion of the shading competition, the stand dynamics feature the growth of a single dominant tree that (with time) suppresses all of the other trees on a plot. When this dominant tree dies, there is a burst of regeneration, growth, and establishment; also, one particularly fast-growing tree becomes the new dominant tree. Repetition creates a cyclical birth-growth-death-birth cycle called gap-phase replacement. In arbitrarily large plots, the burst of regeneration and growth that is associated with the death of a large tree is diminished. This is a consequence of the fact that in a large modelled plot, the portion of the total leaf area associated with a single large tree is relatively small; the change in shading on the forest floor with the death of a single large tree also is small. When extremely large plots are used as the basis of a gap model, the resultant modelled stand dynamics do not feature gap-phase replacement.

Unlike shading competition, in which a dominant tree has a negative effect only on subordinate trees but neither on itself nor on taller trees, competition for resources acts on all of the other trees in the modelled stand; it also can self-limit the growth of a tree. In an arbitrarily small plot, if a dominant tree uses all or most of some available nutrients, its growth will slow and stop. If the modelled plot is too small, even a dominant tree will be unable to grow as large as it should; in fact, it will be unable to grow at all. The resource competition in a gap model sets a lower limit on the size of the modelled plot. The inclusion of gap-phase replacement in the model sets an upper limit on the plot size. Table 3.1 (from Shugart and West [1979]) illustrates the results of experimenting with the size of the plot used in the gap model. In this case, the two tests indicate that if gap-phase replacement is an observed phenomenon in a given forest, then the size of the model plots should be between 0.04–0.08 ha. In practice, the appropriate plot size seems to be smaller in wet tropical forests. This may be due to the fact that the steeper sun angles at low latitudes allow light to reach the forest floor in relatively small gaps. Generally, the most appropriate size of a plot to use for a gap model is that of a typical large-canopy dominant tree for the forest in question.

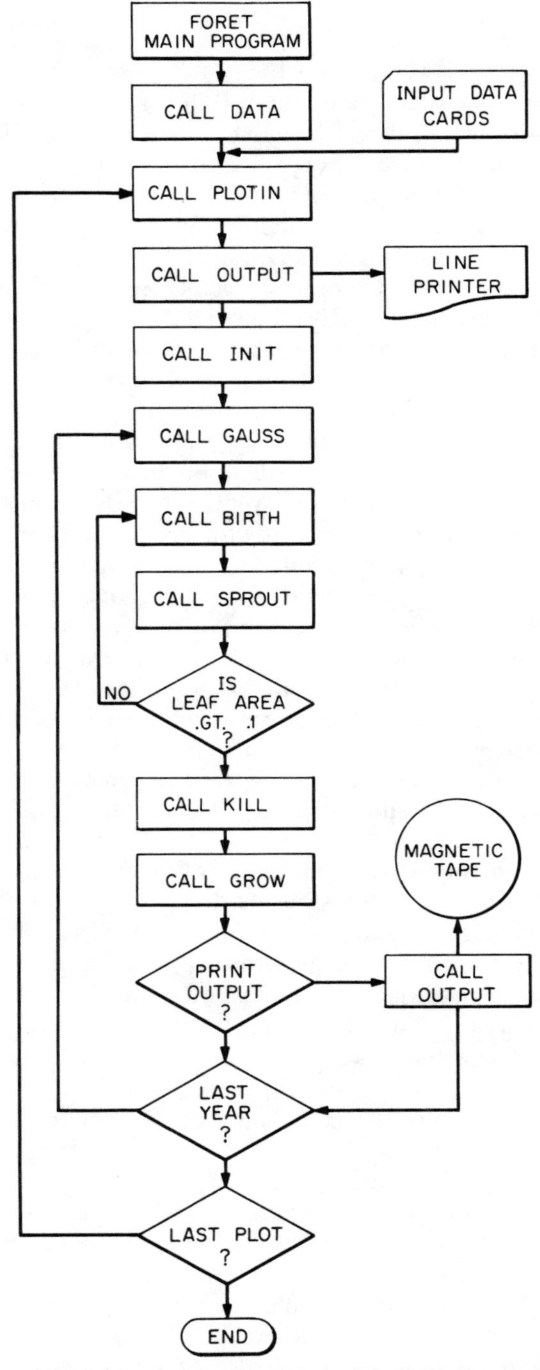

Figure 3.7. Flow diagram of the FORET model. Each rectangle represents a computer subroutine. (From Shugart and West [1977].)

Table 3.1. Results of experiments with different spatial scales for the plot size used in a gap model

Plot size (ha)	Test 1 Can a dominant tree grow to a correct maximum size?	Test 2 Does gap-phase replacement occur?
0.01	No	Yes
0.02	No	Yes
0.04	Yes	Yes
0.08	Yes	Yes
0.20	Yes	Rarely
0.40	Yes	No

From Shugart and West (1979).

Conclusions

In this chapter, the rationale, assumptions, and general formulations that are common to all gap models have been outlined. It is the general intent, in developing gap models, to use simple, and easily parameterized equations to approximate the mechanisms that cause a forest to change (e.g., as manifested by growth, birth, and the death of individuals) on a small plot of land. In practice, the gap model formulations are solved by using a computer program to compute the relationships that were discussed in this chapter. Because the FORET model (Shugart and West [1977]) is the basis for most of the models used in this book, and since its program details have not been published elsewhere, a listing of the program is included as Appendix 1. For those who are acquainted with the FORTRAN IV computer language, this appendix can be studied to obtain an impression of the internal workings of a gap model in greater detail. The general flow of operations in the model (which correspond roughly to this chapter's subheadings) is shown in Figure 3.7. Model parameters for this and all other gap models are listed by species in, and by, model in the following chapter. Given an explanation of how gap models work, the next concern is how well these models work.

Chapter 4

Performance of Gap Models

The first question that usually comes to mind after constructing a model is, "How well does it work?" This question arises with any model, and it is appropriate to discuss model validation in general before illustrating the behavior of gap models. Since models are complex hypotheses, the evaluation of a model's performance is the hypothesis-testing step in that iterative process known as the scientific method. In practice, the testing of ecosystem models has come to be known as model "validation." The term "validation" should not be interpreted to indicate that one model is valid while another is invalid (Goodall [1972]). Validity in this sense is a relative term. A model may be relatively valid for one purpose, but not for another.

Mankin, et al (1977) and Cale, et al (1983) used set theory to explain and define several concepts in model validation. The attributes of a model (e.g., its output, what it predicts, and so on) and the attributes of an ecosystem (e.g., its output, how it responds, and so on) can be thought of as sets in a universe of possible attributes. Similarly, the sets of model attributes are denoted M in the universe, U, as shown in the Venn diagram in Figure 4.1. The set of ecosystem attributes are denoted S in the same universe, U. When testing ecological models, one would like to learn about the existence and magnitude of the set Q, which is defined as the intersection of M and S. Ideally, we would like to find that:

Condition 1. $M \cap S \neq 0$. One would like to know that the model and the system share at least some attributes.

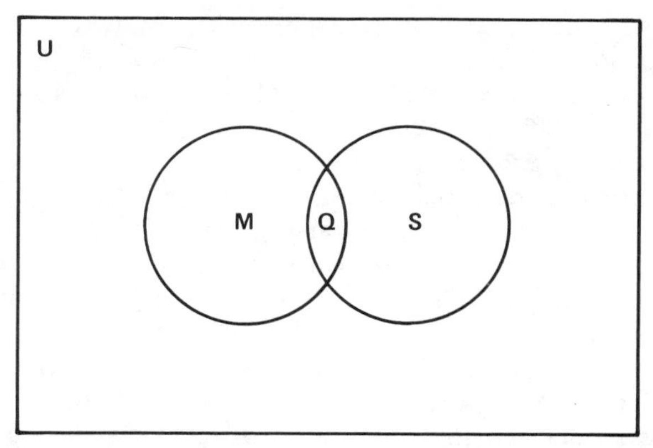

Figure 4.1. A Venn diagram showing the intersection, Q, of the set of ecosystem attributes, S, with the set of model attributes, M, in the universe of possible attributes, U. (From Mankin, et al [1977].)

Condition 2. M = Q. One would like the model to have no attributes that are not of the system.

Condition 3. S = Q. One would like the ecosystem to have no attributes that are not described by the model.

When all three of these conditions are all met, we have a case in which M = S or the model and the ecosystem share all attributes; such a case is impossible (Goodall [1972]). By definition, a model is only a partial representation of a system. The more usual case is for the first condition to be met; a model that shares no attributes with the real system normally is rejected. The second two conditions (2 and 3) normally are only met to a degree. The degree to which a model satisfies Condition 2 is a measure of the reliability of the model (Mankin, et al [1977]). A reliable model rarely would produce predictions that were inappropriate to the ecosystem. The degree to which a model satisfies Condition 3 is termed "the adequacy of the model" (Mankin, et al [1977], and Cale, et al [1977]). An adequate model contains in its set of attributes more of the ecosystem behavior of interest than a less adequate model. At times, models are constructed to maximize either reliability or adequacy. For example, linearizations of dynamic responses of ecosystems in the form of linear models represent the system's behavior at a small point in set S (Patten [1975]). These linear models presumably have a very high reliability, but it is not known how adequate they are. As another example, models often are constructed to take on extreme parameter values with the intent of bracketing the system's possible behavior. Models that are used to compute a radiological dose usually are designed to overestimate the actual values. These models have a very high adequacy, but they only may be minimally reliable in their ability to predict exactly what a system will do. In the more typical cases, models would have a degree of reliability (condition 2) and a degree of adequacy (Condition 3).

Model testing can be divided into two basic types of procedures (Shugart and West [1980]):

(1) *Verification procedures*, in which a model is tested to determine whether it can be made consistent with some set of observations. As an extreme example, the F test of linear regression is a verification procedure in which simple models are tested for their statistically significant agreement with observations after the regression parameters have been adjusted over an infinite range. Usually, ecological models are verified by using structures and parameters that are constrained by biological considerations. Very often, these constraints are so strong that the agreement between the model and the ecosystem can be viewed as a test of the internal consistency of the state of knowledge about an ecosystem. Model verifications are a set of methodologies designed to ensure that a model will satisfy condition 1: $M \cap S \neq 0$.

(2) *Validation procedures*, in which a model is tested on its agreement with a set of observations that are independent of those observations used to structure the model and to estimate its parameters. The degree of independence between the model and the observations before the validation test can be as important as the degree of agreement between the model and the observations following the test. Regular independent testing of a model's output against real system behavior tends to establish the degree of reliability of the model. One would like to determine if things that are predicted by the model can be relied on to occur in nature. Validation procedures, thus, are largely involved with Condition 2 (described above).

A special case of model validation involves applications in which results from a model are judged for their usefulness in providing specific insights into either basic or applied problems of interest. The applications of models can involve some of the most powerful model validations, because the model predictions that have a high likelihood of being tested by future observations or experiments. Often, model applications emphasize an aspect of ecosystem behavior that can be predicted by the model, but can not be easily measured in nature. These applications draw heavily on the prior performance of the model when subjected to both validation and verification procedures for their credibility. Many applications focus on the ability of the model to adequately describe a specific system behavior. In these applications, the degree to which the model meets Condition 3 is of considerable importance.

The procedure used to develop gap models has been the use of comparisons with the general composition of a forest to verify the model structure (i.e., new computer subroutines, additional types of parameters, and so on). These verifications were made under the constraint that most of the parameters used in gap models simply cannot be manipulated to make the model fit the known patterns of a region's forests. For example, the height of a tree at a maximum diameter for a species used to compute the b_2 and b_3 parameters is not an arbitrary variable that can be changed to make the model perform better. If it is given that a model agrees with the general pattern of forest composition over successional time scales (200–1000 years), it is then appropriate to attempt to validate it against independent data (if such data were available). Information that actually can be used to test a gap model can be difficult to obtain. The tendency in developing gap models has been to reserve the more detailed data sets for model validations.

The remainer of this chapter is a sampler of the performance of gap models. The initial results that are presented are from verifications in which the models have been compared to the general compositional patterns of the forest they simulate. In these cases, the models are described and the species parameters are tabulated. The compositional pattern of the particular forest is discussed in a brief section, and references that described the pattern of a given forest are provided. These initial model tests are verifications and are used here to describe the models and the forests that they simulate. However, sources are provided for those readers who are interested in more detail. The detailed output from a gap model is an annual tally of the size (diameter breast height) and species of each tree on a simulated plot that is approximately the size of a sample quadrat (approximately 0.10 ha). The output, in this respect, resembles primary field data and must be generally expressed in a summarized form for purposes of clarity. Because the models are stochastic, output from the model varies and usually is averaged over 50–100 simulated plots. The detail in the model output creates the potential for a variety of tests of the model against data; some of these more detailed tests will be discussed in the later sections of this chapter. Chapter 9 provides three examples of model applications and also can be consulted for examples of model performance.

In this chapter, "verified," "validated," and "applied" will be defined in the restricted senses as types of model tests. The same kind of model prediction can be a verification (when the model is constrained to match actual data), a validation (when the model output and the data are independent), or an application (when the data are not available, but predictions are needed for some reason).

The objective of the remainder of this chapter is to present a sampling of the diverse array of tests that have been applied to gap models. The initial tests largely will be model verifications and will center on the ability of the models to capture the broad compositional patterns in a variety of forested ecosystems. The later tests largely will be model validations and will focus on the detailed structure and dynamics of forests.

General Patterns of Compositional Dynamics Simulated by Gap Models

It is an important test of a model to be able to simulate general patterns of an ecosystem under the constraint that all of the parameters in the model are realistic. The task of matching known patterns in nature is considerably more difficult when the parameters of a model (whether statistical or otherwise) are constrained. However, both the structure and the emphasis of gap models make it appropriate to have a high level of realism in the model parameters. In the construction of most gap models, the initial model verifications involve determining a model's ability to reproduce the general features of a forest pattern while restricted to using model parameters that are reasonable. These model tests are convenient introductions to the spectrum of forests that have been treated with gap models.

Australian Montane Eucalyptus Forests

The BRIND model (Shugart and Noble [1981]) simulates the dynamics of 18 arborescent species (Table 4.1) that are found on southeast-facing slopes at elevations ranging from 900–2,400 meters in the Brindabella Range, Australian Capital Territory (A.C.T.). Nominal locations of the simulated stands are in the Cotter catchment near Canberra, A.C.T. All of the forests are dominated by species in the genus *Eucalyptus*. Figure 4.2 shows simulations that are used to verify the model for the 1600 meter elevation—alpine ash (*Eucalyptus delegatensis*) zone (Costin [1954], Pryor and Moore [1954], and Anon [1973]). The two cases shown are for wildfire with a fire return probability of 0.020 years^{-1} (Anon [1973]) and the absence of wildfire. Fire intensity is simulated as a function of climatic conditions and fuel loads by using climate records for the area (Anon 1973) combined with fire equations (Noble, et al, [1980]) derived from the empirical fire meters of McArthur (1967) and Luke and McArthur (1978).

Without fire in the model, *E. delegatensis*, which is a fast-growing tall eucalyptus, completely dominates the stands and comprises almost 98% of the stands' biomass by year 100 (Fig. 4.2). Of the remaining biomass, most is associated with the understory shrubby trees, *Olearia argophylla* and *Bedfordia salicina*. The dominance of *E. delegatensis* can be attributed to its superior growth rate in relation to any of its possible competitors, and also to its great potential height (61 meters, Hall, et al [1970]). *Eucalyptus delegatensis* is quite sensitive to fire; when fire is added to the model, the dynamics are greatly altered.

With periodic fire (50-year return frequency) (Fig. 4.2), there is a systematic reduction of *E. delegatensis* over the 600-year period of the simulation. *Eucalyptus delegatensis* regenerates from seeds that are released

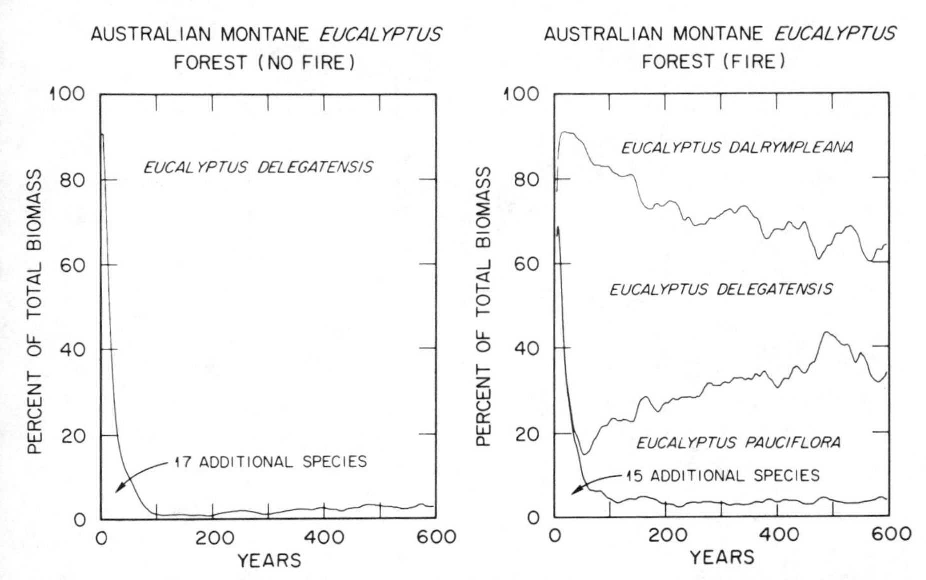

Figure 4.2. Compositional dynamics of the BRIND model (Shugart and Noble [1981]) for the 1600-meter. Alpine Ash Zone of the Brindabella Range in Australia. Each graph is the average response for 50 simulated 1/12-ha forest stands. Forests are initiated (year 0) with four individuals of all species to initialize the seed bank in the model. The area associated with each species on each graph indicates the proportion of total biomass associated with the species. (From Shugart, et al [1981a].)

after a fire that also kills the mature trees. If two fires occur close enough in time (approximately 20 years apart), there are no seed-bearing trees following the second fire and the species is eliminated for lack of a seed source. As *E. delegatensis* declines with recurrent fires, the mountain gum (*E. dalrympleana*) and the snow gum (*E. pauciflora*) become more dominant. Both of these latter species are quite fire tolerant because of ecological attributes that confer increased fire survival (e.g., lignotuberous resprouting, epicormic buds, and the ability to survive total leaf scorching).

The responses of the model in Figure 4.2 (as well as more extensive experiments with different fire probabilities and at different altitudes in Shugart and Noble [1981] and discussed below) are consistent with replacement patterns of vegetation in the Brindabella Range in particular (Costin [1954], Pryor and Moore [1954], and Anon [1973]) and the wet sclerophyll forests of southeastern Australia and Tasmania in general (Ashton [1970]). Both the BRIND model and the forests of the region appear to have multiple stable vegetation types, when regularly perturbed by fire.

Table 4.1. Basic parameters used for species attributes in the BRIND model (Shugart and Noble 1981). H_{max} and D_{max} are maximum height and diameter, respectively. B2 and B3 are the parameters of Eq. 3.2. AGE is species longeviety (AGE_{max} in Eq. 3.16). G is the growth-rate parameter (Eq. 3.6). $DEGD_{max}$ and $DEGD_{min}$ scale the temperature response (see Eq. 3.11). FRSTMX is the number of days with frost tolerated by the species. SDSIZE is the size of the seeds and SDLIFE is the seed longeviety. These latter two parameters are used to load a seed bank submodel. SPRIND is the tendency to sprout and SPRIMN is the minimum diameter of sprouts. SWITCH (1), (2) and (3) are for species with seeds requiring fire to trigger germination, with well-dispersed seeds and with enhanced germination after fire (respectively). FORM is the percentage of the trees height that is in leaf canopy and is used to compute fire-scorch. Scientific binomials follow Burbidge and Gray (1976).

	H_{max} (cm)	D_{max} (cm)	B_2 [a]	B_3 [a]	Age (years)	G [a]	$DEGD_{max}$ (°C day)	$DEGD_{min}$ (°C day)	FRSTMX (days)	SDSIZE (cm)	SDLIFE (yr)	SPRIND	SPRTMN (cm)	SWITCH (1)	SWITCH (2)	SWITCH (3)	FORM [d]
Acacia dealbata	1219[e]	30[e]	71.01	1.1649	30[f]	360	2000	800	120	3	50[g]	1[f,h]	3	T	F	T	0.10
″ falciformis	1067[h]	30[e]	61.01	1.008	20[h]	470	2300	1400	90	3	50[g]	1	3	T	F	T	0.10
″ implexa	1219[e]	30[e]	71.01	1.1649	20[h]	525	3200	1400	90	3	50[g]	1[f]	2	T	F	T	0.10
″ melanoxylon	2134[e]	61[e]	65.50	0.5373	100[h]	190	6256[i]	835[i]	70[j]	6	50[g]	1[h]	3	T	F	T	0.33
″ pycnantha	1219[f]	25[k]	85.21	1.6774	20[f,k]	510	2300	1400	90	3	50[g]	0[l]	3	T	F	T	0.10
Banksia marginata	914[m]	30[k]	51.01	0.8368	50	160	2400	400	150	3	5[k]	1[m]	—	T	T	T	0.10
Bedfordia salicina	914[k]	30[k]	51.01	0.8368	50	225	2300	1400	90	3	1[k]	1[n]	0[n]	F	T	F	0.10

Eucalyptus dalrympleana	3658[j]	122[j]	57.75	0.2368	400[e]	80	2583[j]	1050[j]	100[j]	12	5[o]	1	0	F	F	F	0.50[j]
" delegatensis	6096[j]	213[j]	55.86	0.1309	220[e]	235	2042[j]	1231[j]	100[j]	21	5[o]	0[p]	—	F	F	T	0.66[j]
" dives	2438[j]	76[j]	60.40	0.3964	300[q]	70	3913[j]	1501[j]	80[j]	8	5[o]	1[r]	0[j,s]	F	F	F	0.33[j]
" fastigata	4572[j]	183[j]	48.50	0.1326	275[e]	145	2493[j]	1682[j]	100[j]	18	5[o]	0[p,s]	—	F	F	F	0.58[j]
" pauciflora(H)[t]	1829[j]	91[j]	37.00	0.2033	125[e,f]	135	2050	509[j]	150[j]	9	8[k]	1[s]	0[s]	F	F	F	0.33[j]
" pauciflora(L)[t]	1829[j]	91[j]	37.00	0.2033	125[e,f]	135	2674[j]	1950	150[j]	9	8[k]	1[s]	0[s]	F	F	F	0.33[j]
" robertsonii	3962[j]	152[j]	50.20	0.1641	220[e]	155	3202[j]	1358[j]	100[j]	15	5[o]	1[s]	0[s]	F	F	F	0.50[j]
" rubida	3048[j]	91[j]	63.67	0.3482	125[f]	210	2854[j]	1592[j]	70[j]	9	5[o]	1[s]	0[s]	F	F	F	0.50[j]
" stellulata	1524[j]	61[j]	45.51	0.3732	250	55	2764[j]	1050[j]	160[j]	6	5[o]	1[s]	0[s]	F	F	F	0.33[j]
" viminalis	3658[j]	122[j]	57.75	0.2368	325[e]	100	2854[j]	1682[j]	60[j]	12	8[k]	1[s]	0[s]	F	F	F	0.58[j]
Exocarpus cupressiformis	762[k]	30	41.01	0.6727	50	140	2300	1400	90	3	1[k]	1[r]	3	F	T	F	0.10
Olearia argophylla	914[k]	30[k]	51.01	0.8368	25	340	2300	1400	90	3	1[k]	1[n]	0[n]	F	T	F	0.10

[a] Derived from other parameters.

[b] Assume that trees can produce viable seeds when the dbh is ~0.1 D_{max}.

[c] Given a value of zero for species with lignotubers.

[d] Ratio of bole/total height of typical tree. Assumed to be 0.1 for shrubby trees.

[e] L. D. Pryor, personal communication.

[f] Pryor 1968.

[g] Ewart (1925) based on pp. 137-139.

[h] Simpfendorfer 1975.

[i] Farrell & Ashton 1978.

[j] Hall et al. 1970.

[k] Ewart 1925.

[l] M. Gill, personal communication.

[m] Holliday and Watton 1975.

[n] Ashton 1970.

[o] Ewart (1925) based on pp. 243-244.

[p] Gill 1975.

[q] Jacobs 1955.

[r] Purdie and Slatyer 1976.

[s] Pryor 1976.

[t] H indicates upper elevation cline-form.
L indicates lower elevation cline-form.

Arkansas Upland Forest

The FORAR (Mielke, et al [1977, 1978]) simulates 33 species (Table 4.2) that are found on upland sites in Union County, Arkansas. Without fire, FORAR model simulations of the mixed oak-pine forests in the uplands of south Arkansas (Fig. 4.3) demonstrate a pattern that agrees with field observations of successional replacement, particularly when the model response is averaged over several stands. For the first 250 years of the simulation, loblolly pine (*Pinus taeda*) dominates the forest stands. As the canopies of these stands break up, the understory hardwoods fill the forest gaps, particularly sweetgum (*Liquidambar styraciflua*) and black cherry (*Prunus serotina*). Southern red oak (*Quercus falcata*) dominates in hardwood stands by year 600. These dynamics are in agreement with forest types in south Arkansas and generally are the patterns that are expected in nonburned mixed pine-hardwood forests throughout the southeast (Oosting [1942], and Johnston and Odum [1956]), particularly in abandoned old fields.

With periodic fires (mean return frequency of 24 years corresponding to the apparent fire occurrence rate for south Arkansas), the pattern (Fig. 4.3) is altered radically. *Pinus taeda* dominates the stands for the entire 600-year simulation. The hardwood species that dominated the older non-burned plots are considerably reduced in abundance and are found only in

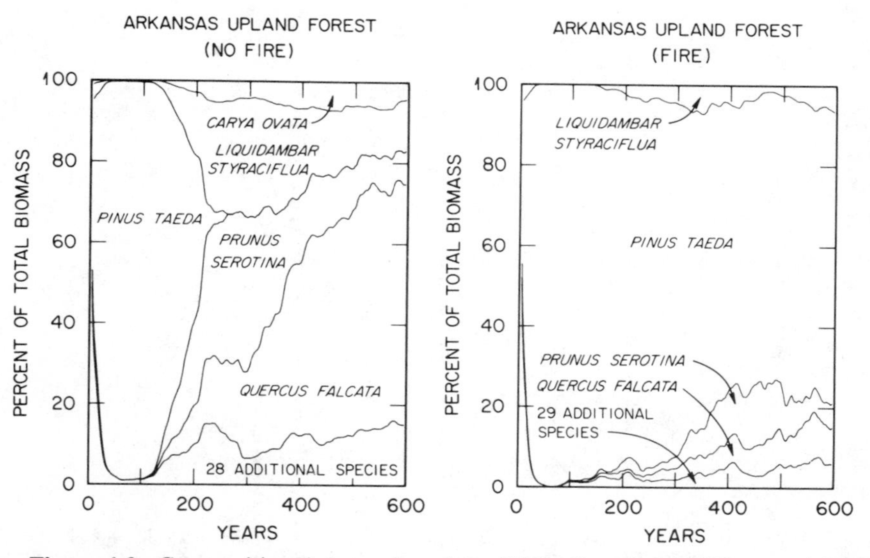

Figure 4.3. Compositional dynamics of the FORAR model (Mielke, et al [1977, 1978]) for the upland sites in Union County, Arkansas. Each of the 50 simulated stands (averaged here) is started from an open stand with no trees. (From Shugart, et al [1981a].)

stands that experience light fires or no fires for a long period (owing to the stochastic nature of the wildfire generating function). This result is consistent with historical observations of extensive (fire-maintained) pine lands in south Arkansas (Lesquereux, M. L., in Owen [1860]).

Australian Subtropical Rain Forest

The KIAMBRAM model (Shugart, et al [1980a] and Shugart, et al [1981b]) simulates the dynamics of 125 species (Table 4.3) that are found in the subtropical rain forest in the vicinity of the New South Wales-Queensland border. Nominal locations of the simulated stands are in the Wiangaree State Forest, New South Wales. The compositional dynamics of the KIAMBRAM model (Shugart et al [1981a]) correspond to those actually documented by Webb, et al (1972) and Hopkins, et al (1977) in subtropical rain forests in the vicinity of the New South Wales-Queensland border. Following a brief period (approximately 15 years) that included the increase and demise of pioneer species (e.g., *Dubosia myoporoides* or *Trema aspera*), both the simulation and the forest (Webb, et al [1972], Hopkins, et al [1977]) had a short-lived appearance of "early secondary" or "nomad" species (van Steenis [1958]) between years 30–60. Typical species in this period would be *Acacia melanoxylon* or *Polyscias elegans*. The decrease in the dominance of these pioneer and early secondary species is indicated by the decrease in the "97 other species" category that is shown in Figure 4.4. For a brief period, "late-secondary" species (Hopkins, et al [1977]), including many of the "22 commercial species" shown in Figure 4.4, have a period of abundance. The simulated mature forest eventually is dominated by *Ackama paniculata*, *Argyrodendron* species and *Ficus* species. The composition of a mature rain forest shows a close similarity to that found in the Wiangaree State Forest, New South Wales (Burgess, et al [1975]).

Southern Appalachian Deciduous Forest

The FORET model (Shugart and West [1977]) simulates 33 species (Table 4.4) in a southern Appalachian deciduous forest (Fig. 4.5). These forests are typified by both the development of a mixed oak (*Quercus* species) forest type and the persistence of the shade-intolerant species tulip-poplar (*Liriodendron tulipifera*) by means of gap-phase replacement (Watt [1947]). The behavior of the model compares favorably with records by Foster and Ashe (1908) regarding forest composition before the chestnut blight (discussed below). The mature forest composition resembles historical descriptions of the southern Appalachian forests (Sargent [1884], Ashe [1897], Pinchot [1906, 1907], Greeley and Ashe [1907], Hall [1907],

Table 4.2. Basic parameters used for species attributes in the FORAR model (Mielke et al. 1977, 1978) using original parameter notation. D_{max} and D_{min} scale the temperature response (see Eq. 3.11). B2 and B3 are parameters for Eq. 3.2. ITYPE is the tolerance type (see Eq. 3.10). AGEMX is the species longeviety (Eq. 3.16). C refers to which of a set of empirical equations is used to compute leaf area. G is the growth parameter (Eq. 3.6). STEND is the sprouting tendency. SPT_{min} and SPT_{max} are the minimum and maximum diameters for sprouting. S(1), (2), (3), (4), and (5) are for species requiring mineral soil, leaf litter, wetter than average years, lowered wildlife population and absence of disease for successful regeneration, respectively. SOILM is a drought index. Scientific binomials follow Little (1971).

Species	D_{max}[3]	D_{min}[3]	B_3	B_2	ITYPE	AGEMX	C[8]	G	STEND	SPT_{min}	SPT_{max}	S (1)	S (2)	S (3)	S (4)	S (5)	SOIM
Acer rubrum	13395	1810	0.173	52.58	1[8]	150[2]	1	240.2	2[1]	12.0[8]	150[11]	F[1]	T[1]	T	F[10]	F[8]	0.5
Carpinus caroliniana	10947	2097	0.228	33.56	1[2]	100	1	129.9	2	6.0	70[11]	F	F	F	F[10]	F	0.4
Carya cordiformis	9461	3686	0.306	69.21	3[1]	300[8]	1	117.2	2[5]	12.0[8]	110[11]	T[8]	F[2]	T[8]	F[1]	F[8]	0.8
Carya glabra	12652	3686	0.168	49.65	1[1]	300[8]	1	111.3	2[5]	12.0[8]	145[11]	T[1]	F[2]	T[8]	F[10]	F[8]	0.5
Carya laciniosa	8718	5081	0.318	72.46	1[1]	300	1	123.2	2[5]	12.0	90[11]	T	F	T	F[10]	F[5]	0.4
Carya ovata	12652	4105	0.312	69.66	1[1]	300[2]	1	116.5	2[5]	2.0[1]	110[11]	T[8]	F[2]	T[8]	F[10]	F[8]	0.7
Carya texana	9461	5526	0.318	57.56	1	300	1	79.6	2[5]	12.0	110[11]	T	F	T	F[10]	F[5]	0.5
Carya tomentosa	10947	3686	0.190	49.41	3[1]	300[8]	1	98.6	2[5]	12.0[2]	130[11]	T[1]	F[2]	T[8]	F[10]	F[8]	0.8
Castanea ozarkensis	7756	5526	0.085	21.72	1	60	1	250.9	2[12]	12.0	125[11]	T	F	T	F[10]	T[5]	0.8
Celtis laevigata	12652	5526	0.336	71.26	1[1]	350	1	112.7	2[1]	6.0[1]	115[11]	T[1]	F	F	F[10]	F	0.6
Cornus florida	10947	3686	0.536	40.81	1[1]	100[8]	1	88.7	3[1]	12.0	35[11]	F[8]	F[1]	T[1]	F[10]	F[8]	0.9
Diospyros virginiana	13395	5526	0.084	35.84	1[1]	150[8]	1	234.1	2[1]	12.0[8]	210[8]	F[1]	F[2]	T[8]	F[10]	F[8]	0.3
Fagus grandifolia	10204	2097	0.151	46.20	1[1]	400[2]	1	80.7	2[1]	6.0[8]	30[8]	F[1]	F[2]	T[8]	F[10]	F[8]	0.5
Fraxinus americana	10947	2414	0.080	34.43	3[1]	300[8]	1	113.3	2[1]	6.0[1]	20	F[1]	T[2]	T[8]	F[10]	F[8]	0.9
Ilex opaca	10947	5526	0.156	42.61	1[2]	200[2]	1	133.5	1[5]	6.0	35[11]	F[8]	F	F	F[10]	F	0.5
Juglans nigra	8499	3686	0.074	36.37	3[1]	250	1	161.5	1[1]	6.0[1]	40	T[8]	F[2]	T[8]	F[10]	F[8]	0.7

Juniperus virginiana	10204	2966	0.236	57.75	3[1]	300[1]	2	107.1	0[1]	12.0[1]	120[11]	F[1]	T[1]	F[1]	F[10]	F[8]	0.9
Liquidambar styraciflua	10947	5526	0.191	55.20	3[1]	300[2]	1	119.4	2[1]	12.0[1]	80[1]	F[1]	F[1]	F[1]	F[10]	F[8]	0.7
Morus rubra	13395	3686	0.061	22.15	1[4]	75[5]	1	260.6	0[5]	12.0[1]	180[11]	T[5]	F[5]	T	F[10]	F	0.3
Nyssa sylvatica	12652	3686	0.202	37.00	1[5]	300[8]	1	54.8	1[1]	60.0[1]	90[1]	F[8]	T[8]	F[8]	F[10]	F[8]	0.7
Ostrya virginiana	10204	1810	0.266	49.50	1[7]	100	1	219.1	1	6.0	90	F	F	F	F[10]	F	0.4
Pinus echinata	9461	5526	0.290	70.75	3[1]	300	2	128.4	2[1]	6.0[1]	20[1]	F[1]	T[1]	F[1]	F[10]	F	1.0
Pinus taeda	10947	6391	0.189	60.81	3[1]	350[2]	2	174.6	0[1]	12.0[1]	160[11]	F[1]	T[1]	F[1]	F[10]	F[8]	1.0
Prunus serotina	10947	3899	0.083	35.57	3[2]	250[1]	1	139.5	2[1]	6.0[1]	210[1]	F[1]	T	F[8]	T[10]	F[8]	0.7
Quercus alba	10204	2966	0.074	36.37	1[1]	400[9]	1	100.7	2[1]	12.0[1]	40[1]	T[1]	F[1]	F[1]	T[10]	F[8]	0.8
Quercus falcata	10947	5526	0.078	33.57	1[6]	400	1	82.9	2[1]	12.0[1]	30[1]	T[8]	F[8]	F[8]	F[10]	F[8]	0.9
Quercus marilandica	10204	5081	0.111	24.86	3[5]	400	1	36.3	2	12.0	40	F[5]	F	F	F[10]	F	1.0
Quercus shumardii	10947	5081	0.199	65.19	3[2]	400	1	115.9	2[1]	12.0[1]	160[1]	F[1]	F[1]	F	F[10]	F	0.7
Quercus stellata	10947	5526	0.195	47.75	3[1]	400	1	67.9	2[1]	12.0[1]	40[1]	T[1]	F[1]	F[8]	T[10]	F[8]	1.0
Quercus velutina	9461	3313	0.097	41.57	1[2]	200[9]	1	100.1	2[1]	12.0[1]	40[1]	F[1]	T[1]	F[8]	F[10]	F[8]	0.8
Sassafras albidum	10947	3686	0.619	75.51	3[2]	200[8]	1	108.1	3[1]	6.0[1]	60[1]	T[1]	F[1]	F[8]	F[10]	F[8]	0.8
Ulmus alata	10947	5526	0.230	51.81	3[1]	125[5]	1	212.5	1	6.0	110	F	F[5]	F	F[10]	F	0.4
Ulmus americana	12652	1522	0.082	39.35	3	300[1]	1	141.8	1[1]	6.0[1]	240[1]	F[1]	F[1]	T[1]	F[5]	T[10]	0.4

[1] From Fowells (1965)
[2] From Harlow and Harrar (1969)
[3] From Little (1971) and U.S. Geological Survey (1965)
[4] From Wigginton (1964)
[5] From Carl Amason (personal communication, 1975)
[6] From Baker (1949)
[7] From Peattie (1950)
[8] From Shugart and West (1977)
[9] From Northeastern Forest Experiment Station (1971)
[10] From Martin et al. (1971)
[11] From Pardo (1973)
[12] From Moore (1960)
[13] All other values not otherwise referenced were developed by Mielke et al. (1978)

Table 4.3. Basic parameters used for species attributes in the KIAMBRAM model (Shugart et al. 1980, Shugart et al. 1981b). D_{max} is the maximum diameter and H_{max} is the maximum height of the species. B_2 and B_3 are parameters derived from Eq. 3.2. Age is the species longeveity (Eq. 3.16). G is the growth parameter (Eq. 3.6). TOL is the shade tolerance. P1 and P2 refer to classes of seed phenology and seed longeveity used in a seed-bank submodel. SWITCHES for seed dispersal and germination are (1) bird-dispersed species, (2) wind-dispersed species, (3) gravity-dispersed species, (4) species requiring mineral soil, (5) species that require an elevated microsite for germination and (6) stranglers.

Species	D_{max} †	H_{max} †	B_2	B_3	Age	G	TOL	P1‡	P2‡	SWITCH 1	2	3	4	5	6
Acacia melanoxylon R. Br.	91	2438	50.34	0.275	60	363	3	1	4	T	F	F	T	F	F
Ackama paniculata Engl.	91	3962	83.67	0.458	300	248§	1	2	1	F	T	F	F	T	F
Acmena australis (C. Moore) L. Johnson	91	3048	63.67	0.348	200	134	1	2	4	T	F	T	F	F	F
Acmena smithii (Poir) Merr. et Perry	30	1829	111.01	1.821	80	200	1	1	2	T	F	F	F	F	F
Acronychia baueri Schott.	30	1829	111.01	1.821	125	128	3	2	2	T	F	F	F	F	F
Acronychia oblongifolia (A. Cunn. ex Hook.) Endl. ex Heunk.	25	1524	109.21	2.150	90	149	3	1	2	T	F	F	T	F	F
Acronychia pubescens C. T. White	25	1524	109.21	2.150	90	149	3	2	2	T	F	F	F	F	F
Acronychia suberosa C. T. White	30	2743	171.01	2.805	60§	251§	3	2	2	T	F	F	F	F	F
Ailanthus triphysa (Dennst.) Alston	61	3048	95.51	0.783	100	264	3	1	3	F	T	F	F	F	F
Akania lucens (F. Muell.) Airy.Shaw	23	1219	94.68	2.071	138§	79§	1	2	2	T	F	F	F	F	F
Alangium villosum (Bl.) Wangerin	23	1219	94.68	2.071	82§	132§	1	2	2	T	F	F	F	F	F
Alphananthe philippinensis Planch.	61	2438	75.51	0.619	200	107	1	2	2	T	F	F	F	F	F
Alphitonia excelsa (Fenzl.) Benth.	38	2438	120.81	1.585	73§	288§	3	1	2	T	F	F	T	F	F
Araucaria cunninghamii Ait. ex D. Don	152	4877	62.20	0.204	300	140	3	1	2	F	T	F	F	F	F
Argyrodendron actinophyllum (F. M. Bail.) H. L. Edlin	107	4267	77.43	0.363	300	179§	1	2	1	F	T	F	F	F	F
Argyrodendron trifoliolatum F. Muell.	122	4267	67.75	0.278	300	437§	1	2	1	F	T	F	F	F	F
Arytera distylis (F. Muell. ex Benth.) Radlk.	15	1219	142.02	4.659	60	176	1	2	2	T	F	F	F	F	F
Arytera divaricara F. Muell.	61	1219	142.02	4.659	60	176	1	2	2	T	F	F	F	F	F
Austromyrtus acmenioides (F. Muell.) Burret	30	1829	111.01	1.821	75	214	2	2	3	T	F	F	F	F	F
Austromyrtus lasioclada (F. Muell.) L. S. Smith	30	1829	111.01	1.821	125	128	2	2	3	T	F	F	F	F	F
Baloghia lucida Endl.	30	1829	111.01	1.821	65§	244§	1	1	2	F	F	T	F	F	F
Beilschmiedia elliptica C. T. White and W. D. Francis	46	3048	127.34	1.393	100	262	1	1	2	T	F	F	F	F	F

Bosistoa euodiformis F. Muell.	25	1219	85.21	1.677	100	110	2	2	1	F	F	T	F	F	F
Brachychiton acerifolium F. Muell.	91	3658	77.00	0.421	92§	341§	2	2	2	T	F	F	F	F	F
Cinnamomum oliveri F. M. Bail.	38	1829	88.81	1.165	100	445§	1	2	2	T	F	F	F	F	F
Cinnamomum virens R. T. Bak	61	2743	85.51	0.701	213§	112§	1	2	2	T	F	F	F	F	F
Citriobatus lancifolius F. M. Bail.	8	457	84.04	5.515	20	207	3	2	3	T	F	F	F	F	F
Citronella moorei (F. Muell. ex Benth.) Howard	152	3658	46.20	0.152	300	107	1	2	1	F	T	F	F	F	F
Claoxylon australe Baill.	15	762	82.02	2.691	50	140	1	2	2	F	F	T	F	F	F
Clerodendrum floribundum R. Br.	15	762	82.02	2.691	30	234	1	2	2	T	F	F	F	F	F
Cryptocarya erythroxylon Maid. and Betche	122	3658	57.75	0.237	387§	82§	1	2	2	T	F	F	F	F	F
Cryptocarya foveolata C. T. White and W. D. Francis	61	3048	95.51	0.783	300	88	1	2	2	T	F	F	F	F	F
Cryptocarya glaucescens R. Br.	38	2134	104.81	1.375	121§	154§	1	2	2	T	F	F	F	F	F
Cryptocarya microneura Meissn	36	1829	95.15	1.338	175	92	1	2	2	T	F	F	F	F	F
Cryptocarya obovata R. Br.	91	3962	83.67	0.458	210§	163§	1	2	3	T	F	F	F	F	F
Cryptocarya triplinervis R. Br.	15	610	62.02	2.035	25	233	1	2	2	T	F	F	F	F	F
Cupaniopsis foveolata (F. Muell.) Radlk.	38	1219	56.81	0.746	160	71	2	2	2	T	F	F	F	F	F
Cupaniopsis serrata (F. Muell.) Radlk.	38	1219	56.81	0.746	75	151	2	1	2	T	F	F	F	F	F
Daphnandra micrantha (Tul.) Benth.	61	3658	115.51	0.947	150§	209§	1	2	1	F	T	F	F	F	F
Decaspermum fruticosum Forst.	25	1219	85.21	1.677	75	147	2	2	3	T	F	F	F	F	F
Dendrocnide excelsa (Wedd.) Chew	183	3658	38.50	0.105	100	322	3	1	4	T	F	F	T	F	F
Denhamia pittosporoides F. Muell.	30	1524	91.01	1.493	75	181	1	2	2	T	F	F	F	F	F
Diospyros australis (R. Br.) Hiern.	15	914	102.02	3.347	50	164	1	2	2	T	F	F	F	F	F
Diospyros pentamera (Woolls and F. Muell.) Woolls and F. Muell ex Hiern.	76	3962	100.40	0.659	185§	182§	1	2	2	T	F	F	F	F	F
Diploglottis australis (G. Don) Radlk.	61	3048	95.51	0.783	100	264	3	2	2	T	F	F	F	F	F
Doryphora sassafras Endl.	122	3658	57.75	0.237	108§	292§	2	2	1	F	T	F	F	F	F
Drypetes australasica (J. Muell.) Pax and K. Hoffm.	61	2743	85.51	0.701	200	120	1	2	2	T	F	F	F	F	F
Duboisia myoporoides R. Br.	30	1219	71.01	1.165	25	446	3	1	4	T	F	F	T	F	F
Dysoxylum fraseranum (A. Juss.) Benth.	152	3962	50.20	0.165	220§	156§	2	2	2	T	F	F	F	F	F
Dysoxylum muelleri Benth.	152	3962	50.20	0.165	200	173	2	2	2	T	F	F	F	F	F
Dysoxylum rufum (A. Rich) Benth.	30	1829	111.01	1.821	110	146	2	2	2	T	F	F	F	F	F
Ehretia acuminata R. Br.	76	2743	68.40	0.449	178§	135§	2	2	2	T	F	F	F	F	F
Elaeocarpus kirtonii F. Muell. ex F. M. Bail.	91	3048	63.67	0.348	150	178	1	2	2	T	F	F	F	F	F
Elaeocarpus obovatus G. Don	91	3658	77.00	0.421	200	158	1	2	2	T	F	F	F	F	F
Elaeodendron australe Vent.	23	1219	94.68	2.071	55	199	1	1	2	T	F	F	F	F	F
Elattostachys nervosa (F. Muell.) Radlk.	30	1524	91.01	1.493	70	194	1	2	2	T	F	F	F	F	F

Table 4.3. (continued)

Species	D_{max} †	H_{max} †	B_2	B_3	Age	G	TOL	P1 ‡	P2 ‡	SWITCH 1	2	3	4	5	6
Elattostachys xylocarpa (A. Cunn. ex F. Muell.) Radlk.	23	1219	94.68	2.071	55	199	1	2	2	T	F	F	F	F	F
Emmenosperma alphitonioides F. Muell.	91	3659	77.00	0.421	264§	120§	1	2	3	T	F	F	F	F	F
Endiandra crassiflora C. T. White and W. D. Francis	38	1524	72.81	0.955	125§	110§	1	2	2	T	F	F	F	F	F
Endiandra discolor Benth.	91	3658	77.00	0.421	300	106	1	2	3	T	F	F	F	F	F
Endiandra muelleri Meissn.	38	1219	56.81	0.746	110	103	1	2	2	T	F	F	F	F	F
Euodia micrococca F. Muell.	30	2438	151.01	2.477	70	298	3	1	4	T	F	F	T	F	F
Ficus obliqua Forst. f. var. obliqua	305	4877	31.30	0.051	300	142	1	1	3	T	F	F	F	T	T
Ficus virens Ait.	183	3048	31.84	0.087	225	121	1	1	4	T	F	F	F	T	T
Ficus watkinsiana F. M. Bail.	183	4572	48.50	0.133	300	132	1	1	3	T	F	F	F	T	T
Flindersia australis R. Br.	152	3962	50.20	0.165	150	230	2	2	2	F	T	F	F	F	F
Geijera salicifolia Schott.	46	2438	100.67	1.101	70	304	3	1	1	T	F	F	F	F	F
Geissois benthamii F. Muell.	91	3048	63.67	0.348	128§	209§	1	2	1	F	T	F	F	T	F
Gmelina leichhardtii (F. Muell.) F. Muell. ex Benth.	122	3962	62.75	0.257	300	115	2	2	3	T	F	F	F	F	F
Guilfoylia monostylis (Benth.) F. Muell.	20	762	61.52	1.514	45	161	1	2	2	T	F	F	F	F	F
Guioa semiglauca (F. Muell.) Radlk.	38	1524	72.81	0.955	70	197	1	2	2	T	F	F	F	F	F
Halfordia kendack (Montri.) Guill.	76	2743	68.40	0.449	149§	90§	1	2	2	T	F	F	F	F	F
Harpullia pendula Planch. ex F. Muell.	61	2438	75.51	0.619	160	134	1	2	2	T	F	F	F	F	F
Helicia glabriflora F. Muell.	30	1524	91.01	1.493	153	88	2	2	2	T	F	F	F	F	F
Hymenosporum flavum (Hook.) F. Muell.	41	2134	98.26	1.209	90	208	2	2	2	F	T	F	F	F	F
Jagera pseudorhus (A. Rich.) Radlk.	38	1524	72.81	0.955	75	184	1	2	2	T	F	F	F	F	F
Litsea reticulata (Meissn.) F. Muell.	152	3658	46.20	0.152	444§	72§	1	2	2	T	F	F	F	F	F
Lomatia aborescens Fraser and Vickory	23	1219	94.68	2.071	60	182	3	2	2	T	F	F	F	F	F

Macaranga tanarius (L.) J. Muell.	46	1219	47.34	0.518	40	286	3	1	3	T	F	F	T	F	F
Mallotus philippensis (Lam.) J. Muell.	30	1524	91.01	1.493	90	151	1	1	2	T	F	F	T	F	F
Melia azedarach L. var. australasica (A. Juss.) C.DC.	122	4572	72.75	0.298	200	197	3	1	3	T	F	F	F	F	F
Melicope octandra (F. Muell.) Druce	20	1829	166.52	4.097	57§	274§	2	2	3	F	T	T	F	F	F
Mischocarpus pyriformis (F. Muell.) Radlk.	46	1829	74.01	0.809	110	149	1	2	2	T	F	F	F	F	F
Neolitsea cassia (L.) Kosterm.	30	1524	91.01	1.493	75	181	3	2	2	T	F	F	F	F	F
Neolitsea dealbata (R. Br.) Merr.	15	610	62.02	2.035	50	116	3	1	3	T	F	F	T	F	F
Notelaea longifolia Vent.	30	914	51.01	0.837	75	116	1	1	2	T	F	F	F	F	F
Olea paniculata R. Br.	61	3048	95.51	0.783	200	132	1	2	2	T	F	F	F	F	F
Oreocallis pinnata (Maid, and Betche) Sleumer	91	2438	50.34	0.275	60	363	2	2	2	F	T	F	F	F	F
Orites excelsa R. Br.	61	3048	95.51	0.783	96§	274§	3	2	1	F	T	F	F	F	F
Pennantia cunninghamii Miers	91	2438	50.34	0.275	150	145	1	2	2	T	F	F	F	F	F
Pentaceras australe (F. Muell.) Hook. f. ex Benth.	38	1829	88.81	1.165	70	232	3	1	4	F	T	F	F	F	F
Phaleria chermsideana (F. M. Bail.) C. T. White	15	762	82.02	2.691	40	175	2	2	2	T	F	F	F	F	F
Pithecellobium grandiflorum Benth.	23	1524	121.35	2.654	40	333	3	1	3	F	F	T	F	F	F
Pittosporum rhombifolium Ait.	23	914	68.01	1.488	75	114	3	1	3	T	F	F	F	F	F
Pittosporum undulatum Vent.	30	1219	71.01	1.165	45	248	3	1	2	T	F	F	F	F	F
Planchonella australis (R. Br.) Pierre	61	3048	95.51	0.783	68§	387§	1	2	2	T	F	F	F	F	F
Polyosma cunninghamii J. J. Benn.	20	1219	106.52	2.621	141§	76§	1	2	2	T	F	F	F	F	F
Polyscias elegans (C. Moore and F. Muell.) Harms.	76	2743	68.40	0.449	80	302	3	1	3	T	F	F	T	F	F
Polyscias murrayi (F. Muell.) Harms.	30	1524	91.01	1.493	45	302	3	1	3	T	F	F	T	F	F
Premna lignum-vitae (A. Cynn. ex Schau.) Pieper	91	3962	83.67	0.458	100	114	1	2	2	T	F	F	F	F	F
Pseudocarapa nitidula (Benth.) Merr. and Perry	46	2743	114.01	1.247	110	214	1	2	2	T	F	F	F	F	F
Pseudoweinmannia lachnocarpa (F. Muell.) Engl.	107	3962	71.72	0.336	300	114	1	2	1	F	T	F	F	F	F
Quintinia seiberi A.DC.	61	2134	65.51	0.537	120§	157§	1	2	1	F	T	F	F	T	F
Quintinia verdonii F. Muell.	61	2134	65.51	0.537	100	190	1	2	2	F	T	F	F	T	F
Randia benthamiana F. Muell.	15	762	82.02	2.691	49§	177§	1	2	2	T	F	F	F	F	F
Rhodamnia argentea Benth.	91	3048	63.67	0.348	280	95	1	2	2	T	F	F	F	F	F
Rhodamnia trinervia (Sm.) Bl.	76	2743	68.40	0.449	200	121	1	2	2	T	F	F	F	F	F
Rhodomyrtus psidioides (G. Don) Benth.	30	1219	71.01	1.165	55	203	3	2	3	T	F	F	F	F	F
Rhodosphaera rhodanthema (F. Muell. ex Benth.) Engl.	46	2134	87.34	0.955	180	104	1	2	3	T	F	F	F	F	F
Rhysotoechia bifoliolata (F. Muell.) Radlk.	41	1219	53.26	0.655	160	71	1	2	2	T	F	F	F	F	F

Table 4.3. (continued)

Species	D_{max} †	H_{max} †	B_2	B_3	Age	G	TOL	P1‡	P2‡	SWITCH 1	2	3	4	5	6
Sacopteryx stipitata (F. Muell.) Radlk.	23	1219	94.68	2.071	125	87	1	2	2	T	F	F	F	F	F
Scolopia brownii F. Muell.	61	2743	85.51	0.701	160	150	2	2	2	T	F	F	F	F	F
Sloanea australis (Benth.) F. Muell.	61	3048	95.51	0.783	117§	225§	1	2	2	T	F	F	F	F	F
Sloanea woolsii F. Muell.	122	4267	67.75	0.278	84§	437§	1	2	2	T	F	F	F	F	F
Stenocarpus salignus R. Br.	61	2743	85.51	0.701	118	202	1	2	1	F	T	F	F	F	F
Stenocarpus sinuatus Endl.	76	3048	76.40	0.501	200	133	1	2	1	F	T	F	F	F	F
Streblus brunonianus (Endl.) F. Muell.	23	1219	94.68	2.071	100	109	1	2	2	T	F	F	F	F	F
Symplocus stawellii F. Muell.	61	2743	85.51	0.701	145	165	1	2	2	T	F	F	F	F	F
Synoum glandulosum (Sm.) A. Juss.	61	2134	65.51	0.537	100	190	2	2	2	T	F	F	F	F	F
Syzygium coolminianum (C. Moore) L. Johnson	23	914	68.01	1.488	55	155	1	2	2	T	F	F	F	F	F
Syzygium crebrinerve (C. T. White) L. Johnson	91	3658	77.00	0.421	200	158	1	2	4	T	F	F	F	F	F
Toona austrais (F. Muell.) Harms.	183	4267	45.17	0.123	104§	355§	2	2	2	F	T	F	F	F	F
Trema aspera (Brongn.) Bl.	10	457	63.03	3.102	20	218	3	1	2	T	F	F	T	F	F
Wilkiea huegeliana (Tul.) A.DC.	15	762	82.02	2.691	78§	88§	1	2	2	T	F	F	F	F	F
Zanthoxylum brachyacanthum F. Muell.	15	914	102.02	3.347	44§	182§	1	2	2	T	F	F	F	F	F

†Taken from Francis (1970) and from observations in Lamington National Park.

‡Based on three years' observation of marked trees along two transects in primary and disturbed forests at O'Reilly's in Lamington National Park, Queensland.

§Parameter computed from data collected during experiment D1/8.1 at Wiangaree Forest. Information provided by the New South Wales Forestry Commission.

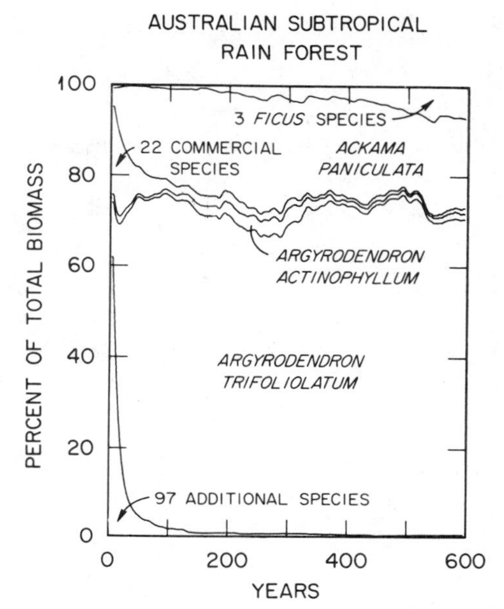

Figure 4.4. Compositional dynamics of the KIAMBRAM model (Shugart, et al [1980a, 1981b]) as the average of 50 1/20-ha plots. (From Shugart, et al [1981a].)

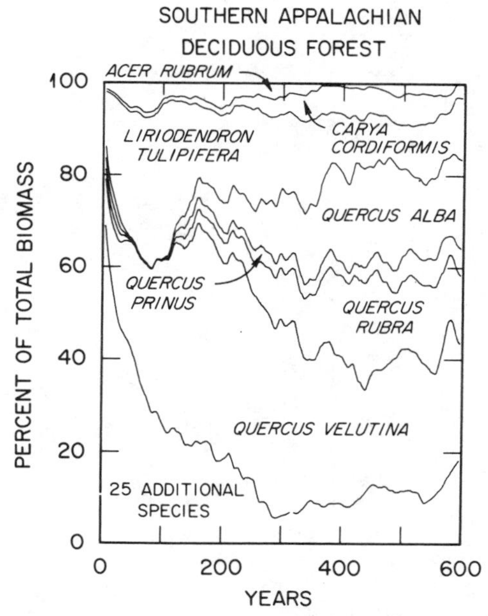

Figure 4.5. Compositional dynamics of the FORET model (Shugart and West, [1977]) as the average of 50 1/12 ha plots. (From Shugart, et al [1981a].)

Table 4.4. Basic parameters used for species attributes in the FORET (Shugart and West 1977) model

Species	$DEGD_{max}$	$DEGD_{min}$	B_3	B_2	T_c[†]	AGEMX	F[‡]	G	SPRTND	SPRTMN	SPRTMX	[§]Reproduction 1	2	3	4
Acer rubrum	13395	1810	0.173	52.58	1	150	1	238.5	3	12	200	F	T	T	T
Acer saccharum	7366	1522	0.119	52.11	1	300	1	167.6	3	12	80	F	F	T	T
Aesculus octandra	7366	5526	0.132	57.00	2	100	1	242.4	1	12	200	T	F	T	F
Carya cordiformis	9461	3686	0.306	69.21	3	300	1	116.8	1	12	200	T	F	T	F
Carya glabra	12652	3686	0.168	49.65	2	300	1	110.8	1	12	200	T	F	T	F
Carya ovata	10947	3686	0.312	69.66	2	300	1	116.1	1	12	200	T	F	T	F
Carya tomentosa	12652	4105	0.190	49.41	3	300	1	98.1	1	12	200	T	F	T	F
Castanea dentata	8499	3686	0.038	23.10	1	300	1	108.1	3	12	200	T	F	F	F
Cercis canadensis	9461	3686	0.952	59.03	1	100	1	87.4				F	T	T	F
Cornus florida	10947	3686	0.536	40.81	1	100	1	88.7	3	12	200	F	F	T	F
Diospyros virginiana	13395	5526	0.084	35.84	1	150	1	231.7	2	12	200	F	F	T	F
Fagus grandifolia	10204	2097	0.151	46.20	1	400	1	80.4	2	6	30	F	F	T	F
Fraxinus americana	10947	2414	0.080	34.43	3	300	1	112.6	2	6	20	F	T	T	T
Juglans nigra	8499	3686	0.074	36.37	3	250	1	160.4	1	6	20	T	F	T	F
Juniperus virginiana	10204	2966	0.236	57.75	3	250	2	106.7				F	T	F	F
Liquidambar styraciflua	10947	5526	0.191	55.20	3	300	1	119.4	2	12	80	F	T	F	T
Liriodendron tulipifera	10947	3686	0.044	32.35	3	300	3	174.8	2	12	200	F	T	F	T

Nyssa sylvatica	12652	3686	0.202	37.00	3	300	1	56.2	1	80	200	F	T	F	F
Oxydendron arboreum	8499	5526	0.348	63.57	1	200	1	133.4	3	12	200	F	F	F	T
Pinus echinata	9461	5526	0.290	70.75	3	300	2	127.9	2	6	20	F	T	F	F
Pinus strobus	6391	731	0.196	71.84	2	400	2	142.9				F	T	T	F
Pinus virginiana	7366	5526	0.421	77.00	3	300	2	105.6				F	T	F	F
Prunus serotina	10947	3899	0.083	35.57	3	250	1	138.6	3	12	200	F	F	F	T
Quercus alba	10204	2966	0.074	36.37	2	400	1	100.3	2	12	40	T	F	F	F
Quercus coccinea	8499	4105	0.133	39.00	2	400	1	66.6	2	12	80	T	F	F	F
Quercus falcata	10947	5526	0.078	33.57	2	400	1	82.6	2	12	30	T	F	F	F
Quercus prinus	7756	3686	0.064	27.28	2	267	1	102.2	2	12	40	T	F	F	F
Quercus rubra	8499	731	0.079	28.70	2	400	1	61.8	2	12	40	T	F	F	F
Quercus stellata	10947	5526	0.195	47.75	2	400	1	67.6	2	12	30	T	F	F	F
Quercus velutina	9461	3313	0.097	41.57	2	400	1	99.7	2	12	40	F	T	F	F
Robinia pseudoacacia	7366	5526	0.125	38.20	2	200	1	135.8	3	12	80	F	T	F	T
Sassafras albidum	10947	3686	0.619	75.51	3	200	1	107.5	3	12	200	T	F	F	F
Tilia heterophylla	8499	5526	0.396	60.40	2	150	1	144.5	3	12	80	F	T	T	F

†T_c is the tolerance class of each species. Classes 1 and 2 are considered shade-tolerant and class 3 is considered intolerant.

‡F is a classification that dictates which of 3 empirical curves from Sollins et al. (1973) is used to calculate biomass as a function of DBH for use in equation (9). Class 1 is deciduous trees other than Liriodendron; class 2 is coniferous trees and class 3 is Liriodendron.

§Reproduction switches are used in the BIRTH subroutine and take values of T (true) or F (false). Switch 1 is T if the species requires leaf litter for successful recruitment. Switch 2 is T if the species requires mineral soil. Switch 3 is T if the species recruitment is reduced by hot years. Switch 4 is T if the species is a preferred food of deer or small mammals.

and Ashe [1911]). Results from the FORET model will be discussed in some detail in the chapters that follow.

Mississippi River Floodplain Forest

The FORMIS model (Tharp [1978]) simulates the forest associated with different flooding regimes and soil types for the forests of the Mississippi River floodplain. The model has a repertoire of 33 species, (mostly deciduous hardwoods) (Table 4.5) many of which are endemic to this and a few other large southern river floodplains. The model results for high-flood regimes capture the apparent patterns around oxbows, sloughs, and other standing water systems as they are presently found in the floodplain. As flood frequency increases, there is a transition from cypress (*Taxodium*) swamp (Fig. 4.6) to Tupelo (*Nyssa*) swamp, which has its origins in the differing silvics of those species (Fowells [1965]). *Taxodium* can grow under conditions of year-round flooding, but it must be established on dry land. Thus, in very wet situations (e.g., an oxbow lake), *Taxodium* regeneration is dependent on an extended dewatered period that is due to drought combined with reduced river flooding. The episodic *Taxodium*

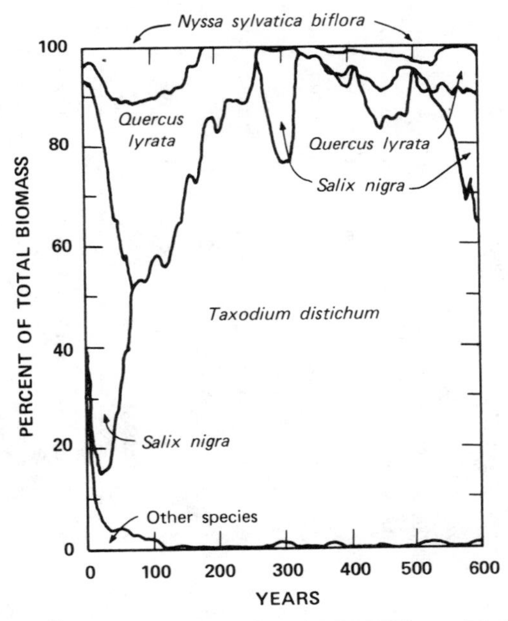

Figure 4.6. Compositional dynamics of the FORMIS model (Tharp [1978]) at relatively high durations of flooding for forests in the floodplain of the Mississippi River. Cypress swamp at 120 days/year of inundation.

regeneration is inferior to the regular *Nyssa* sprouting regeneration, as an adaptive strategy, when the dewatered condition occurs infrequently.

Northern Hardwood Forest

The JABOWA model (Botkin, et al [1972]) simulates forest dynamics that are associated with 13 species (Table 4.6) found in northern hardwood forests. Nominal locations are along the altitudinal gradient in the Hubbard Brook Watershed (Bormann and Likens [1979a,b]) in the White Mountains of New Hampshire. The compositional dynamics simulated by the model (Fig. 4.7) have patterns that are very similiar to that occur in this watershed to such a degree that Bormann and Likens (1979a,b) used the model as a unifying paradigm to relate the apparent pattern of forests in that watershed with the processes that caused those patterns. The longer term (greater than 600 years) response of the model matches several earlier observations as to the position of the species on the elevational gradient (Chittenden [1905]).

Puerto Rican Montane Rain Forest

Doyle (1981) developed the FORICO model to simulate the lower montane rain forest of the Luquillo Experimental Forest in northeastern Puerto Rico; it is locally known as the "tabonuco" (*Dacryodes excelsa*) forest (Wadsworth [1951, 1957]) after the dominant tree. The model simulates the dynamics of 36 tree species (Table 4.7) found on 1/30-ha plots. Successional dynamics of six of the more important species (Fig. 4.8) agree with the pattern of species replacement and codominance that is found in the actual forest, insofar as it is known at differet periods following disturbances (Crow [1980], Crow and Grigal [1980]). The model tends to underestimate the abundance of the palm, *Euterpe globosa*, which is a frequent understory species (particularly on boggy sites).

Southern Wetlands Vegetation

The SWAMP model developed by Phipps (1979) for the White River National Wildlife Refuge near DeWitt, Arkansas simulates the growth of 25 species (Table 4.8) on 1/25-ha plots in the same sort of forest ecosystem that was simulated by the FORMIS model (Tharp [1978]), which was previously discussed. Phipps initialized his simulator with actual plot data from sites that had been disturbed by lumbering, and he noted the change in basal area of trees by layers (Fig. 4.9) as these plots recovered. This sort of model test is structural, as contrasted to the general compositional

Table 4.5. Basic parameters used for species attributes in the FORMIS (Tharp 1978) model

Species	$DEGD_{max}$	$DEGD_{min}$	B_3	B_2	T_c [a]	AGEMX	G	SPRTND	SPRTMN	SPRTMX	FLOOD	Reproduction [b]				
												1	2	3	4	5
Acer negundo - Boxelder	9900	1600	0.0642	23.50	1	75	280	1	6	40	0.167	T	T	F	F	F
Acer rubrum - Red maple	13390	2300	0.1725	52.60	1	150	241	3	12	200	0.0	F	T	F	F	F
Acer saccharinum - Silver maple	9000	16000	0.0773	33.00	2	125	259	1	6	50	0.167	T	T	F	F	F
Carya aquatica - Water hickory	11100	5800	0.2556	54.58	2	250	108	1	6	70	0.750	T	T	T	T	F
Carya cordiformis - Bitternut hickory	9170	3470	0.3412	83.25	1	300	149	1	12	200	0.0	F	T	F	F	F
Carya illinoensis - Pecan	9900	4300	0.1598	58.50	1	250	189	1	6	25	0.167	F	T	T	F	F
Carya tomentosa - Mockernut hickory	10820	3470	0.3916	71.67	1	300	99	1	12	200	0.0	F	T	T	F	F
Celtis laevigata - Sugarberry	12560	4820	0.3844	70.34	1	200	146	2	6	115	0.0	F	T	T	F	F
Cornus drummondi - Dogwood	10700	1600	0.9039	41.36	1	50	117	1	6	15	0.750	F	T	F	T	F
Diospyros virginiana - Common persimmon	13390	4820	0.0840	35.86	1	150	232	2	12	200	0.0	F	T	T	F	F
Foresteria acumeneda - Swamp-privet	9600	5300	1.8600	76.14	2	50	170	1	6	20	4.999	F	T	F	T	F
Fraxinus pennsylvania - Green ash	9900	1600	0.2060	50.25	2	150	188	1	6	50	0.333	F	T	F	F	F
Gleditsia aquatica - Waterlocust	9900	4300	0.3479	63.67	2	150	178	1	6	25	0.333	F	T	F	F	F
Gleditsia triacautuus - Honeylocust	9900	4300	0.1234	45.17	2	150	248	1	6	25	0.333	F	T	F	F	F
Liquidambar styraciflua - Sweetgum	10820	4820	0.1908	58.20	2	250	158	2	12	80	0.0	F	T	F	F	F

Morus rubra - Red mulberry	13390	3470	0.1753	37.43	1	100	192	0	0	0	0.0	F	T	F	F	F
Nyssa aquatica - Water tupelo	9300	6000	0.0961	35.17	2	300	99	1	6	50	4.999	T	T	T	T	F
Nyssa sylvatica biflora - Swamp tupelo	11200	5700	0.2367	57.75	2	300	107	3	6	50	4.999	T	T	T	T	F
Planera aquatica - Planertree	9300	4800	1.6900	85.56	2	60	185	1	6	25	0.750	F	T	F	T	F
Platanus occidentalis - American sycamore	9900	3500	0.0388	28.42	2	500	93	2	6	50	0.167	T	T	F	F	F
Populus deltoides - Eastern cottonwood	10000	1600	0.0388	28.42	2	150	309	1	6	30	0.333	T	T	F	F	T
Quercus falcata pagodaefolia - Cherrybark oak	10000	4600	0.0907	38.72	2	300	177	1	6	12	0.0	T	T	T	F	F
Quercus lyrata - Overcup oak	9600	5300	0.1436	43.80	2	250	175	1	6	12	0.750	T	T	T	T	F
Quercus michauxii - Swamp chestnut oak	10200	5000	0.0508	27.89	2	300	166	1	6	12	0.0	F	T	T	F	F
Quercus nigra - Water oak	10700	5500	0.1098	40.17	2	250	191	2	6	12	0.167	F	T	T	F	F
Quercus nuttalli - Nuttall oak	9500	6100	0.2367	57.75	2	250	182	1	6	12	0.333	F	T	T	F	F
Quercus phellos - Willow oak	9600	4800	0.1143	41.84	2	250	198	1	6	12	0.167	F	T	T	F	F
Quercus shumardii - Shumard oak	10820	4520	0.0899	43.87	2	300	158	2	12	160	0.0	T	T	F	F	F
Salix nigra - Black willow	11200	2900	0.0694	33.88	2	70	534	3	6	30	0.750	T	T	F	T	T
Taxodium distichum - Baldcypress	11800	5300	0.0331	24.25	2	400	100	0	0	0	4.999	F	T	F	T	F
Ulmus americana - American elm	12560	2200	0.0421	28.27	1	300	142	2	6	240	0.0	F	T	F	F	F
Ulmus crassifolia - Cedar elm	11000	1600	0.8093	74.01	1	150	109	1	6	50	0.333	F	T	F	F	F

[a] T_c is the shade tolerance class of each species. Class 1 is considered shade-tolerant; class 2 is considered intolerant.

[b] Reproduction switches are used in the Birth subroutine and take values of T (true) or F (false). Switch 1 is T if the species requires mineral soil. Switch 2 is T if species cannot germinate when flood frequency exceeds flood tolerance. Switch 3 is T if the species is heavy-seeded (edible). Switch 4 is T if the species requires two-year drought conditions for germination. Switch 5 is T if the establishment of the species is restricted by leaf litter.

Table 4.6. Basic species attributes used in the JABOWA (Botkin et al. 1972a,b) model. G = growth constant. C = leaf area constant. AGEMX = maximum age (years). ITYPE is the tolerance to shade, 1 = intolerant, 2 = tolerant. D_{max} = maximum known diameter (cm). H_{max} = maximum known height (cm). B_2 and B_3 are constants in the equation $H = 137 + B_2D - B_3D^2$ relating height to diameter. D_{min} and D_{max} are minimum and maximum degree-days. W_{min} and W_{max} are minimum and maximum values for the index of evapotranspiration.

	G[5]	C[6]	AGEMX	ITYPE	D_{max}/H_{max}	B_2	B_3	D_{min}	D_{max}	W_{min}	W_{max}
Acer saccharum	170	1.57[1]	200[3]	2	152.5(3)/4011[3]	50.9	.167	2000	6300	300	
Fagus grandifolia	150	2.20[1]	300[3]	2	122(3)/3660[3]	57.8	.237	2100	6000[7]	300	
Betula alleghaniensis	100	1.486[1]	300[3]	1	122(3)/3050[3]	47.8	.196	2000	5300	250	
Fraxinus americana	130	1.75	100	2	50/2160[3]	80.2	.802	2100	10700	320	
Acer spicatum	150	1.13[1]	25	2	13.5/500	53.8	2.00	2000	6300	320	
Acer pennsylvanicum	150	1.75	30	2	22.5/1000	76.7	1.70	2000	6300	320	
Prunus pennsylvanica	200	2.45[2]	30	1	28(2)/1126[2]	70.6	1.26	1100	8000	190	
Prunus virginiana	150	2.45	20	1	10/500	72.6	3.63	600[3]	10000	155	
Abies balsamea	200	2.5	80[3]	2	50/1830[3]	67.9	.679	1100	3700	190	
Picea rubens	50	2.5	350[3]	2	50/1830[3]	67.9	.679	600	3700	190	
Betula papyrifera	140	0.486	80	1	46/1830[3]	73.6	.800	1100	3700	190	600[8]
Sorbus americana	150	1.75	30	2	10/500	72.6	3.63	2000	4000	300	
Acer rubrum	240	1.75	150[3]	2	152.5[3]/3660[3]	46.3	0.152	2000	12400	300	

(0): Values not otherwise referenced were developed by Botkin et al. (1972a,b).
(1): Whittaker 1970(Personal communication).
(2): Marks 1971.
(3): Harlow & Harrar 1941.
(4): Climatological ranges in growing degree days, obtained by matching northern and southern
(5): Growth constants adjusted for reasonable growth of individual tree in full sun with climate and soil factors equal to 1 (values of G will give 2/3 of maximum diameter at 1/2 maximum age starting from an 0.5 cm stem).
(6): Actual leaf area in square meters is $\sim CD^2/45$ for D in cm.
(7): Northern strain.

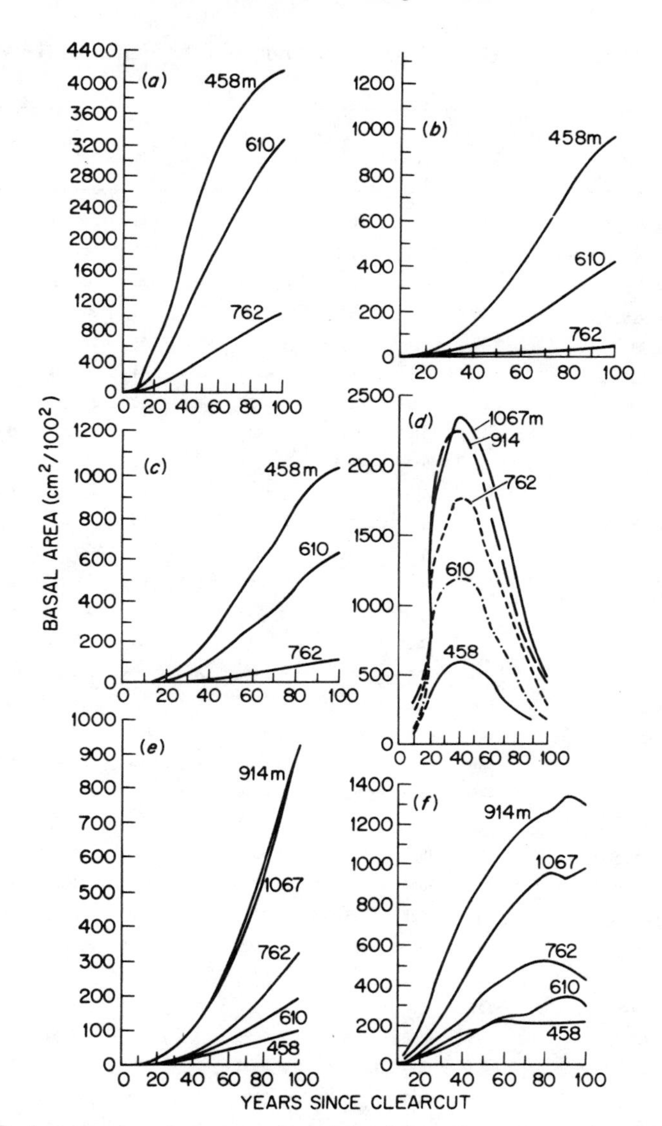

Figure 4.7. Average basal area as a function of time since clear-cut, and elevation for each of the six major species simulated by the JABOWA (Botkin, et al [1972a]) model. (a) Yellow birch; (b) Beech; (c) Sugar maple; (d) White birch; (e) Red spruce; and (f) Balsam fir. Each line represents the average for 100 plots at a single elevation with identical site conditions, including a deep well-drained soil. (From Botkin, et al [1972a]).

Table 4.7. Basic parameters for species attributes in the FORICO (Doyle 1981) model

Species name[a]	D_{max}[a]	H_{max}[a]	B_2	B_3	TOL[b]	AGE[c]	G[d]	S_{min}	S_{max}
Alchornea latifolia	46	1524	60.30	0.6554	2	58	240.19	0.05	0.12
Alchorneopsis portoricensis	46	1524	60.30	0.6554	2	56	246.40	0.05	0.12
Buchenavia capitata	122	2438	37.72	0.1545	2	120	182.85	0.05	0.12
Byrsonima coriacea	46	1829	73.56	0.7996	3	46	202.78	0.12	1.00
Casearia arborea	15	914	103.60	3.4533	3	47	175.37	0.12	1.00
Casearia sylvestris	10	457	64.00	3.2000	2	44	99.00	0.05	0.12
Cecropia peltata	61	2134	65.47	0.5366	3	50	271.12	0.12	1.00
Cordia boringuensis	13	610	72.76	2.7988	1	43	131.99	0.004	0.75
Cyathea arbores	13	914	119.53	4.5976	3	17	461.89	0.12	1.00
Dacryodes excelsa	152	3043	38.30	0.1259	1	205	132.34	0.004	0.05
Didymopanax morototoni	46	1829	73.56	0.7996	3	52	267.65	0.12	1.00
Drypetes glauca	15	914	103.60	3.4533	1	56	147.19	0.004	0.05
Eugenia stahlii	30	1829	112.80	1.3800	1	103	154.85	0.004	0.05
Euterpe globosa	20	1524	138.70	3.4675	1	22	587.34	0.004	1.00
Guarea ramiflora	15	762	83.33	2.7777	1	41	169.56	0.05	0.12
Guarea trichiloides	91	2286	47.23	1.2095	2	121	169.66	0.004	0.05
Hirtella rugosa	8	610	118.25	7.3906	1	32	164.62	0.004	0.05
Homalium racemosum	61	2134	65.47	0.5366	2	107	178.41	0.05	0.12
Inga laurina	46	2134	86.82	0.9437	1	75	229.10	0.05	0.12
Inga vera	46	1524	60.30	0.6554	2	65	271.38	0.05	0.12
Laetia procera	30	2286	143.26	2.3877	3	86	228.60	0.05	0.12
Linociera domingensis	30	1829	112.80	1.8800	1	65	247.67	0.05	0.12
Manilkara bidentata	122	3048	47.72	0.1955	1	157	171.76	0.004	0.05
Matayba domingensis	46	1829	73.56	0.7996	1	70	232.60	0.004	0.05
Miconia prasina	10	762	125.00	6.2500	2	26	257.17	0.05	1.00
Miconia tetrandra	30	614	92.46	1.5411	1	53	257.17	0.05	0.12
Ocotea leucoxylon	25	1524	110.96	2.2192	2	35	387.09	0.05	0.12
Ocotea moscnata	76	2438	60.55	0.3983	1	76	57.03	0.004	0.05
Ormosia krugii	61	1829	55.47	0.4547	2	84	197.39	0.05	0.12
Palicourea riparia	8	457	80.00	5.0000	2	29	142.81	0.05	1.00
Psychotria berteriana	10	610	94.60	4.7300	3	28	195.68	0.05	1.00
Sapium laurocerasus	61	1829	55.47	0.4547	2	167	99.00	0.05	0.12
Sloanea berteriana	91	3048	63.97	0.3515	1	118	226.08	0.004	0.05
Tabebuia heterohyila	46	1524	73.56	0.7996	3	79	206.75	0.05	0.12
Tetragastris balsamifera	46	2438	100.04	1.0874	1	169	125.87	0.004	0.05
Trichilia pallida	15	914	103.60	3.4533	1	254	32.23	0.004	0.05

[a]Taken from Little and Wadsworth (1964).

[b]Tolerance classes for each species drawn from Smith (1970).

[c]Computed from Eq. 3.7.

[d]Derived from tree growth data provided by the U.S. Forest Service Institute of Tropical Forestry (see Crow and Weaver 1977).

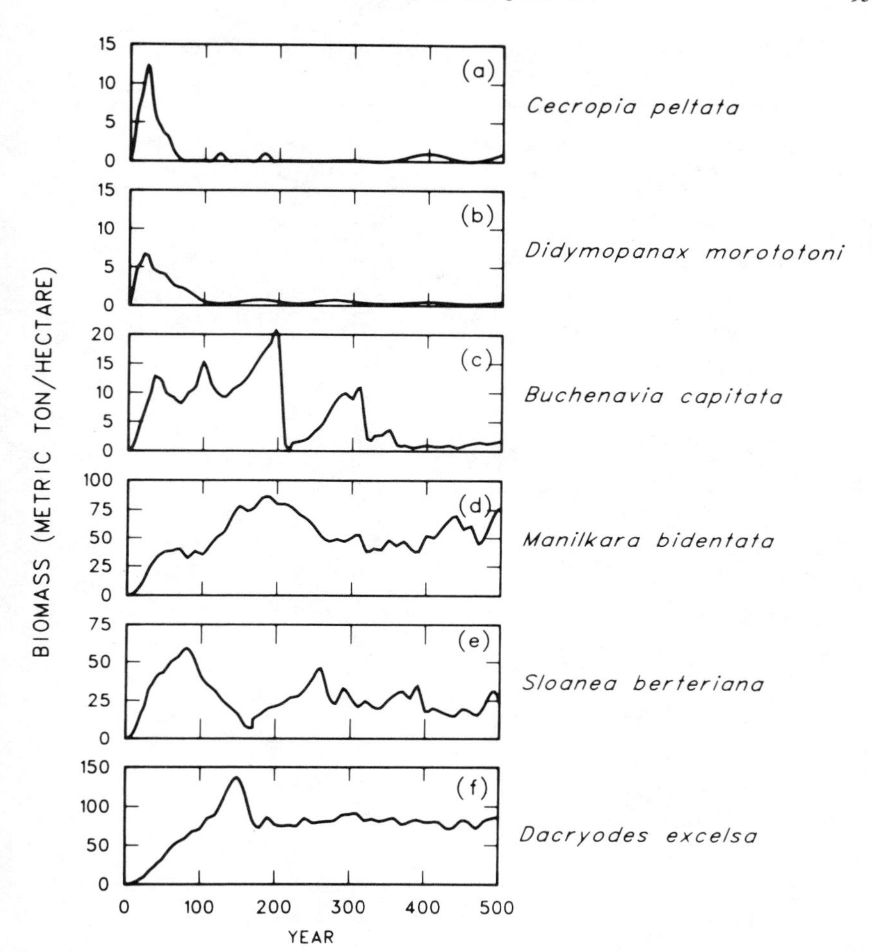

Figure 4.8. Successional dynamics of six dominant species of the Puerto Rican montane rain forest as simulated by the FORICO model: (a) *Cecropia peltata*; (b) *Didymopanax morototoni*; (c) *Buchenavia capitata*; (d) *Manilkara bidentata*; (e) *Sloanea berteriana*; and (f) *Dacryodes excelsa*. (From Doyle, [1981].)

model tests that have been discussed up to this point in this chapter. This case is included to complete the list of the diverse array of gap models. Other structural tests on gap models will be discussed below.

Extended Compositional Tests on Gap Models

The verifications described above test whether or not a gap model captures the qualitative compositional dynamics of the forests in some location when the model parameters are constrained to realistic values. Even

Table 4.8. Basic species attributes used in the SWAMP (Phipps 1979) model to simulate tree growth

Species	Shade tolerance[a]	B[b]	W[c]	Flooding[d]
Acer rubrum	1	500	0.60	0-30%
Carya aquatica	2	390*	0.30	10-40%
C. ovata	2	400	2.10	2-0 years
C. tomentosa	2	400	2.10	2-0 years
Celtis laevagata	1	350	0.60	5-35%
Diospyros virginiana	2	120*	1.50	0-30%
Fraxinus pennsylvanica	2	500	0.90	10-30%
Gleditsia aquatica	1	400	0.30	30-40%
G. triancanthos	2	400	1.50	0-15%
Liquidambar styraciflua	1	450	1.20	10-20%
Morus rubra	2	300	1.80	0-15%
Planera aquatica	1	350	0.30	30-40%
Prunus serotina	1	300	2.10	8-0 years
Quercus alba	2	500	2.40	5-0 years
Q. falcata	2	1380*	1.80	2-0 years
Q. lyrata	2	490*	0.60	2 years-40%
Q. nigra	2	500	1.50	10-15%
Q. nuttallii	2	1050*	0.90	2 years-30%
Q. phellos	2	700	1.20	0-20%
Q. shumardii	2	700	1.50	3 years-10%
Q. stellata	2	300	2.40	8-0 years
Taxodium distichum	2	775	0.15	30-50%
Ulmus alata	2	200	0.90	0-15%
U. americana	1	450	1.50	0-20%
U. crassifolia	1	375	1.20	5-30%

[a]Shade tolerance: 1 = shade tolerant, 2 = shade intolerant. In part after Baker (1950) and in part after Botkin et al. (1972a,b).

[b]B = basic growth rate, in mm^2/π (see Phipps 1979). * = entirely from data collected at White River.

[c]W = estimated optimum depth to water table in metres.

[d]Flooding range (duration in percent of time, and/or flood frequency in years) for which species is expected to reproduce. Adapted from Bedinger (1971).

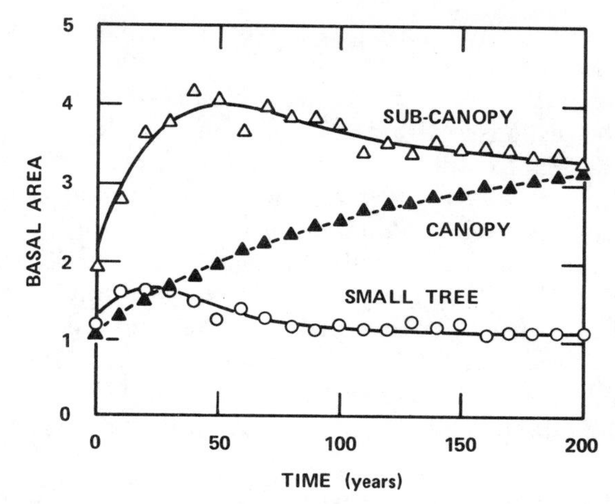

Figure 4.9. Basal area of sites recovering from lumbering disturbance over 200 years as simulated by the SWAMP model by layers. Basal area is in square centimeters, for small tree and subcanopy layers ($\times$ 10^3) and for canopy layer ($\times$ 10^4) (From Phipps [1979].)

these general tests are useful for rejecting inappropriate models and for providing useful information about the forests. The tests usually involve the pattern of species replacement, the dynamics or abundance of major species and the distribution of abundance among species in a "mature" forest. In some cases, the prediction of old-growth forest composition can be used as a validation of the models (as is alluded to in the JABOWA model (Botkin, et al [1972a]) and used in the KIAMBRAM model (Shugart, et al [1980a]) because the species parameters have been estimated from the silvics of the species and are independent of the old-forest stand data.

Extended compositional tests involve a model's ability to simulate forests under experimental conditions. It is rare to have access to data that was developed by direct experimentation on a forest. For this reason, most of the extended compositional tests use inference to shape natural events into an experimental context.

For example, in testing the FORET model, Shugart and West (1977) used the chestnut blight as a "before and after" experimental condition for a model validation. The fungus parasite, *Endothia parasitica*, was introduced into New York on imported Asiatic chestnut nursery stock (Walker [1957]). Within 30 years of its discovery (in 1904), this blight had destroyed virtually all of the chestnut (*Castanea dentata*) timber in the eastern United States. The FORET model parameters were estimated from easily available sources and were developed in such a manner that the parameters for chestnuts all could be estimated without running the model to obtain a curve fit. The model was verified by noting its ability to simulate the present-day forests in the southern Appalachians (with chestnut absent as a canopy tree species). The chestnut species was then "introduced" to the model, the model was run (100 1/12-ha plots for 600 years), and the response was tested against independent data on the composition of the mature southern Appalachian forest before the chestnut blight. The model was verified on the "experimental" condition (the present-day forests) and validated against the "control" (the mature forests of southern Appalachia with chestnuts). One result of this test is shown in Table 4.9, where the simulated pattern of forest composition is compared to historical data collected by Foster and Ashe (1908) for Tennessee forests. The FORET model was developed for Anderson County; thus, it should agree with the central part of Table 4.9. The information in the rest of the table is included to provide an impression of the regional

Table 4.9. Reproduction of a table from Foster and Ashe (1908) with the inclusion of results from the FORET model. The original table legend read, "Composition of forests in which chestnut oak forms more than 10% of the mixture—Southern portion of the region—Virgin growth" (Trees 10 in and over in diameter breast high).

	Slope type					Ridge type	
	Average of 738 acres†		Average of 262 acres*			Average of 62 acres	
Species	Number of trees	Percent	Number of trees	Percent	Percent simulated by FORET model	Number of trees	Percent
Chestnut oak	6•02	16•91	7•46	16•98	10	12•54	24•34
Black oak	5•02	14•11	6•33	14•41	15	8•04	15•60
Chestnut	13•65	38•35	8•73	19•88	20	14•67	28•47
White oak	1•56	4•38	6•09	13•87	15	0•93	1•81
Shortleaf pine	1•82	5•11	0•60	1•37	1		
Scrub pine	0•85	2•39	0•33	0•75	1		
Hickory	0•82	2•30	3•14	7•15	5		
Gums	1•01	2•84	2•37	5•40	4		
White pine	1•67	4•69	0•20	0•46	3		
Yellow poplar	0•71	2•00	1•84	4•19	10	1•01	1•96
Maples	0•55	1•55	3•55	8•08	12	4•51	8•77
Hemlock	0•24	0•67	0•12	0•27	1		
Brasswood	0•11	0•31	0•42	0•96	1		
Birch	0•07	0•20	0•24	0•55	1	0•15	0•29
Buckeye	0•02	0•06	0•13	0•29	1		
Cherry	0•01	0•03	0•04	0•09	1		
Beech			1•20	2•73	4	0•14	0•27
Locust			0•28	0•64	1		
Other species	1•46	4•10	0•85	1•93	1	9•53	18•49
Total	35•59	100•00	43•92	100•00		51•52	100•00

†Polk and Monroe Counties, Tennessee.

*Scott, Campbell, and Anderson Counties, Tennessee. Harlan County, Kentucky and Lee County, Virginia.

variability of forests. The Spearman rank correlation (Siegel [1956]) between model results and the empirical values given for Scott, Campbell, and Anderson Counties was 0.83 (it is significant at the 0.99 level). In addition to this direct comparison with a rather unusual and good quantitative historical summary of forest composition, the model also conforms to the more qualitative descriptions of the southern Appalachian forests which are found in historical accounts of the region (Sargent [1884], Ashe [1897, 1911], Pinchot [1906, 1907], and Greeley and Ashe [1907]).

Environmental gradients also can be used to test the response of a forest dynamics model. The JABOWA model (Botkin, et al [1972a]) was validated on its ability to predict the elevation of the transition from hardwood to coniferous forest with elevation in the Hubbard Brook Watershed. Tharp's (1978) FORMIS model was verified against predictions of the composition of forests in zones of the Mississippi River floodplain, with different flood intensities and different soils. Figure 4.10 shows an altitudinal zonation of the *Eucalyptus* forests in the Australian Alps. Shugart and Noble (1981) used the BRIND model to predict this altitudinal

Figure 4.10. An elevational zonation of *Eucalyptus* forests in the Brindabella Range (Australian Capital Territory). Three different zones are shown: (1) Next to the lake in the foreground is a mixed peppermint forest (*E. viminalis*, *E. robertsonii*, and *E. radiata* mixtures); (2) In the middle ground is alpine ash forest (*E. delegatensis*); and (3) In the background at the top of the mountain is snow gum (*E. pauciflora*).

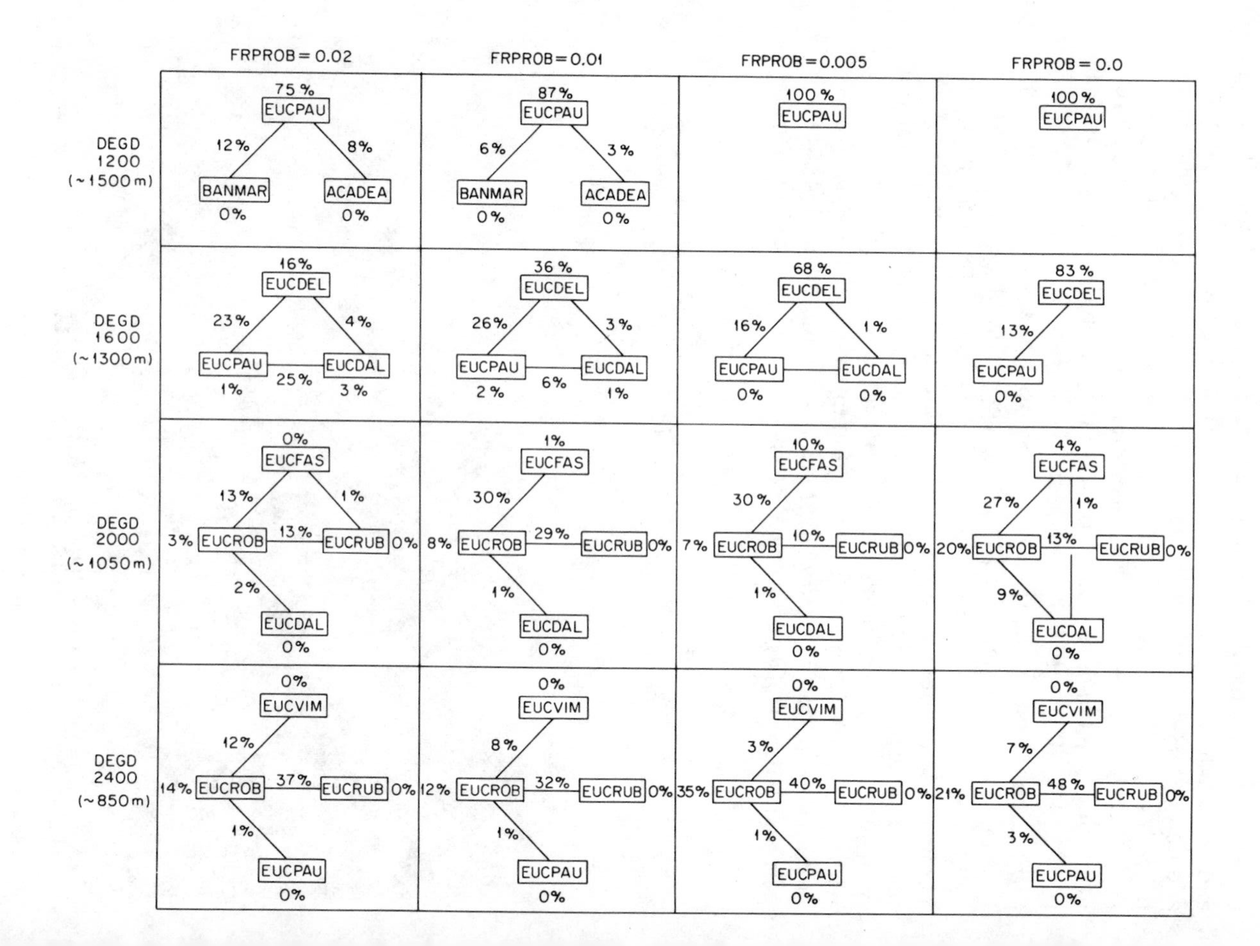
FRPROB = 0.02
FRPROB = 0.01
FRPROB = 0.005
FRPROB = 0.0
DEGD 1200 (~1500 m)
DEGD 1600 (~1300 m)
DEGD 2000 (~1050 m)
DEGD 2400 (~850 m)
75 % EUCPAU
12 %
8 %
BANMAR 0 %
ACADEA 0 %
87 % EUCPAU
6 %
3 %
BANMAR 0 %
ACADEA 0 %
100 % EUCPAU
100 % EUCPAU
16 % EUCDEL
23 %
4 %
EUCPAU 1 %
25 %
EUCDAL 3 %
36 % EUCDEL
26 %
3 %
EUCPAU 2 %
6 %
EUCDAL 1 %
68 % EUCDEL
16 %
1 %
EUCPAU 0 %
EUCDAL 0 %
83 % EUCDEL
13 %
EUCPAU 0 %
0 % EUCFAS
13 %
1 %
3 % EUCROB
13 %
EUCRUB 0 %
2 %
EUCDAL 0 %
1 % EUCFAS
30 %
8 % EUCROB
29 %
EUCRUB 0 %
1 %
EUCDAL 0 %
10 % EUCFAS
30 %
7 % EUCROB
10 %
EUCRUB 0 %
1 %
EUCDAL 0 %
4 % EUCFAS
27 %
1 %
20 % EUCROB
13 %
EUCRUB 0 %
9 %
EUCDAL 0 %
0 % EUCVIM
12 %
14 % EUCROB
37 %
EUCRUB 0 %
1 %
EUCPAU 0 %
0 % EUCVIM
8 %
12 % EUCROB
32 %
EUCRUB 0 %
1 %
EUCPAU 0 %
0 % EUCVIM
3 %
35 % EUCROB
40 %
EUCRUB 0 %
1 %
EUCPAU 0 %
0 % EUCVIM
7 %
21 % EUCROB
48 %
EUCRUB 0 %
3 %
EUCPAU 0 %

zonation and the effects of varying fire frequencies as a matrix of differing forest types (Fig. 4.11). At high elevations (about 1500 meters, Fig. 4.11), the model simulates forests that are dominated by *E. pauciflora* (snow gum) either in pure stands or (with high wildfire probabilities) with a shrubby understory of *Acacia dealbata*, *Banksia marginata*, or mixtures of the two. Such forests correspond well with those referred to as snow gum forests in Figure 4.10, At elevations of about 1300 meters (Fig. 4.11), the model predicts pure stands of *E. delegatensis* (as seen in Fig. 4.10); or less frequently, it stands dominated by *E. dalrympleana* and *E. pauciflora*. *E. delegatensis* dominance (due to its superior growth rate) is asserted with a diminution of the probability of wildfire. At 1,050 meters (Fig. 4.11), the model predicts the occurrence of brown barrel (*E. fastigata*) stands that are found at this altitude. Furthermore, the model predicts that *E. rubida*, although often present in mixtures with *E. robertsonii*, does not gain sufficient dominance to account for 90% of the biomass on an average stand. This is consistent with the observed inability of *E. rubida* to show dominance on southeast slopes, even in the face of an adequate seed source in the Brindabella Range; here, the species only dominates on northwest slopes (Anon., [1973]). Similarly, the rarity of *E. dives* in the simulations is consistent with the natural restriction of this species to dry sites. One might expect *E. viminalis* to be more in evidence (and *E. rubida* less so) than the model indicates at this elevation. At 850 meters, the model simulates stands with mixtures of *E. robertsonii*, *E. rubida*, *E. viminalis*, and (rarely) a low altitude form of *E. pauciflora*. Of these species, only *E. robertsonii* is able to show clear dominance at sites. The forest zonations (their location, their response to fire, and the details of their composition) show community types that are very much in evidence in the Brindabella Range (Anon [1973]) and in the higher elevations of southeastern Australia in general (Costin [1954]).

Simulating changes along an altitudinal zonation is somewhat analogous to simulating the long-term temporal change in a forested landscape in response to a change in climate. Solomon, et al (1980, 1981) applied the

---

Figure 4.11. Summary forest-type constellations for one- and two-species-dominated forest types simulated by the BRIND model at four different altitudes (expressed by the heat sum, DEGD) and four different wildfire probabilities (FRPROB). The value written over a box is the percentage of instances the species named in the box contributed 90% or more of the stand's total biomass. The value near lines between boxes is the percentage of instances in which the two named species contributed 90% or more of a stand's total biomass. Species mnemonics are taken from the first three letters of the scientific binomial: ACADEA = *Acacia dealbata*, BANMAR = *Banksia marginata*, EUCDAL = *Eucalyptus dalrympleana*, EUCDEL = *E. delegatensis*, EUCFAS = *E. fastigata*, EUCPAU = *E. pauciflora*, EUCROB = *E. robertsonii*, EUCRUB = *E. rubida*, EUCVIM = *E. viminalis*. (From Shugart and Noble [1981].)

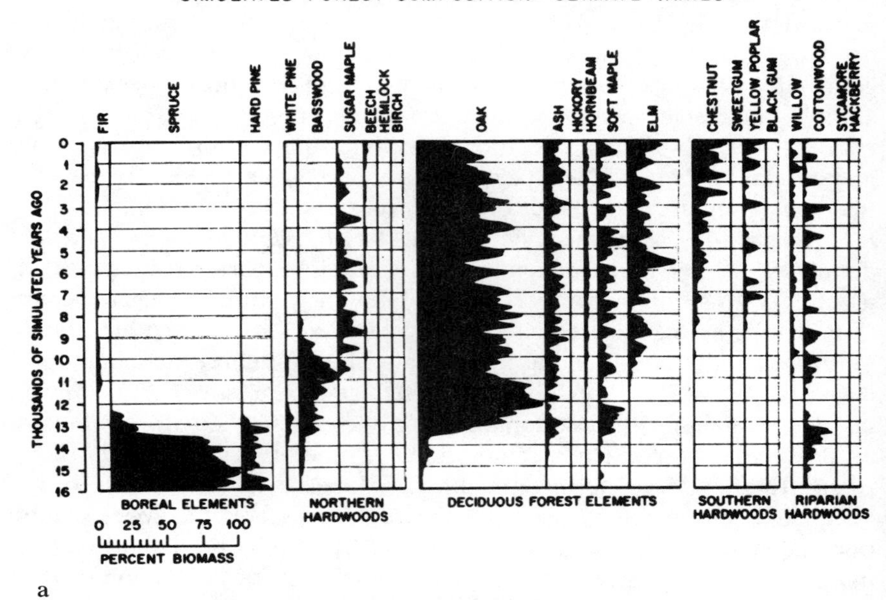

a

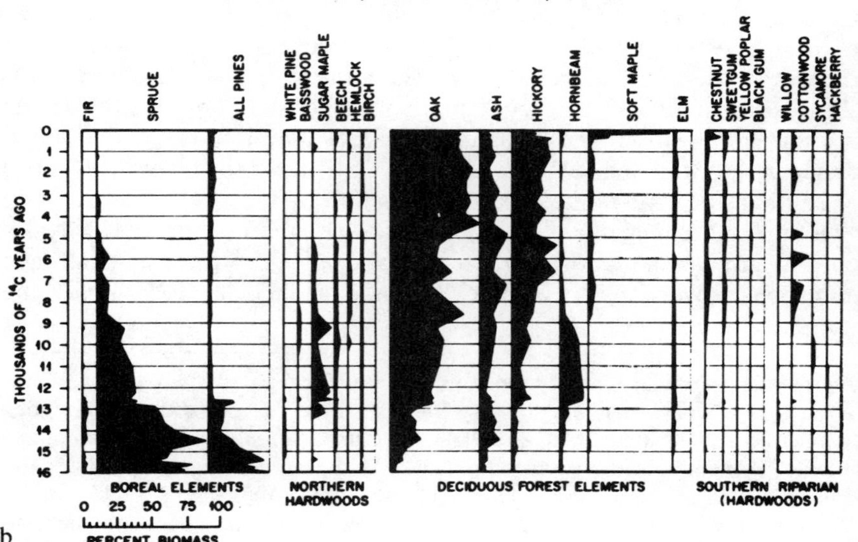

b

Figure 4.12. (a) Pollen percentage measured at Anderson Pond (White County, Tennessee) for arboreal taxa (Delcourt [1979]). (b) Biomass percentage composition simulated by the FORET model. (From Solomon, et al [1981].)

FORET model to predicting the pattern of forest composition indicated by the fossil pollen in a sediment core that was taken from Anderson Pond in White County, Tennessee (Delcourt [1979]). The temporal pattern of arboreal pollen for the past 16,000 years (as evidenced by this core) (Fig. 4.12a) was found to be in agreement with that predicted by the FORET model (Fig. 4.12b) when the model was driven with climate variations that were considered by Delcourt (1979) to be appropriate for eastern Tennessee for the past 16,000 years. Solomon, et al (1980) computed the correlation coefficient between simulated biomass and observed pollen values for 22 comparable taxa at 500-year intervals. The correlation was significant in all 22 cases, and the r^2 values averaged about 0.50 (indicating r values of approximately 0.70) across the entire simulation. Such correlations are quite good for data as variable as pollen data—particularly when given that no effort was made to include the surface heterogeneity (soils, topography, and so on) of the airshed that supplies Anderson Pond with its pollen. Simulations that incorporate these details are presented in Chapter 9.

Extended tests on the ability of a model to predict forest composition can be reasonably effective ways of gaining confidence in the model's reliability. Often, the ability of a model to predict a forest response under an altered condition is as close to actual experimental validation of long-term predictions as one can reasonably expect to obtain. Usually, the experimental context of these tests is developed in the test premise. For example, Doyle (1981), in validating his FORICO model of Puerto Rican tabonuco forest, hypothesized that the influence of hurricanes on Carribean forests was important enough that (without this frequent disturbance) the forest pattern predicted by tree silvics alone would not be appropriate to the island. Doyle tested this hypothesis by actually determining the pattern of dominance and diversity for this island with and without hurricanes and by comparing these two curves with actual data from Puerto Rico (Fig. 4.13).

Patterns in Forest Structure Simulated by Gap Models

Because gap models keep track of the diameter of each tree on a simulated plot, they provide a detailed set of predictions of forest structures that can be compared with real data as a model test. At the most general level of inspection, one can examine whether or not the pattern of structural stand dynamics is reasonable. This verification procedure (see Fig. 4.14) on forest structure is linked with what is called compositional tests in the first part of this chapter in that the composition of a forest contains some structural implications. The composition of a forest often is ex-

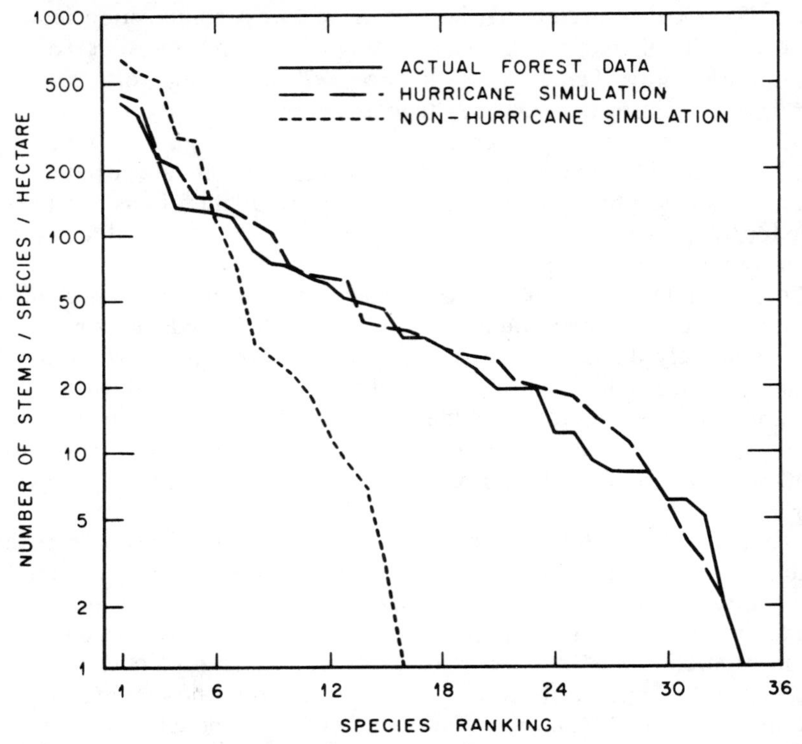

Figure 4.13. Dominance-diversity curves for model simulation with and without hurricane effects and with tabonuco forest data. Species rank in ranges from most abundant (1) to least abundant (36). Species abundance is represented as the total number of stems (above 4 cm dbh) per hectare. (From Doyle [1981].)

pressed by the species that are present as large trees. Figure 4.15 is a diagram of the compositional change of forests in the alpine ash (*E. delegatensis*) zone of the Brindabella Mountains. This figure was developed by inspecting many model simulations of structural and compositional changes simulated by the model. An example of the model output from one such simulation case is illustrated in Figure 4.14. Structural tests are tests of the model's ability to accurately predict statistical distributions of tree sizes.

Doyle (1981) validated the FORICO model by predicting the statistical distribution of diameters of trees for forests of different ages, and then by comparing these distributions to actual data provided by the USDA Forest Service Institute of Tropical Forestry in Rio Piedras, Puerto Rico (Fig. 4.16). One advantage of this sort of model test is that it simultaneously tests the birth, growth, and death formulations used in the models. The predicted curves can differ from the observed in their basic shape due to

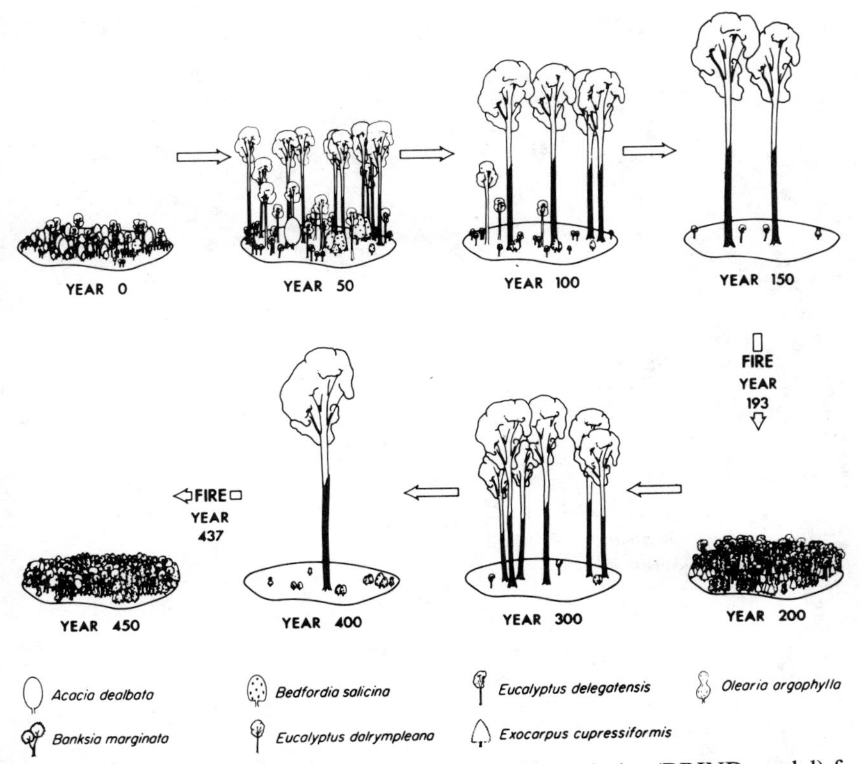

Figure 4.14. 450 years of change on a single simulated plot (BRIND model) for the alpine ash zone of the Brindabella Mountains. The species are drawn to scale by height, and the width of the plot is 32 meters.

several causes. If there is too little regeneration in the model, the initial size classes will be too low; if there is too much, they will be too high. Errors in tree growth rates can distort the rate at which the curve changes, as can errors in the mortality rates. Slow growth and/or high mortality can cause the diameter curve to drop away too rapidly in the larger-diameter classes. The effect of growth suppression that is associated with high mortality shapes parts of the curve associated with the smallest and largest diameter class. This is due to the fact that diameter growth is slower in large trees and that the likelihood of competition causing growth suppression is highest in small trees. In the example (Fig. 4.16a), the model appears to have the appropriate shape for the younger stands, but it regularly predicts too few trees in each size class. For older forests (Fig. 4.16b), the shape and details of the predicted and observed diameter frequency curves are similar; however, there is a tendency for the model to predict the presence of a few more trees in the 20- to 50-cm size range. Doyle used the Kolmogorov test (Siegel [1956]) and found no

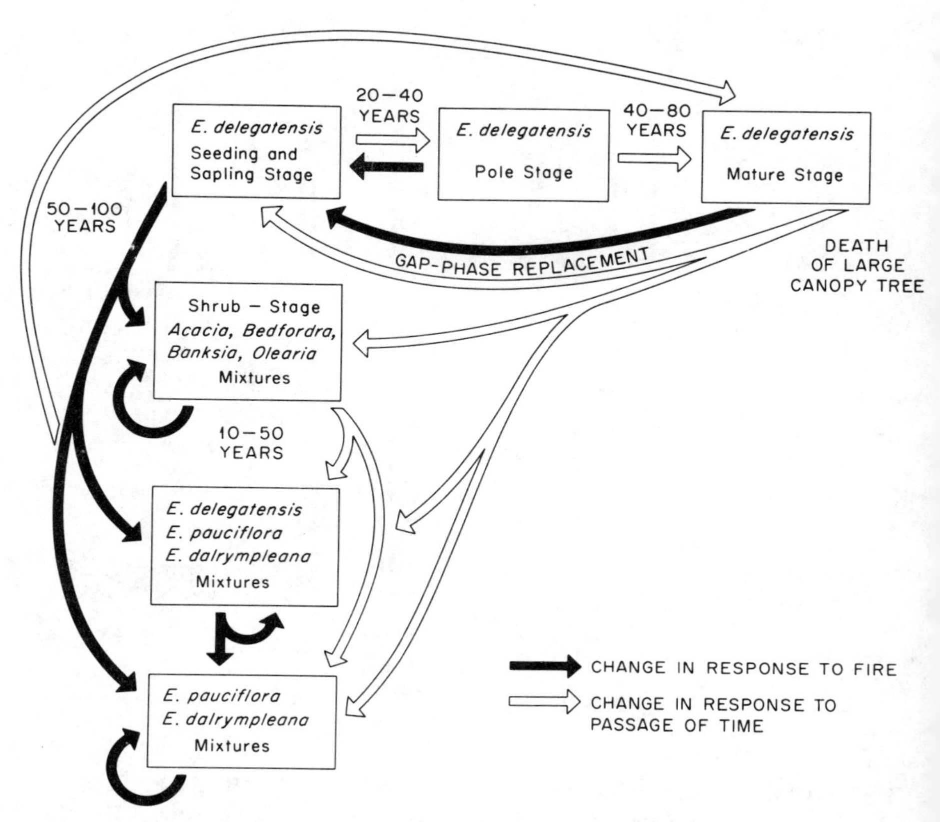

Figure 4.15. Simulated pattern of successional dynamics for forest types on sheltered slopes and southeasterly aspects at growing degree-day values of 1,600 day °C (~1,300 meters in the Brindabella Range). Boxes indicate forest types, light arrows indicate changes expected through time; dark arrows indicate changes expected due to fire. The structural changes in the forest illustrated in Figure 4.14 are two cycles through the top three forest types shown in this diagram. (From Shugart and Noble [1981].)

significant difference between model results (Fig. 4.16) and field data. Both curves (Figs. 4.16a,b) are similar to the degree that if they were data drawn from two different forests of the same type, they would be judged to be within the range of variations expected by chance; note that the horizontal scale in Figure 4.16a is five times less than that in Figure 4.16b. Mielke, et al (1978) used a test similar to that of Doyle as a verification of the FORAR model.

Another related test is the use of the model to predict average diameters, stocking densities, or basal areas of trees in stands of different ages. This test has been used both as a validation and as a verification of different gap models. Linda K. Mann, in modifying the FORAR model to

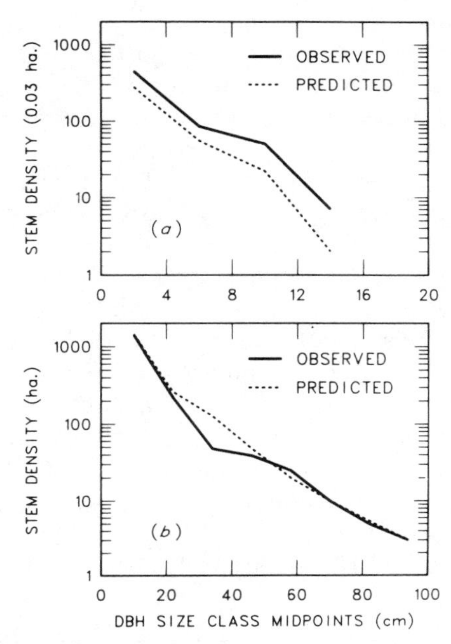

Figure 4.16. Density-diameter distributions between simulated results and field data from stands of different ages: (a) 9-year-old stand, (b) 150-year-old stand. (From Doyle [1981].)

be a forest plantation model for loblolly pine (*P. taeda*), verified the model against yield tables (Smalley and Bailey [1974]) for loblolly pine. By varying the "G" parameter (*see* Equation 3.6), Mann was able to verify the use of the FORAR model as a plantation simulator (Fig. 3.1). In a detailed structural validation on the FORNUT model, Weinstein, et al (1982) accurately (simulated individual tree growth within 5% of that measured) predicted average diameter increments at sites that were dominated by *Liriodendron* over a 13-year remeasurement period. Simulated tree growth on Chestnut oak (*Quercus prinus*) dominated sites were within 7% of measurements. The agreement was not as good (approximately 10–30%) on dry sites.

The independence between data and model outputs sometimes must be traded off against the need to modify a model to create a fair test. For example, Shugart and Noble (1981) used stand yield tables for alpine ash (*Eucalyptus delegatensis*) to validate the BRIND model. After the BRIND model was completed, a series of simulations using the model had been developed and a manuscript on the model had been drafted for publication; a previously unpublished data set collected by A. Lindsay in 1939 in Bago, New South Wales was summarized and published by Borough, et al (1978). This data set was in the form of a yield table for unthinned natural stands of *E. delegatensis* (site index: 40 meter height expected of "dominant" trees at year 50), and it was from a location near enough to the Brindabella Range (the site of the BRIND model development) that the model output should be comparable to the data. It seemed logical to compare Lindsay's data with output from the already developed computer simulations. However, there was one obvious problem with making this comparison. Lindsay's data had been collected from sites that had considerably higher initial stocking densities (about 5,600 stems ha^{-1}) than the example simulation that was available (*see* legend, Fig. 4.17). This difference may have been due to the *E. delegatensis* regeneration in the simulation that competed for sites with lignotuberous *E. pauciflora* regeneration. Regardless of the cause of the difference in initial conditions, this presented a difficulty. Should one adjust the model initial condition to better match the test data? Or should the validation test be conducted under a handicap, but with model/data independence? The latter option was taken. As one would expect, the diameters that were simulated in the understocked model stands was higher than those in Lindsay's yield table (Fig. 4.17). Nonetheless, the general rates of growth of the simulated trees were similar to the growth rates from Lindsay's

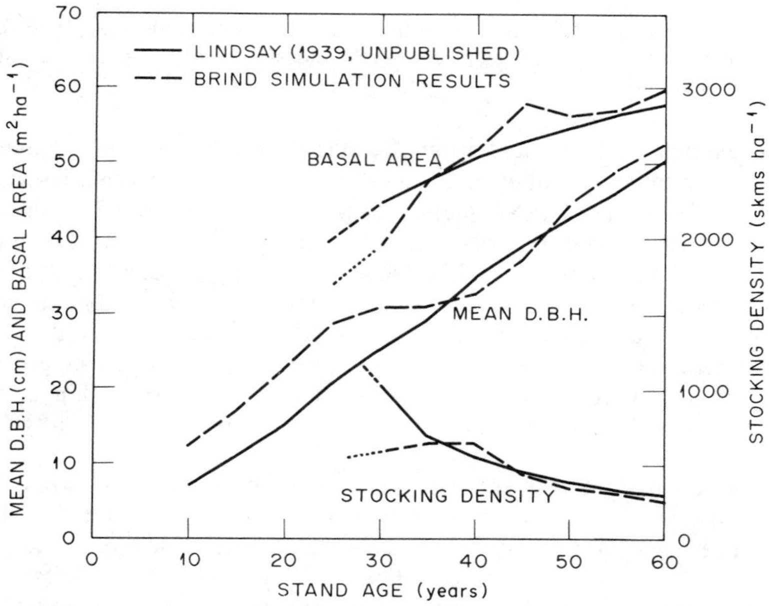

Figure 4.17. A comparison of basal areas, stocking density, and mean diameters from a yield table (Lindsay's 1939 unpublished data summarized by Borough, et al [1978]) with independent output from the BRIND model. Stocking densities are not shown for purposes of clarity at year 10 but are about 5,600 for Lindsay's data and about 2,600 for the example simulation. Solid lines are Lindsay's data; dashed lines are BRIND model output. (From Shugart and Noble [1981].)

table; also, the two DBH curves in Figure 4.17 are parallel for the first 20 years. At year 35, Lindsay's yield table had sufficient stand thinning to cause the yield table to drop to the density of the simulated data. From this point, the yield table and the data (Fig. 4.17) are very similar in the average DBH and stocking densities. The ability of the BRIND model to converge on real data, even when started at an inappropriate initial condition, seemed to constitute a strong test on the model. The model/data agreement that is seen after year 35 in Figure 4.17 is manifested throughout the simulation when the model is started at the appropriate initial stocking densities.

Conclusions

This chapter has presented several example gap models that have been developed for forests in different parts of the world. These models will be used in the remaining chapters to explore the dynamics of forested systems. The performance of the models in terms of reproducing compositional patterns and dynamics in forests, the response of forests to gradients, and the structure of forests has been discussed in this chapter—hopefully with enough detail to provide the reader with an impression of how the models perform. These and similar explorations of model behavior that are found elsewhere are summarized in Table 4.10. Some of the model applications in this table will be discussed in subsequent chapters. *In toto*, gap models appear to be able to mimic a nontri-

Table 4.10. Tests of gap models[a]

Verification

Model can be made to predict a known feature of a forest.

1. Predict forestry yield tables for loblolly pine (*Pinus taeda*) in Arkansas (FORAR).
2. Predict relations of forest types in succession in middle altitudinal zone in the Australian Alps (BRIND) and the Smoky Mountains (FORNUT).
3. Predict response to clear-cut in Arkansas wetlands (SWAMP).
4. Predict forest types changing as a function of flood frequency in Arkansas wetlands (SWAMP) and the Mississippi floodplain (FORMIS).
5. Be consistent with structure and composition of forests in New Hampshire (JABOWA), Tennessee (FORET), Puerto Rico (FORICO), and the floodplain of the Mississippi River (FORMIS).
6. Compare to forest of known age in subtropical rain forests (KIAMBRAM).
7. Predict Arkansas upland forests based on 1859 reconnaisance (FORAR).

Table 4.10. (continued)

Validation

Model independently predicts some known feature of a forest.

1. Predict response of *Eucalyptus* forests to fire (BRIND).
2. Assess effects of the chestnut blight on forest dynamics in southern Appalachian forests (FORET).
3. Predict forestry yield tables for alpine ash (*Eucalyptus delegatensis*) in New South Wales (BRIND).
4. Predict average diameter increment for forests at four different sites in Walker Branch Watershed, Anderson County, Tennessee (FORNUT).
5. Predict frequency of trees of various diameters in rain forests in Puerto Rico (FORICO) and uplands in Arkansas (FORAR).
6. Predict vegetation change in response to elevation in New Hampshire (JABOWA) and the Australian Alps (BRIND).
7. Determine effects of hurricanes on the diversity of Puerto Rican rain forest (FORICO).

Application

Model is used to predict a response of a forest to changed conditions.

1. Predict response of northern hardwood forest to increased levels of CO_2 in atmosphere (JABOWA).
2. Predict response of southern Appalachian hardwood forest to decreased growth due to air pollutants (FORET).
3. Predict response of northern hardwood forest (JABOWA), southern Appalachian forest (FORET), Arkansas upland forests (FORAR), Arkansas wetlands (SWAMP), and Australian subtropical rain forest (KIAMBRAM) to various timber management schemes.
4. Predict changes in a 16,000-year pollen chronology from east Tennessee in response to climate change (FORET).
5. Assess habitat management schemes for endangered species (FORAR), nongame bird species (FORET), and ducks (SWAMP).

[a] BRIND is a model of Australian *Eucalyptus* forests (Shugart and Noble, [1981]); FORAR of Arkansas mixed pine-oak forests (Mielke, et al [1978]); FORET of Tennessee Appalachian hardwood forests (Shugart and West [1977]); FORICO of Puerto Rican tabonuco montane rain forest (Doyle [1981]); FORMIS of Mississippi River floodplain deciduous forest (Tharp [1978]); JABOWA of northern hardwood forest (Botkin, et al [1972a]); KIAMBRAM of Australian subtropical rain forest (Shugart, et al [1981]); and SWAMP of Arkansas wetlands forest (Phipps [1979]).

vial range of the behavior of forest ecosystems. The models, in some cases have been successful in duplicating test data that were independent of the data used to develop the models (model validation as discussed in this chapter).

In my opinion, gap models should be considered a class of computer models that are worthy of exploration because of their implications regarding the larger scale response (in space and time) of forested systems. When considering these models in this more theoretical context, it is important to remember that these models only are representative of a limited set of the potential behavior of forested ecosystems. The worth of gap models will be related to the extent that this restricted set contains forest attributes that are of both theoretical and practical interest. In speaking of the tree-by-tree Markov model that he developed for the Princeton Forest, Horn (see Chapter 2) noted, "Even where the model works, it is intended only to be sufficient as a caricature of reality, rather than necessary as a mechanistic explanation" (Horn [1981]). Horn's caution to regard a model as a "caricature of reality" also is appropriate for gap models and the interpretation of their behaviors.

Chapter 5

Patch Dynamics in Forested Mosaics

If one considers a forested landscape as being a mosaic whose elements are the areas occupied by a dominant canopy tree (Fig. 5.1), it is convenient to define the dynamics at the scale of the mosaic element as patch dynamics, and the dynamics at the scale of the total mosaic as landscape dynamics. This chapter will discuss the important features of patch dynamics, both in general and as simulated by gap models. The dynamics of landscapes will be treated in later chapters, particularly Chapter 6, which parallels this chapter.

Regeneration Cycles in Gaps

Over periods of time that are comparable to the potential lifetime of trees, many important aspects of small patches (about 1/10 ha) change dramatically in response to internal cycles. The most important of these cycles is the periodic change in a patch's biomass, which drives other periodic changes in light profiles, temperature at the forest floor, and nutrient supply. A single patch within a forest is characterized by the growth of a number of small trees until one or two trees grow large enough to gain dominance over the others. After 100 years or more, these dominant trees die and a new cohort of more or less even-aged individual trees begins to grow. These young trees compete, grow, and die until one tree gains dominance and closes the cycle. The changes in biomass of a 1/12-ha patch over several centuries should be represented by a saw-toothed curve (Fig. 5.2). Such a pattern was determined by a detailed reconstruction of a small plot of land (0.35 ha) in the Harvard Forest, Petersham,

a)

Figure 5.1. Regeneration waves. (a) Regeneration waves near Mount Katahdin, Maine. (b) Diagrammatic section through a regeneration wave. (From Sprugel, DG, Bormann, FH, "Natural disturbance and the steady state in high-altitude balsam fir forests. *Science* 211:390–393, © 1981 by the AAS. Original photo courtesy of DG Sprugel.)

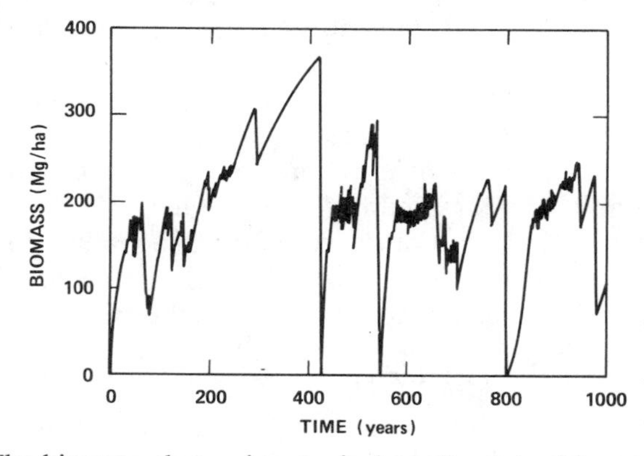

Figure 5.2. The biomass change in a typical small patch of forest (a 1/12-ha plot simulated here by the FORET model) fluctuates in response to the death of canopy trees and growth of their successors. (From Shugart and West [1981].)

Massachusetts (Oliver and Stephens [1977], and discussed later in Chapter 7), and it was implied by earlier studies at nearby sites (Henry and Swan [1974], and Stephens [1956]). The conceptualization that a small patch of forest has an agrading and degrading phase that corresponds to the saw-toothed biomass curve was discussed by Watt (1947). An established idea in ecology for some time has been that the small-scale pattern of a forest is that of a nonequilibrating system with respect to biomass (or the related variables such as tree volume, leaf area, and so on). Early soil scientists held the same concept in viewing tree uprooting as a soil process (Shaler [1891], Van Hise [1904], and Lutz and Griswold [1939], all cited by Stephens [1956] in a brief review of the topic). Watt's (1947) ideas about small-scale patch dynamics have been continued by several recent investigators (Whitmore [1975], Oliver and Stephens [1977], Bormann and Likens [1979a,b], and Shugart and West [1981]).

There are two central hypotheses about the underlying recruitment processes that generate the saw-toothed biomass dynamics (Fig. 5.2). These two hypotheses have been presented by several authors (notably Egler [1954], Spurr and Barnes [1973], Mueller-Dombois and Ellenberg [1974], and Drury and Nisbet [1973]). At one extreme, one can view the recruitment of new trees following a disturbance that is due to the death of a large canopy tree as being drawn from young trees that already were present and were tolerant to the shading from the original canopy dominant. The operation of this process over several birth-growth-death cycles would cause the dominance of gaps to pass to successively more tolerant species whose progeny could survive shading (and other alterations of the patch's environment due to the presence of a large tree). This tendency for an intolerant species to "relay" a site to a more tolerant species prompted Egler (1954) to term this process "relay floristics." This autogenic mechanism is presented, in most classic accounts of forest dynamics, as the process that drives succession (Clements [1916], Oosting [1956], Odum [1959], MacArthur and Connell [1966], and Daubenmire [1968]).

At the other extreme, the establishment of new trees may be a consequence of recruitment from seedlings or sprouts that occupy the site after a disturbance and are able to out-compete any subsequently recruited individuals ("initial floristic composition," Egler [1954]). If the disturbances that initiate this recruitment are outside the influence of the vegetation at a site, the pattern is termed "allogenic succession" (Spurr and Barnes [1973]). The allogenic nature of forest dynamics has been pointed out by several ecologists (Raup [1957], Olson [1958], Henry and Swan [1974], and Oliver and Stephens [1977]).

The importance of autogenesis versus allogenesis in forest dynamics has been debated by ecologists for some time (Egler [1954], and Drury and Nesbit [1973]); there is little unity on the topic. Part of the problem is the lack of direct evidence to exhaustively test either hypothesis. Knowledge of the age distributions of trees in a forest could provide insight into this problem. Unfortunately, tree size is not always a reliable indicator of tree age (Harper [1977]), and the inspection of size distributions of trees can be misleading as indicators of age distributions. It also is difficult to observe forest behavior that is elicited by an unpredictable event (a disturbance) and is manifested by a phenomenon that is difficult to sample (population dynamics of small trees—both seedlings and sprouts).

There are two other fundamental problems with the autogenic/allogenic hypotheses that are coupled with this measurement difficulty. First, the two hypotheses are not necessarily exclusive (*see* Chapter 7). It is quite possible that one species might regenerate itself in an autogenic pattern, while another species might require a disturbance and regenerate in an allogenic succession. Measurements in a forest could produce evidence to support one theory at one point in time and the other at some other time; or, one forest at one location might be autogenic, while another forest at a different location might be allogenic in its successional pattern. Second, many (but not all) of the mechanisms (e.g., trees growing beneath other trees) that have been discussed as driving autogenic succession operate at a spatial scale that is small compared to the disturbances driving allogenic successions (e.g., fires, hurricanes, and floods). Small disturbances, such as the death of an individual tree, are intrinsically autogenic. The magnitude of a tree-fall disturbance depends on the size of the fallen tree. In addition, some of the allogenic disturbances, particularly fire, may be controlled to a degree by such autogenic considerations as fuel load. Heinselman (1981), referring to the coniferous forests of northern North America, noted that wildfire ignition seemed more likely in old stands and that "blowups" of fires into major fires occurred in old stands. Regardless of the degree of allogeneity in large-scale disturbances, the state of a forest has a considerable influence on the nature and intensity of a disturbance at the scale of a patch that is simulated by a gap model.

Small Disturbances in Gap Models

While the biomass dynamics of gap models have a characteristic sawtooth pattern (Fig. 5.2), it is more difficult to generalize the structural and compositional response at the same scale across all of the models. The structure of a forest gap as simulated by the FORET model, can be used as an example (Fig. 5.3). The initial stage of the recovery of the biomass curve following the death of a large tree features the recruitment of several hundred seedlings and sprouts (Fig. 5.3a). This more or less evenaged cohort grows; there is competition within it (Fig. 5.3b) until one individual dominates the rest (Fig. 5.3c). If this individual grows large enough, it eventually shades out virtually all smaller trees. Upon the death of this tree, the biomass on the plot drops sharply and a new wave of regeneration is initiated (Fig. 5.3d).

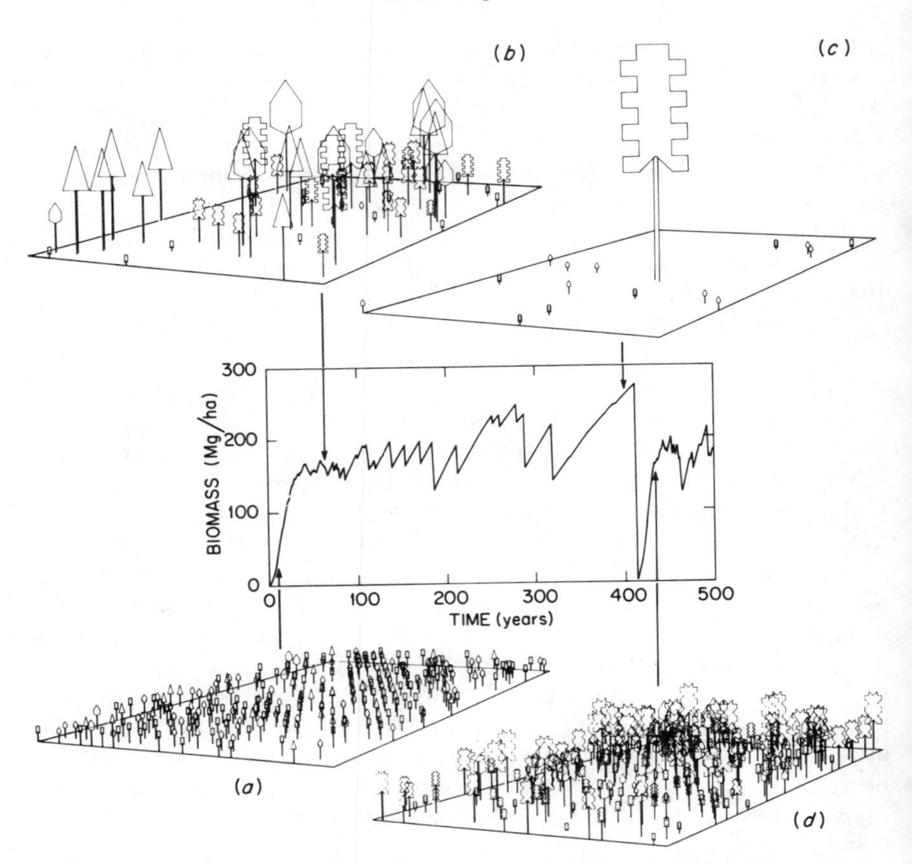

Figure 5.3. The structure of a forest gap simulated by the FORET model for different parts of the gap-scale biomass curve center. The gap-scale biomass curve (mg/ha of trees versus time) for one cycle. (a) Structure of regenerating forests; (b) Canopy competition in an even-aged patch; (c) Dominance of plot by one canopy tree; and (d) Regeneration following the death of the dominant tree. The horizontal scales are twice the vertical scale.

In simulations of the FORET model, there are considerable variations in the pattern of stand regeneration. If the canopy tree is of a species that is relatively small or if (due to either chance or the short-lived character of its species), it dies soon after becoming the dominant tree, then the next successful tree is likely to be recruited from the already established subordinate trees—not from seedling recruits. Sometimes, already established trees of a reasonable size can be overgrown by seedlings if the disparity of growth rates is great enough. In some short-lived species, mortality can open the canopy so that tolerant species can develop underneath them and grow at a slow, but continual, rate toward stand dominance.

Species also can change their successional status in response to growth rate changes that are induced by environmental gradients. For example, yellow-poplar (*Liriodendron tulipifera*), as simulated by the FORET model, can persist as a mature forest species in reasonable numbers by gap-phase replacement (much the same as the case shown in Fig. 5.3). In fact, if it becomes established on a plot, it often is capable of growing past larger individuals of competing species that were established before the gap formed. This model behavior of the species is consistent with the actual ecology of the species (Skeen [1976]). When the growth rate of *Liriodendron* is reduced in the model, the species eventually loses the ability to grow past small established trees to dominate canopy openings, and it is lost as a mature forest species. It is important to note that the regeneration of species in the patches beneath canopy gaps is a complex process that produces a more or less generally similar pattern of biomass response.

The saw-tooth biomass curve and the changes of other aspects of a gap that is associated with this curve cause different aspects of a tree's life history to be emphasized at different stages in the gap development. When the biomass and leaf area are low following the fall of a large tree, the regeneration aspect of the trees is particularly emphasized. As the established seedlings and sprouts close the canopy, the emphasis for survival shifts toward growth and form. Finally, at the peak of the biomass curve, species longevity and the pattern of mortality can influence the next regeneration stage.

Regeneration and Gap Size

The fall of a large tree produces an abrupt change in the forest floor in relation to several important biotic variables. Light is dramatically increased in quantity, and it also is altered in quality with a shift to more radiation in the red end of the spectrum and less in the blue end. With more sunlight, there is an increase in soil temperature and a related decrease in relative humidity and surface soil moisture. In the absence of overhead leaves and branches, the energy of rainfall is delivered to the soil surface. Changes in soil properties after gap formation include increased decomposition and nutrient availability. Mineral soil is exposed with coincidental changes in microtopography when the tree falls. The sudden change in these and other important variables can kill already established seedlings that were adjusted to the microclimate and can simultaneously advantage new seedlings, perhaps of other species. The change in microclimate also may serve to the advantage (or disadvantage) of plants or other life forms with which the seedlings compete. For gaps of different sizes, the magnitude of the microenvironmental change following the death of a tree changes; also, the effect on the regeneration niche space (Grubb [1977]) can be manifested as a change in the species that successfully produces seedlings (Whitmore [1975]). Lyford and MacLean (1966) found that certain species are more likely to germinate on mounds, while others are more likely to germinate in pits that result from tree fall and uprooting.

When a small tree dies, the resultant gap in the canopy is small, and it may be filled by the growth of trees that are present on the plot (Trimble and Tryon [1966]). In this case, the alteration of the forest floor environment is slight; tree species that regenerate under other trees or that sprout are given an advantage. In slightly larger gaps, there may be a regeneration of various species. However, the established saplings would tend to have an edge in the competition for the canopy space. In still larger gaps that are caused by the death of a large canopy tree, the typical gap pattern (Fig. 5.3) applies.

Denslow (1980) noted a spectrum of environmental changes as a function of gap size in tropical forests. She classified rain forest trees into categories according to their ability to colonize gaps of different sizes. Large-gap specialists have highly shade-intolerant seedlings produced from seeds that germinate only in high-temperature and high-light conditions. Small-gap specialists have seeds that germinate in shade, but that require the presence of a gap for growth. Understory specialists do not seem to require gaps for either regeneration or growth. Denslow (1980) pointed out that there was a continuous gradation across these general categories.

Gap models consider many species attributes that Denslow (1980) and others (van der Pijl [1972], Whitmore [1975], Grubb [1977], and Bazzaz and Pickett [1980]) have noted as being important in the differentiation of the gap-size-related regeneration patterns of trees. Such attributes generally are applied as filters to screen the species list against the simulated microenvironment to obtain a list of species that are eligible for establishment in a given year. Such filters (*see* Chapter 3) include light requirements, moisture requirements, dispersal, seed longevity, and seed storage of each tree species. The patterns of temporal variation in established seedlings differ considerably (Fig. 5.4) when even these simple filters are applied to the dynamics of gap formation over periods of several tree generations. If variations among species' establishment that are generated by such simple methods (Fig. 5.4) are realistic, then the determination of more subtle establishment adaptations in nature will require studies of an intensity (and duration) that is far beyond what is typically conducted. Most current reviewers recognize this and tend to discuss regeneration from the pragmatic view that factors influencing the establishment of seedlings can be usefully grouped in broad classes (Kozlowski [1971a,b], van der Pijl [1972], Grubb [1977], and Denslow [1980]).

Even broad classes of regeneration-niche categories combined with

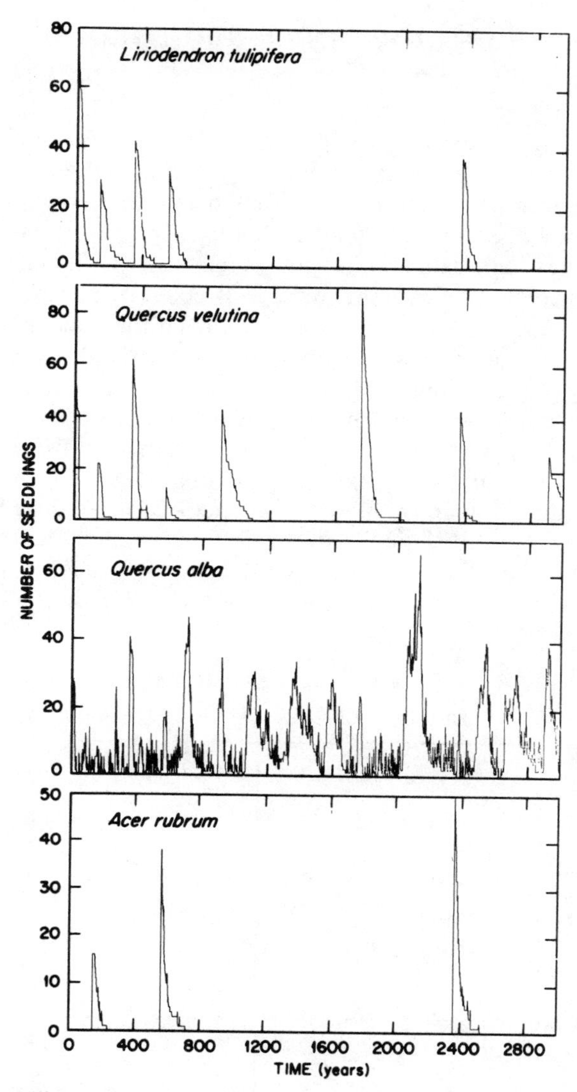

Figure 5.4. Seedling establishment in a forest gap (1/12-ha) for selected species, as simulated by the FORET model for 3,000 years.

stochastic processes and the temporal variability in the forest floor environment allow a maintenance of diversity in gap model simulations. The actual size of the tree that dies is an important determinant of which species are successful in regenerating in the gap created by the death event—a feature of gap models and of actual forests (Whitmore [1975], Denslow [1980], Bazzaz and Pickett [1980], and Opler, et al [1980]).

The Role of the Species in Determining Gap Size

Landscape-scale processes (Chapter 6) such as fires, hurricanes, and landslides, all can create very large open patches that encourage certain species' regeneration. Disturbances that occur at infrequent intervals, even when they operate on a small scale (e.g., species-specific disease or lightning strike), can cause large canopy gaps, if the length of intervals between disturbances allows the average affected tree to grow large. Frequent disturbances can reduce the size of gaps if they occur so frequently that the canopy mortality is high enough to greatly reduce the size of the typical canopy tree. Along with these exogenous influences on gap size, an endogenous variable—species composition—can determine the size of the gaps as well. The stature of the dominant individual tree can influence the gap size distribution as a function of composition much in the same way that the disturbance frequency influences gap size distribution. This is due to the fact that the species of the dominant individual tree indicates its maximum size. Furthermore, some species typically die violently when they are mature trees (e.g., mortality due to windfall). Other species

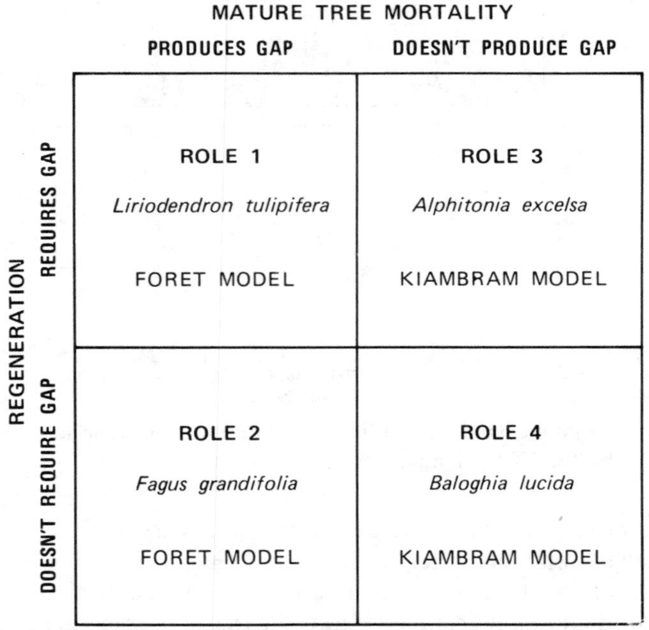

REGENERATION \ MATURE TREE MORTALITY	PRODUCES GAP	DOESN'T PRODUCE GAP
REQUIRES GAP	ROLE 1 *Liriodendron tulipifera* FORET MODEL	ROLE 3 *Alphitonia excelsa* KIAMBRAM MODEL
DOESN'T REQUIRE GAP	ROLE 2 *Fagus grandifolia* FORET MODEL	ROLE 4 *Baloghia lucida* KIAMBRAM MODEL

Figure 5.5. The role of gap-requiring and gap-producing tree life history traits with example species for each of the four roles. *See* Figure 5.6 for the dynamics of forests that are composed solely of each of the example species.

typically die by attrition, so that the standing snag of the tree continues to affect the microenvironment even after the death of the tree. Such differences in the usual pattern of death also influences the degree and the abruptness of change in the forest floor environment.

Because the size of the tree that dies, thus producing a gap, influences regeneration (which influences canopy composition) and since there can be a degree of influence between the canopy composition and canopy gap size, it seems adaptive for species traits to close this causal loop. For example, a tree species that required large gaps for regeneration logically may be expected to grow to a large size so that its mortality would create the conditions needed for its regeneration. Such patterns of mutually adaptive traits determine the roles (Fig. 5.5) of tree species (Shugart, et al [1981]).

Roles of Species in Patch Dynamics

Two very general types of regeneration can be combined with two equally general assessments of the ability of a species to create a gap on mortality that will define four roles (Fig. 5.5). These roles can be used to inspect small-scale forest patch dynamics. It should be made clear that these roles are intended to be high-contrast black-and-white cartoons of multicolored reality. In fact, it may not be particularly useful to think of a given species as having a specific role, except in the context of a given forest ecosystem. For example, a Role 1 species in a given system might behave as a Role 3 species under another condition, if its ability to grow to a large size before death is altered. The effects of species roles on the landscape scale are discussed in Chapter 6. The present listing is intentionally simplified for purposes of discussion (Fig. 5.5), but it does include patch-scale roles that are evidenced across several gap models. The roles that are particularly important at the scale of a small plot of land follow.

Role 1 Species

These require large gaps for regeneration, grow rapidly to a large size, and die abruptly. In nature, such species could be expected to have regularly produced, efficiently dispersed seeds that could be expected to survive for long periods of time. Germination would be triggered by factors associated with the creation of a large gap. An example species for this role is the yellow-poplar (*Liriodendron tulipifera*), as simulated by the FORET model. The species has its best regeneration on mineral soil, with adequate light and moisture. It has wind-dispersed seeds (produced regularly and in large numbers) that can survive up to 7 years in the soil (Fowells

[1965]). This shade-intolerant species grows rapidly, has a considerable height for its diameter, and can eventually reach 50–55 meters in height. Large trees often die from windthrow, and large-standing dead *Liriodendron* are infrequent. The species was well-represented in the original forest of the southern Appalachians, and its survival can be attributed to gap-phase replacement (Skeen [1976], and Runkle [1981]). It is potentially the largest tree in the forests in which it occurs.

The small-scale dynamics of a forest comprised of a single Role 1 species (*Liriodendron tulipifera*, as simulated by the FORET model) (*see* Fig. 5.6a) feature a greatly accentuated saw-toothed biomass curve with very high biomass maxima. The associated number of stems on the small plot (1/12 ha) shows a saw-tooth curve that is oriented in the opposite direction, with large recruitment events occurring when the last tree of the previous cohort dies and a thinning of the new cohort through the remainder of the cycle (Fig. 5.6a).

Role 2 Species

These do not require gaps for regeneration; in some cases, they are disadvantaged by the conditions that are found in a newly formed large gap. They are shade-tolerant and grow at a reasonable rate to a moderately large size. Role 2 species may be quite long-lived and slow to die. Vegetative reproduction can be as important as seedling regeneration (or even more so) in some cases. An example species for this role is the American beech (*Fagus grandifolia*). This species is quite tolerant of shading. It produces large crops of relatively large edible seeds that must be sheltered in leaf litter to have much hope of surviving, because they generally are eaten by squirrels and other forest mammals. Even so, the small plants in many locations are due, to a great degree, to root sprouting. The species does not regenerate particularly well beneath large individuals of its own species. This is evidenced by its involvement in reciprocal replacement with other species (Horn [1971], Fox [1977], Forcier [1975], Woods [1979], and Woods and Whittaker [1981]). Reciprocal replacement occurs when two (or more) species, often both Role 2 species, each have better regeneration in gaps that are produced by the mortality of individuals of the other species. This type of a replacement pattern tends to make two species alternate at a site.

The dynamics of a forest that is comprised of a single Role 2 species (Fig. 5.6b) have a saw-toothed biomass curve. However, compared to the curve for Role 1 species forests (Fig. 5.6a), the curve has reduced biomass maxima. There are more "teeth" on the saw-tooth, and the biomass minima do not plunge as deeply as in the Role 1 curve. The Role 2 species numbers curve is more level. The numbers curves also show a lagged increase following decreases in biomass.

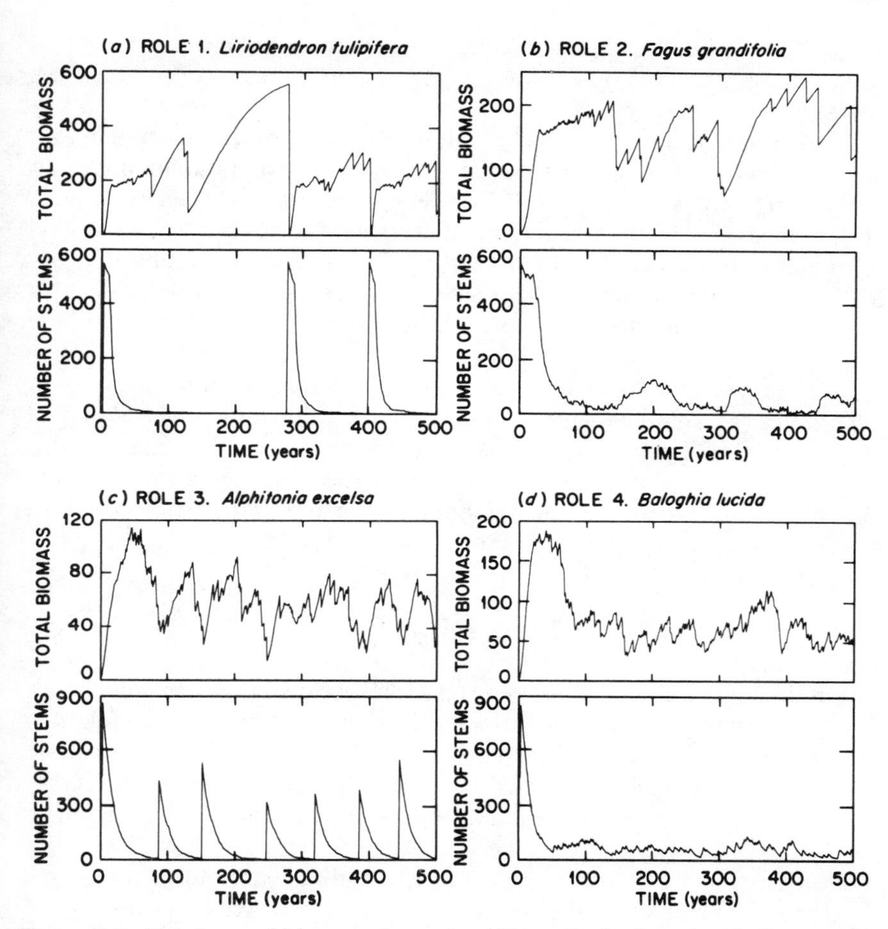

Figure 5.6. Number and biomass dynamics of hypothetical patches in forests of a single species that conform to the roles diagramed in Figure 5.5. (a) Role 1: *Liriodendron tulipifera* on a 1/12-ha plot simulated by the FORET model. (b) Role 2: *Fagus grandifolia* on a 1/12-ha plot simulated by the FORET model. (c) Role 3: *Alphitonia excelsa* on a 1/30-ha plot simulated by the KIAMBRAM model. (d) Role 4: *Baloghia lucida* on a 1/20-ha plot simulated by the KIAMBRAM model.

Role 3 Species

These do not typically produce gaps upon their death, but they do require gaps for successful regeneration. In nature, such trees usually have a very good dispersal ability and are capable of regenerating in large numbers in appropriate regeneration sites—even when these sites are at a considerable distance from a seed source. The trees often have high growth rates and often have short life spans. The actual death of these trees often is due to disease and pest effects. Role 3 species conform reasonably well to what van Steenis (1958) described as a "nomad" species or Whitmore's (1975) "pioneer" species.

An example Role 3 species is *Alphitonia excelsa* of the Australian rain forest. *Alphitonia* is a soft-wooded, fast-growing, and shade-intolerant tree that usually is short-lived. It is widely distributed as seeds in the soils of the forests in which it occurs, and it germinates on exposure of the soils to sunlight. *Alphitonia* is a small tree (maximum diameter greater than 0.5 meters and a maximum height greater than 20 meters). The death of a single tree typically does not create a large enough disturbance to allow regeneration.

The biomass dynamics of a patch in a forest that is composed only of *Alphitonia*, as simulated for a 1/20-ha plot by the KIAMBRAM model (Shugart, et al [1981a]), have a saw-tooth shape with very fine "teeth" (Fig. 5.6c). The regeneration is greater at the time when many of the larger trees have died, and the numbers curve varies with these periods of high regeneration.

Role 4 Species

These do not require gaps for regeneration; because they are small, they do not generate gaps upon their death. Small trees of this sort are found in the subcanopy of many rain forests (Richards [1952], and Whitmore [1975]). Because Role 4 species are small and often occur in very diverse forests, their habits often are not as well known as those of species with particularly Roles 1 and 2. In the KIAMBRAM model parameterization, they often showed moderate growth rates and appeared to be long-lived for trees of their size. Role 4 species may reproduce by root sprouting and are capable of regenerating under intact forest canopy; however, regeneration is better under a sparse canopy.

An example Role 4 species is *Baloghia lucida*, as simulated on a 1/20-ha plot by the KIAMBRAM model. *Baloghia lucida*, sometimes called scrub bloodwood in Australia, occurs in rain forests along the eastern coasts of Australia as well as on Norfolk and Lord Howe Islands and New Caledonia. It is a small tree (height of 20 meters, diameter of 0.3 meters; Francis [1970]) with relatively hard wood that suitable for flooring and small turnery. The tree can be quite abundant (e.g., at Lamington National Park in northern New South Wales) as a subcanopy tree in mature rain forests.

The patch-scale biomass dynamics of a hypothetical forest that is composed of a Role 4 species (simulated here by *Baloghia lucida* an a 1/20-ha patch with the KIAMBRAM model) features a very finely toothed biomass curve (Fig. 5.6d). The numbers curve is relatively invariant, as one would expect in a stable mixed-age stand. Among the species of the same role in a given forest, the details of the entire regeneration process probably are very important in explaining species co-existence and survival. Furthermore, different combinations of the successfully regenerating species at a site (due to either mechanism or random effects) can (in themselves) maintain high diversities of functionally similar species in the gap model simulations.

One may use the four patch-scale roles of forest trees to organize larger patterns among the adaptive traits of a set of species. For example, the Leguminose tree (*Tachigalia versicolor*—as described in its life history by Foster [1977], appears to be a logical extreme of the Role 1 adaptive syndrome. *Tachigalia* is a very large, multibranched canopy tree species that is found in evergreen and semideciduous lowland forests of Panama, southeast Costa Rica, and northwest Columbia (Foster [1977]). The species apparently blooms only once in a tree's lifetime, and the tree dies within ~ 1 year of releasing a wind-dispersed fruit. Several trees in an area bloom (and die) synchronously, but not all trees flower at the same time. Foster (1977) notes, "Because most, if not all, tropical canopy species require an opening or 'light-gap' in the forest canopy for successful maturation, a tree fall should strongly augment the opportunities for seedlings in the vicinity of the parent. . . . *T. versicolor* saplings are often found growing in the openings created by the fall of a canopy adult, or in adjacent openings." Whitmore (1975, particularly pages 69–73), uses essentially the same categories that are shown in Figure 5.6 (based on regeneration strategy, shade tolerance, and size of trees). He lists many example species from the Indo-Malaysian rain forest system. Roles could be a useful way of organizing adaptive traits of trees. However, the present interest is in how these roles interact to form ecosystem dynamics at the scale of the forest patch.

Tree Roles and Forest Ecosystem Dynamics

Different forests may be comprised of different representations of the four roles. For example, if the maximum known tree height is used as an index of tree size (and thus of gap-generation on mortality), the Australian subtropical rain forest (as parameterized for the KIAMBRAM model) appears to have several species in each of the four roles (Fig. 5.7a,b). Similar categorizations of parameters from an extended version of the FORET model (Solomon, et al [1980]) indicates that Role 4 species are not in evidence in the eastern forest of North America and that (relative to the Australian forest) there is a diverse array of Role 1 species (Fig. 5.7c,d). The hurricane-prone Puerto Rican montane rain forest, as simulated in the FORICO model, seems to have no species in the Role 1 category (Fig. 5.7c,d). It is evident from Figure 5.7 that the diversity of

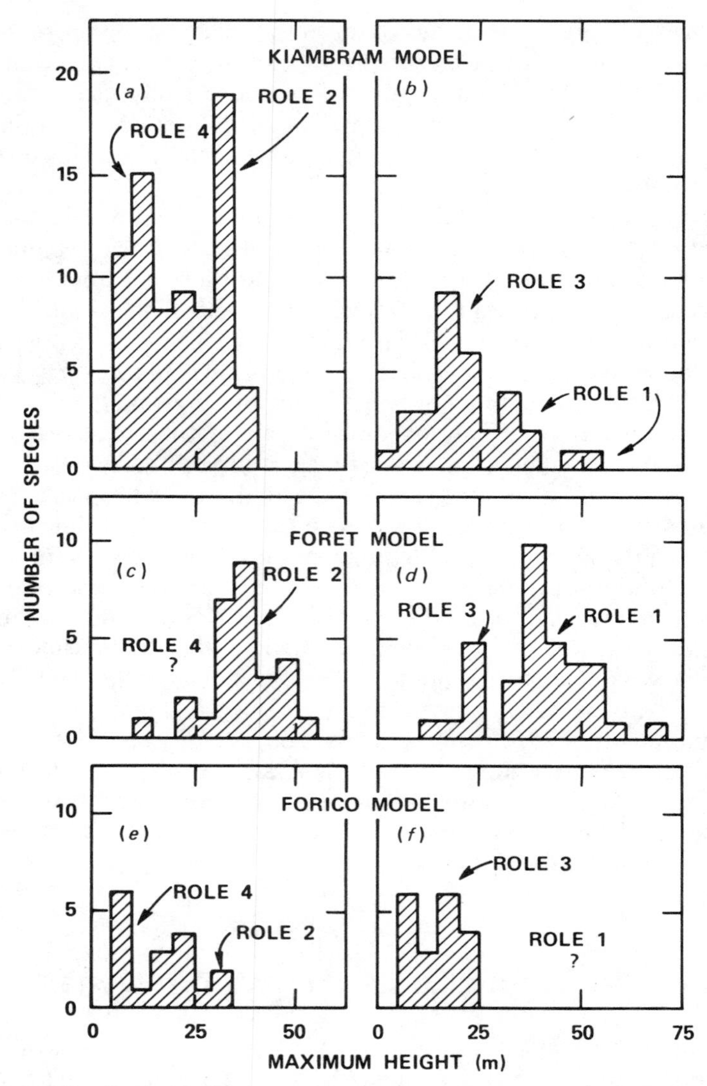

Figure 5.7. Distributions of species by maximum size and regeneration habit for species' parameter lists from three gap models. Roles 1, 2, 3, and 4 are from descriptions in Figure 5.5 and the text. (a) Species listed as shade-tolerant in the KIAMBRAM model of Australian subtropical rain forests; (b) Species either listed as shade-intolerant or that required mineral soil or a raised germination site for regeneration from the KIAMBRAM model; (c) Species listed as shade-tolerant in Solomon, et al (1980) FORET model parameterization for North America; (d) Species listed as shade-intolerant or that required for mineral soil for regeneration in the FORET model; (e) Species listed as tolerant in the FORICO model of Puerto Rican montane rain forest; (f) Species listed as shade-intolerant or intermediate in the FORICO model.

species that are classed by size and regeneration habit often is bimodal and certainly tends not to be uniform. The diversity of species within a role probably is a useful index of the functional parts of a forest ecosystem; however, within a role, a single well-adapted species can have as much influence as a suite of poorly adapted species. For example, the Puerto Rican montane rain forest does not appear to have a great abundance of Role 2 species (Fig. 5.7e). However, one Role 2 species, *Dacryodes excelsa*, is the dominant tree in this particular forest. In fact, the "*Dacryodes-Sloanea*" association (Beard [1944]) is distributed throughout the Lesser Antilles and is named after two successful Role 2 species (*Sloanea berteriana* is the other Role 2 species in Fig. 5.7e). As one might expect in a frequently disturbed ecosystem, there is also an absence of Role 1 species in Puerto Rico (Fig. 5.7f).

Successful species of different roles could be expected to influence other species on a small patch in different ways. For example, a large mature tree of Role 1 would, on its death, be expected to create a large gap that would encourage its own regeneration as well as that of the other Role 1 and Role 3 species in a community. The death of a small individual of Role 1 would create a smaller gap that would tend to favor the regeneration of Role 2 and Role 4 species. The other roles could be expected to influence regeneration among themselves as a complex interaction of ecological roles, and also of size or spatial scales. One would expect (Fig. 5.8) Role 2 and Role 4 to generally be mutually encouraging, Role 1 and Role 3 to prosper when the mortality of canopy trees creates large gaps; and Role 2 and Role 4 to increase in cases where the mortality pattern generates smaller gaps. The interactions among roles are such that species of any role could be expected to regenerate in the gap left by the mortality of species of the same or any other role with a fair amount of stochasticity; the pathways in Figure 5.8 are the most probable.

Forcier (1975) and Woods and Whittaker (1981), in studying the replacement pattern in hardwood forests in northeastern United States, found the tendency for the species replacements to form a closed cycle (with the possibility of reciprocal replacement—*see also* Runkle [1981] and Fox [1977]) among the Role 2 species. This closed cycle (Fig. 5.9a,b) is connected with a gap generation/gap colonization cycle between Role 2 and Role 3 species. Horn (1975a,b) (*see also* Chapter 2) computed transition probabilities for the trees in a somewhat more diverse forest in New Jersey. By multiplying Horn's transition probability matrix with the vector of expected canopy trees at equilibrium, one obtains a diagram of the expected percentages of transitions among roles in the equilibrium forest (Fig. 5.9c). This agrees with the pattern that is predicted from the interactions of species roles (Fig. 5.8) in an environment where the gaps are of intermediate size.

There are two important consequences of considering the trees that comprise a forest as occupying these roles. The first is a practical scien-

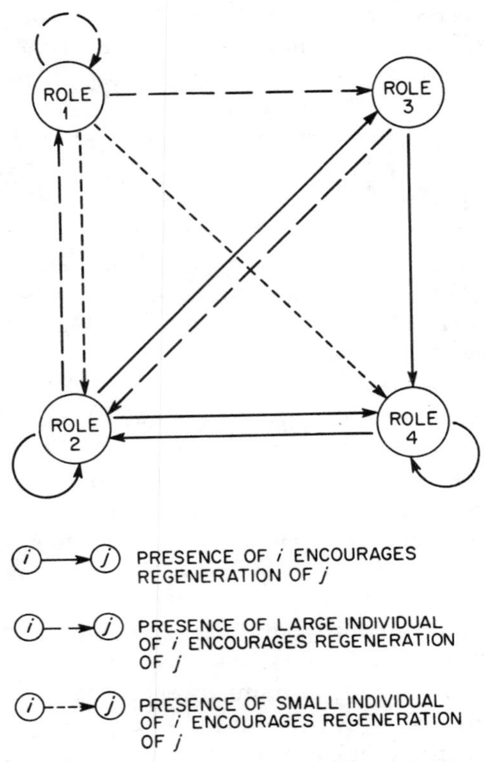

Figure 5.8. The directions of positive or strongly positive interactions among four roles of tree species with respect to their influence on regeneration.

tific consideration. By identifying functional roles for trees, one can begin a categorization that conveys information about aspects of forest dynamics. To say that a forest is comprised mostly of Role 2 trees with a few Role 3 trees conveys a notion of how that forest might behave over time. A Role 2 tree-dominated forest (Fig. 5.6) has a small-scale dynamic response that is different from that of a Role 1 tree-dominated forest.

The use of the four roles as building blocks to construct hypothetical forest's dynamics for a small patch may require a detailed understanding of the life histories of the species that are involved. For example, Shugart, et al (1980b) investigated the behavior of a hypothetical two-species mixture of beech (*Fagus grandifolia*—a Role 2 species) and yellow-poplar (*Liriodendron tulipifera*—a Role 1 species) under the influence of a slowly changing climate. Both of the species could regenerate under all conditions in this climate-change gradient and both of the species monotonically increased their rates of growth toward the warmer climates. Under cool climates (Fig. 5.10), the growth rates were slow, the trees all tended to be smaller, and the Role 2 species dominated the stands (as evidenced

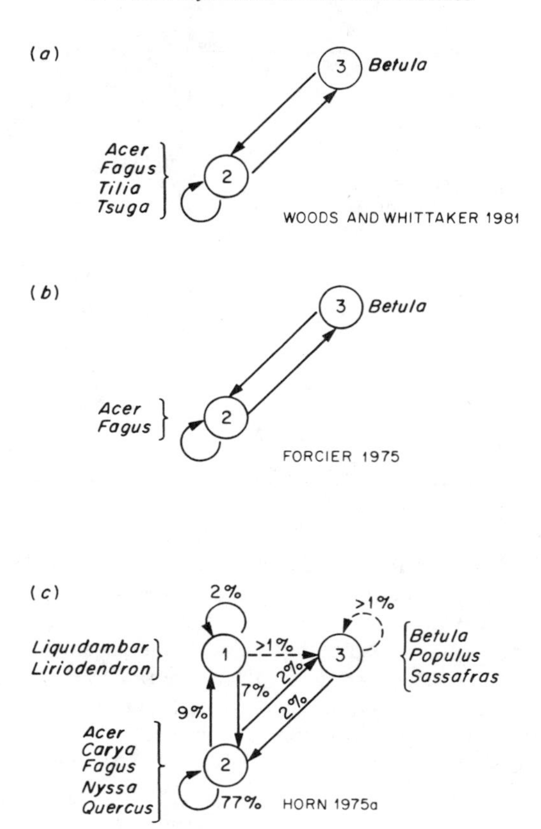

Fig. 5.9. Species role/interaction diagrams for three studies conducted in the northeastern United States, (a) Woods and Whittaker [1981]; (b) Forcier [1975]; (c) Horn [1975a]. The percentages in (c) are computed by multiplying Horn's (1975a) transition probability matrix times the vector of expected composition.

by the low percentage of *Liriodendron*, which imply a high percentage of *Fagus* at the lower degree-day values [Fig. 5.10]). Similarly, under warmer conditions, the trees grew larger and the Role 1 *Liriodendron* dominated the stand. However, the stand dominance for intermediate degree-day values demonstrated an apparent hysteresis. As to which of two possible forests (either *Liriodendron*- or *Fagus*-dominated) actually occurred depended on the direction of change of the climate. Along with indicating the possible complexity of constructing the dynamics of a mixed-species forest (even when the roles of the component species are sharply differentiated) the example (Fig. 5.10) also links the simulation gap-model dynamics with applications to nonlinear differential equations in forests (Yamamura [1976], and Smith [1980]). This also leads one to speculate on the roles that are outlined here and on the theoretical consid-

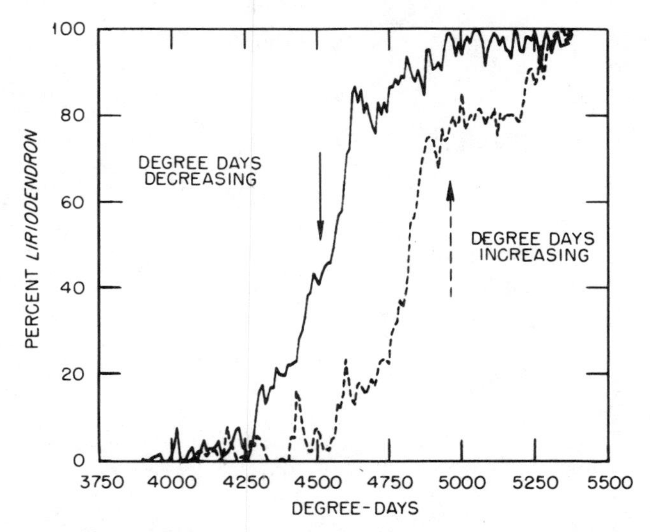

Figure 5.10. Percentage of total stand biomass that is attributable to *Liriodendron tulipifera* in a two-species mixture with *Fagus grandifolia* under gradually changing degree-day (F day) values. Each curve is the mean value of five simulated 1/12-ha stands as the annual degree-day value is increased (dashed line) or decreased (solid line) between 3,800 and 5,300 at the rate of 0.2 degree-day year^{-1}. (Reprinted by permission of the publisher from Shugart, H. H., et al: Environmental gradients in a simulation model of a beech-yellow poplar stand. Math. BioSci. 50:167, 1980. Elsevier Science Publishing Co.)

erations of competition, limiting similarity, and system stability that are largely derived from animal population studies (May [1973], DeAngelis [1975], and Lawlor and Smith [1976]).

This second, more theoretical, consideration of tree roles is speculative, but it also is a potentially important point for understanding the organization of mixed-species forests. In the most general outline, communities often are thought to be strongly structured by competition among their members (Pianka [1981]). These competitive interactions have a negative effect on the populations concerned, and natural selection should favor individuals within the populations whose attributes minimize these effects. One would expect to find that cooccurring species are different from one another, or that there is a limit to the similarity of species (MacArthur and Levins [1967], May [1973], and Lawlor and Smith [1976]) that may even produce regularities in the pattern of differences among species assemblages (Schoener [1974]; *see* Simberloff and Boecklen [1981] for a cautionary review). According to the competitive exclusion principle (Gause [1934]), one expects both coexisting species to differ in some way (Cole [1960], and Hardin [1960]) and the most similar species to be the most competitive with one another.

These considerations often are codified in the traditional alpha matrix (e.g., Levins [1968] of competitive interactions that is derived from the Lotka-Volterra equations of the form:

$$\frac{dN_i}{dt} = \frac{r_i N_i (k_i - N_i - \sum_{j=1}^{n} \alpha_{ij} N_j)}{k_i} \tag{5.1}$$

where N_i is the number (biomass) of the i-th species, r_i is the i-th species intrinsic rate of increase, k_i is the i-th species "carrying capacity," and a_{ij} is the competition coefficient of the effect of species j on species i.

The Lotka-Volterra equations often are thought of as a second-order Taylor series approximation to whatever complexities connect the species in a community (May [1973]); they are studied as approximations to a more complex reality. This approach has legions of precedents in other sciences, but there is a concern that the Lotka-Volterra equations may be too great simplification of nature (Neill [1974]). Whatever the case, the use of the equations has greatly enriched the available literature on competition and has come to be the traditional approach to mathematical investigations of consequences of competitive interactions.

The Lotka-Volterra equations are formulated on the direct effects of one species on another (the a's which are negative); however, one can have indirect effects mediated through other members of the community (Levine [1976]). Of these indirect effects, the ones of interest here are those that cause competitive mutualism (Pianka [1981]). In competitive mutualism, two competitors with moderate competitive effects on one another both have strong negative effects on a third strong competitor (Lawlor [1979, 1980], and Travis and Post [1979]). In inhibiting this mutual competitor, the two species help one another.

The interaction between trees is complex in relation to the Lotka-Volterra equations, largely because the competitiveness of an individual is strongly related to its size. Size is not a state variable in the Lotka-Volterra formulation, so the comments regarding tree competition in relation to the traditional competition theory should be regarded as generalizations. Competition among trees at the patch-scale involves two sorts of interactions. Trees of different roles compete in a way that the rules of the game may be altered by the competition. At one extreme, the competitive interactions may be in the context of episodic recruitment and ingrowth that are released by the death of a single, giant Role 1 species. At the other extreme, the competitive interactions may occur in a background of regular recruitment and ingrowth in a small-scale, mixed-size forest due to dominance by smaller Role 2 and Role 4 species. Figure 5.8 is a diagram of the interactions in determining such rules of the game among the 4 roles. Within a set of rules and within a role, competition of the sort abstracted in the Lotka-Volterra equations may influence the pattern of adaptions according to classic concepts of resource division (Schoener

[1974]) or limiting similarity (May [1973]). Grubb (1977) emphasized the patterns of regeneration as the niche dimensions in plants of similar forms (or roles).

Considering the direct interactions among the four roles as they might be encoded in the Lotka-Volterra equations for interactions among groups of species of different roles, the pattern in Figure 5.8 includes a diverse array of interactions that include not only competition, but competitive mutualism (Lawlor [1979, 1980]), mutualism (Goh [1979], and Travis and Post [1979]), and a nonsymmetrical positive/negative interaction that (in a Lotka-Volterra formulation) resembles predation (or parasitism) as well. For example, Role 2 species enforce a set of conditions (rules) so that gap-regeneration species are favored on the death of a large Role 2 trees. Role 3 species colonize these gaps and create conditions that are conducive to Role 2 regeneration. This reciprocal mutualistic cycle is found, in practical forestry, in the use of nurse crops of Role 3 trees to promote the regeneration and growth of commercially important Role 2 trees.

Tregonning and Roberts (1979) stated; "It is not well understood how large complex systems (such as ecosystems) come into being or how they persist over long periods of time, although recent studies (Gardner and Ashby [1970], May [1972], Roberts [1974], Gilpin [1975], Tregonning and Roberts [1979]) have clarified certain aspects of the associated stability problems."

Tregonning and Roberts noted that the occurrence of such systems presents no theoretical problem if the variables are adjusted for their mutual interactions in advance. Tregonning and Roberts (1979) investigated the problem in which randomly assembled communities following the Lotka-Volterra equation form are extremely unlikely to be homeostatic (e.g., a 20-species assemblage has a probability of less than 10^{-6}). Their initial observation that variable preadjustment would obviate the problem relates to the idea of roles in forest communities. If the trees in a forest could be reasonably assigned to roles such as those described in this chapter, then one could reasonably expect that at least some of the parameters of interaction among species could be thought of as being adjusted in advance. A set of adaptions of a Role 3 species (numerous, well-dispersed seeds, and fast growth) to one Role 1 species would be adjusted for the interaction with another Role 1 species.

Conclusions

This chapter has focused on the dynamics of small patches of a large forest mosaic. The biomass dynamics of gaps were found to be a sawtoothed curve that is associated with the death of large trees and the subsequent regeneration and growth. The nonequilibrium nature of small patches of forest relates directly to the succession theory and also allows the categorization of tree species into function roles according to the way they respond to, and interact, with the cyclical changes of a patch. These roles, which are viewed as a great simplification of reality, have a diverse set of dynamic behaviors that are manifested as changes in the small-scale numbers and biomass dynamics. The roles also interact with one another to create a complex array of dynamics and types of competitive interactions. This chapter ends with speculation on the relation between trees of differing roles and classic competition theory. In the next chapter, these results will be extended to the scale of forested landscapes and their dynamics.

Chapter 6

The Biomass Response of Landscapes

In Chapter 5, the small-scale dynamics of a patch of forest was studied by inspecting the response of a single simulation of a gap model. The analogous response of a landscape could be obtained by assuming that the landscape is a mosaic tessellated with small patches and with each patch being represented by a gap model. The potential complexity of characterizing a landscape in such a way is great. It includes: (1) All of the small-scale interactions of regeneration, growth, and death discussed in Chapter 5; (2) The synchrony (or lack thereof) among the several patches' dynamics that go together to become the forest landscape dynamics; (3) Interactions across space that involve, for example, seed supply on different patches; (4) The gradients of soil fertility that might exist across the landscape; and (5) The possible changes in micrometeorology variables across a large forest clearing. Most ecologists recognize these complexities and attempt to formulate an idealized landscape biomass dynamic that can be applied to a general situation. This ideal case is the response of a forest over a landscape that is large enough to be logically represented by hundreds of patches the size of the plot simulated by a gap model. This response usually is studied by considering a landscape following deforestation by either natural or anthropogenic causes. One usually assumes that there is adequate regeneration over a relatively homogeneous and generally benign substrate.

The Biomass Response of Homogeneous Landscapes

One straightforward model of the biomass dynamics of a homogeneous forest is,

$$\frac{dM}{dt} = c_1 - c_2 M \tag{6.1}$$

where M is the landscape biomass (kg ha^{-1}), c_1 is the input (a constant, kg ha^{-1} t^{-1}), and c_2 is a loss rate (t^{-1}).

This model implies a constant input of biomass (e.g., constant productivity) and that the loss of biomass (due to respiration, branch pruning, death, and so on) is a constant proportion of the biomass. Such a model would generate biomass dynamics that would increase the value both monotonically and without inflection to a constant asymptote. This curve is denoted "concave" (Fig. 6.1). Several ecologists who worked in very different forests (Tiren 1927, Burger 1953, Kira and Shidei 1967, Odum 1969, Albrektson 1980) all noted that this concave shape is the one to be expected in succession. One of the aspects of this model, that of constant (or near constant) gross annual production, was reviewed by Peet (1981). He found that this feature was reported by Ovington (1957) for *Pinus sylvestris* plantations, by Ovington and Madgwick (1959) for *Betula verrucosa* stands, by Van Cleve (1973) for *Alnus* stands, and by Meeuwig (1979) for Pinyon-juniper stands. Odum and Pinkerton (1955), Margalef (1963), Olson (1963), Yoda, et al (1965), and Odum (1969) all provide discussions relating to the second aspect of this model, which is the constant proportion of biomass lost over time. There is abundant evidence that the net production of biomass (Equation 6.1 *in toto*) decreases with stand age (*see* Peet [1981], Meyer [1937], Rodin and Bazilevich [1967], and Attiwill [1979]).

Another model that also features a net production component that approaches zero as the forest's biomass approaches an asymptote is the biomass analogue to the logistic equation:

$$\frac{dM}{dt} = c_1 M - c_2 M^2$$

$$= c_1 M \left(1 - \frac{c_2}{c_1} M\right) \tag{6.2}$$

where M is the landscape biomass (kg ha^{-1}), c_1 is a constant (t^{-1}), c_2 is a constant (kg^{-1} ha t^{-1}), and c_2/c_1 is the reciprocal of the asymptotic biomass value (analogue to the carrying capacity).

This equation is typically used for population models (May [1981]), but DeAngelis (1975) used a related formulation for changes in mass in food webs near equilibrium. Gap models (*see* Equation 3.1) use a similar equation for the change in mass of a single tree. The basic rationale is that production is a linear function of mass (c_1, M in Equation 6.2), and that as the mass approaches some value (namely c_1/c_2), the respiration increases to balance the production. In graphic form, the logistic (Eq. 6.2) differs from the concave (Equation 6.1), mostly by the presence of an inflection of the increasing part of the curve (Fig. 6.1). In actual practice, the concave and logistic curves often are regarded as ecologically equivalent. For example, Peet (1981) discusses both as a single type of monotonically

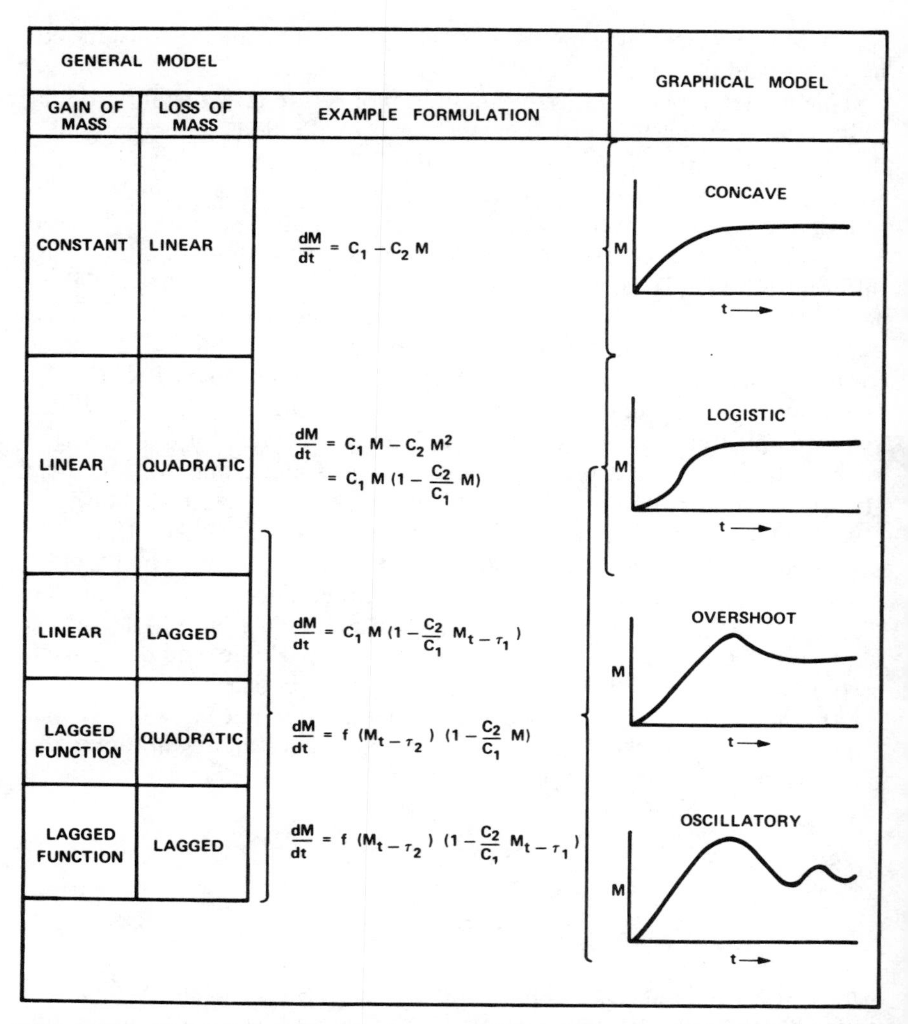

Figure 6.1. General models for change of biomass for an abstracted forested landscape. Conceptually the models differ as to the nature of the gain and loss of phytomass from the forest (constant, linear, quadratic, or lagged lossand gain-functions). Example formulations that have the indicated features are provided. Graphical models (M = biomass; t = time) of the biomass dynamics also are shown and grouped with models that could generate the response.

increasing biomass curve of a ''logistic fashion,'' which (given the variability of the data) probably is appropriate. May (1981) points out that the logistic could be regarded as the first terms of a Taylor-series expansion of the actual dynamics of an ecological system, but the same certainly could be said for the concave (Equation 6.1) formulation. Both the logistic and the concave curves draw support from observations of asymptotic biomass accumulations that were reported for several forests (e.g., Forcella and Weaver [1977], MacLean and Wein [1976], Switzer, et al [1966], and Zavitkowski and Stevens [1972]).

The principal alternative to the monotonic logistic or concave formulations originates from the observation that the biomass of a forest, over time, may overshoot the ultimate asymptotic value; it then may either return to the asymptote or oscillate about it with ever-decreasing amplitude. Bormann and Likens (1979a,b) reported the expectation that the landscape biomass dynamics for hardwood forests in New England should overshoot the asymptote by involving a "shifting-mosaic steady state" concept of landscape dynamics. This concept was developed using the JABOWA (Botkin, et al [1972a,b]) gap model by Bormann and Likens (1979a,b). Watt (1925, 1947) presents a conceptual development of essentially the same ideas, as do Cooper (1913), Aubreville (1933, 1938), Stearns (1949), and Goodlett (1954) in a somewhat less explicit manner. Bormann and Likens (1979a) identify the mechanisms of the overshoot curve (Fig. 6.1) as, "we think the idea of peak biomass prior to the steady state may have wide applicability to terrestrial ecosystems, and the maximum attainable biomass accumulation may be the result of synchrony imposed upon the system by either nature or man. In northern hardwoods, peak biomass is achieved because even-agedness after clear-cutting imposes synchrony on all of the patches comprising the ecosystem. As a consequence, most of the patches, performing in lockstep, accumulate biomass for about a century and a half. Thereafter, ecosystem biomass drops as patches become more asynchronous and all aged." Because Bormann and Liken's (1979a,b) concept of the biomass dynamics involves synchrony of the mortality of the first generation of trees, a simple model that incorporates this effect is a time-lagged logistic curve:

$$\frac{dM}{dt} = c_1 M \left[1 - \frac{c_2}{c_1} M_{(t - \tau 1)}\right] \tag{6.3}$$

where $M_{(t-\tau)}$ indicates the phytomass at some $t - \tau_1$ time earlier and other parameters, as in Equation 6.2.

The function was introduced by Hutchinson (1948) and was recently discussed by May (1975, 1981). May (1981) points out that the actual delay function may be considerably more elaborate than the constant lag indicated in Equation 6.3.

Peet (1981) noted that the overshoot behavior also could feature an oscillatory approach to the asymptote (Fig. 6.1), and he noted some apparent cases of such dynamics (Ilvessalo [1937], and Siren [1955]). Peet extended Bormann and Liken's (1979a,b) general concept in emphasizing that the regeneration in forests often was a lagged function due to suppression of regeneration by the population of trees by resource pre-emption (Monsi and Oshima [1955], Connell and Slatyer [1977]) or:

$$\frac{dM}{dt} = f(M_{[t-\tau 2]}) \left(1 - \frac{c_2}{c_1} M\right) \tag{6.4}$$

where $f(M_{[t-\tau 2]})$ is a lagged regeneration function. Combining the two effects in Equations 6.3 and 6.4, one obtains a general function:

$$\frac{dM}{dt} = f(M_{[t-\tau 2]}) \left(1 - \frac{c_2}{c_1} M_{[t-\tau 1]}\right) \tag{6.5}$$

that could be taken as an example of the model verbalized by Peet (1981). All of the behavior shown as graphic models in Figure 6.1 can be obtained from Equations 6.3, 6.4, and 6.5. These models also can provide a variety of other behaviors, which range from regularly oscillatory to seemingly chaotic changes in the biomass dynamics (May [1981]).

Some Examples of Forest Biomass Response

A theoretical consequence of idealizing the biomass response of a forested landscape to becoming an equilibrium system (asymptotic behavior when biomass gain and loss are equal) is to use internal time delays to represent regeneration delay or synchronous mortality (or both). Equation 6.5 is such a general model that any one of the dynamics postulated for real landscapes (Fig. 6.1) is a special case of the model. A complication is that the delays added to the model actually are surrogates for the age (or size) structure that interacts with the numbers dynamics of forests to cause the biomass dynamics. Both Sullivan and Clutter (1972) and Suzuki and Unemura (1974a,b) incorporated these effects directly in their forest dynamics models; but, these models currently are restricted to single-species stands.

Since Peet's (1981) parsimonious concept of the biomass dynamics of forests is capable of imitating the idealized response of a forest, it is appropriate to see if actual forest changes are regular in a fashion that may constrain the range of Equation 6.5. Figure 6.2 contains the biomass or volume dynamics for a variety of natural forests and forest plantations. The appropriate parameter changes that will make the general model agree with these dynamics vary from case to case. In Finnish *Pinus silvestris* forests (Fig. 6.2a), the volume responses seem to be simple logistics (Equation 6.2), with regular changes in the value of the asymptote (c_2/c_1, in Equation 6.2) as a function of forest type. Peet's (1981) volume data for *Pinus taeda* forests, which vary one from another by mostly initial density (Fig. 6.2b), demonstrate a variety of responses that may be produced by adding a time lag to the general model as some increasing function of

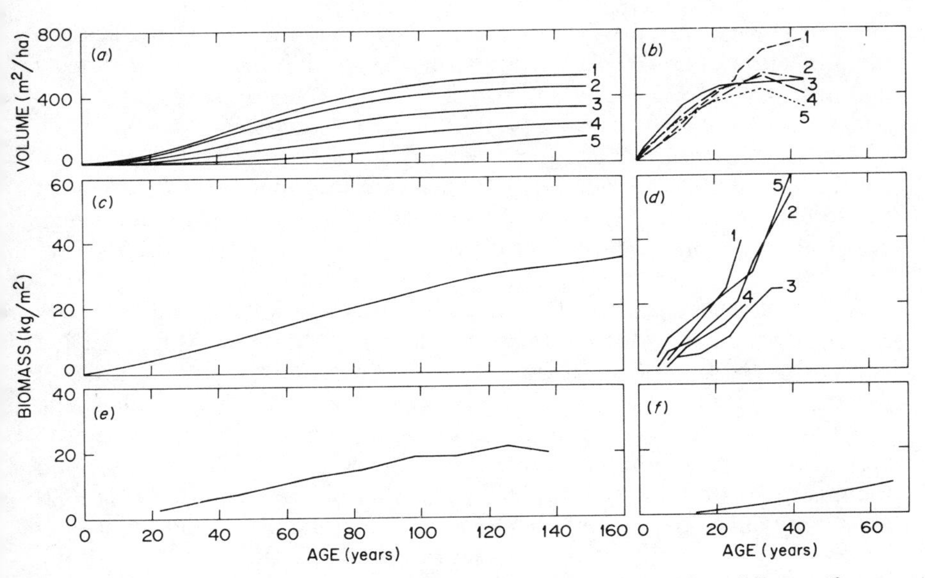

Figure 6.2. The form of the biomass response of a forest for different forest systems. (a) Change in volume (a biomass index) for *Pinus silvestris* forest in Finland. Original data from Ilvessalo (1920) for a forest of southern Finland grouped by vegetation type by Delvaux (1971) and plotted by vegetation type 1) *Oxalis-myrtillus* type; 2) *Myrtillus* type; 3) *Vaccinium* type; 4) *Calluna* type; and 5) *Cladina* type. Figure redrawn from Delvaux (1971). (b) Volume change for *Pinus taeda* forests on old fields in North Carolina (redrawn from Peet [1981]). Forest response varies by initial density of trees and by site index. 1) Site index (SI) = 90 m; Initial Density (D) = 25 trees per 0.1 ha; 2) SI = 86, D = 51; 3) SI = 83, D = 431; 4) SI = 81, D = 234; and 5) SI = 77, D = 1173. (c) Biomass changes following fire for oak-pine forest on Long Island, New York. (Figure redrawn from Whittaker [1970] based on data in Whittaker and Woodwell [1968, 1969]). (d) Biomass change from several fast-growing plantations. 1) Teak (*Shorea robusta*) plantations in the Gorakhpur Forest Division, India (data collected by Q. Foruqi, from DeAngelis, et al [1981]); 2) Sal (*Shorea robusta*) plantations in the Gorakhpur Forest, India (data collected by Q. Foruqi, from DeAngelis, et al [1981]); 3) Sal (*Shorea robusta*) plantations Lacchmipur Range, Gorakhpur Forest, India (data collected by D. Satyauarayana, from DeAngelis et al. [1981]); 4) *Pinus radiata* plantations in Chile (data from Matte [1971]); and 5) *Pinus radiata* plantations in Chile from sandy sites (data from Matte [1971]). (e) *Picea abies* forests of the Myrtillosum type on sand moraines in Karelia, USSR (data collected by Kazimirov and Morozova, from DeAngelis, et al [1981], *see also* Kazimirov and Morozova [1973]). (f) *Populus grandidentata* stands on poor sites, northern lower Michigan. (Redrawn from Cooper [1981].)

density. Whittaker's (1970) biomass data for mixed pine-oak forests on Long Island, New York, appear as a logistic form (Fig. 6.2c) as do the biomass dynamics recorded by Kazimirov and Morozova (1973) for spruce forests in Karelia, USSR (Fig. 6.2e). The latter authors feel that the irregular pattern toward the end of the spruce biomass dynamics is a manifestation of an oscillatory return to one equilibrium (*see* Kazimirov and Morozova [1973]). This opinion is based on other augmenting studies and observations. Data from several different plantations of fast-growing trees show an assortment of concave and logistic responses, but are not of sufficient length to provide much insight into the asymptotic behavior of the forests (Fig. 6.2d). Finally, Cooper's (1981) long data set on a single stand of aspen on a poor site (sandy soil) in Michigan (Fig. 6.2f) shows initial dynamics of a logistic form.

Taken as a whole, there is a range of biomass responses that actually are recorded for forested landscapes. While some of these responses have been recorded over short periods of time in relation to the time scale of forests, they can be taken as evidence against proposing any single shape as being *the* expected form of the landscape biomass response. There do appear to be regularities in the patterns shown in relation to a general theory (Peet [1981]). The range of behaviors that are illustrated (Fig. 6.2) does not include chaotic behavior (May [1976]) or short-period oscillatory behavior. The pattern of the biomass dynamics appears to be well-behaved in a mathematical sense, and there is a possibility of predicting the biomass response of a given forest from case-specific ecosystem attributes. These attributes include loss and gain rates and the amount and nature of the delay* in these rates.

Idealized Landscape Dynamics from Gap Models

Gap models explicitly incorporate the regeneration delays, age-structure effects, and mortality responses that are implicit in the delay functions in the general model (Fig. 6.1). Therefore, it is reasonable to ask whether or not gap models can be used to project patterns of landscape dynamics—at least in specialized cases. In what have been called idealized landscape dynamics in this chapter, a forest landscape that is composed of many patches is altered by a simultaneous disturbance event and the biomass response is determined. If there are no strong interactions among the patches that make up the forest mosaic, the biomass response can be simulated by summing the responses of several patches. An individual patch's dynamics are simulated by a gap model. This Monte Carlo simulation of a landscape is based on the assumption that each patch is indepen-

* Gutierrez and Fey (1980), in a very general model of succession designed for grasslands, list 30 model parameters of which 14 are related to delays. They also recognize that the considerable range of dynamics that are possible from delay-difference equations is not found in actual successional sequences.

dent of the others. This restriction will be relaxed and spatial interactions among the patches will be discussed in the latter part of this chapter.

The mosaic nature of landscapes is an old and well-established paradigm in forest ecology (e.g., Aubreville (1933, 1938) for rain forests, Watt [1925] for beech forests; and Poole [1937] for temperate rain forest). The idea that this mosaic could be the basis of an understanding of landscape dynamics was eloquently developed by Watt (1947) and it recently has generated renewed interest (Wiens [1976], Levin [1976], and Bormann and Likens [1979a,b]). Gap models have been used to construct the landscape dynamics for several different forests (Shugart and West [1977, 1981], and Bormann and Likens [1979a,b]). In such applications, there are two problems of interest:

1. What is a landscape's transient response following a single large-scale disturbance?
2. What is the nature of a landscape at a steady state (if a steady-state exists)?

Before discussing these responses in mixed-species forested landscapes, the theoretical responses of monospecies landscapes as simulated by gap models will be developed.

Dynamics of Monospecies Landscapes

Some forested systems are strongly dominated by a single species so that a monospecies landscape is not necessarily a theoretical entity without real analogues. The expected biomass dynamics for a monospecies landscape (Fig. 6.3) are for the total landscape biomass to increase as trees grow and compete. The maximum value of biomass is obtained at the time when each patch of the mosaic is simultaneously dominated by one or two large trees. The forest, at this time of maximum biomass (and minimal numbers due to thinning from overstory competition), is even-aged with a deep closed canopy and an open understory. As this even-aged initial cohort begins to die more or less synchronously across the landscape, there is enhanced regeneration and survival in canopy gaps. If the mortality is synchronized across all of the patches that make up the landscape, then the gap-phase regeneration of the second cohort also is synchronized. A high degree of such synchrony produces an oscillatory return of the biomass dynamics to an equilibrium; a lesser degree produces a less oscillatory transient.

The rate of initial increase of the landscape biomass transient is mostly slowed by the death of suppressed trees, and it is a function of initial stocking densities at each patch. There generally is a stocking level that will amplify the biomass overshoot response, which amounts to an optimum between having too few trees to provide a maximum potential bio-

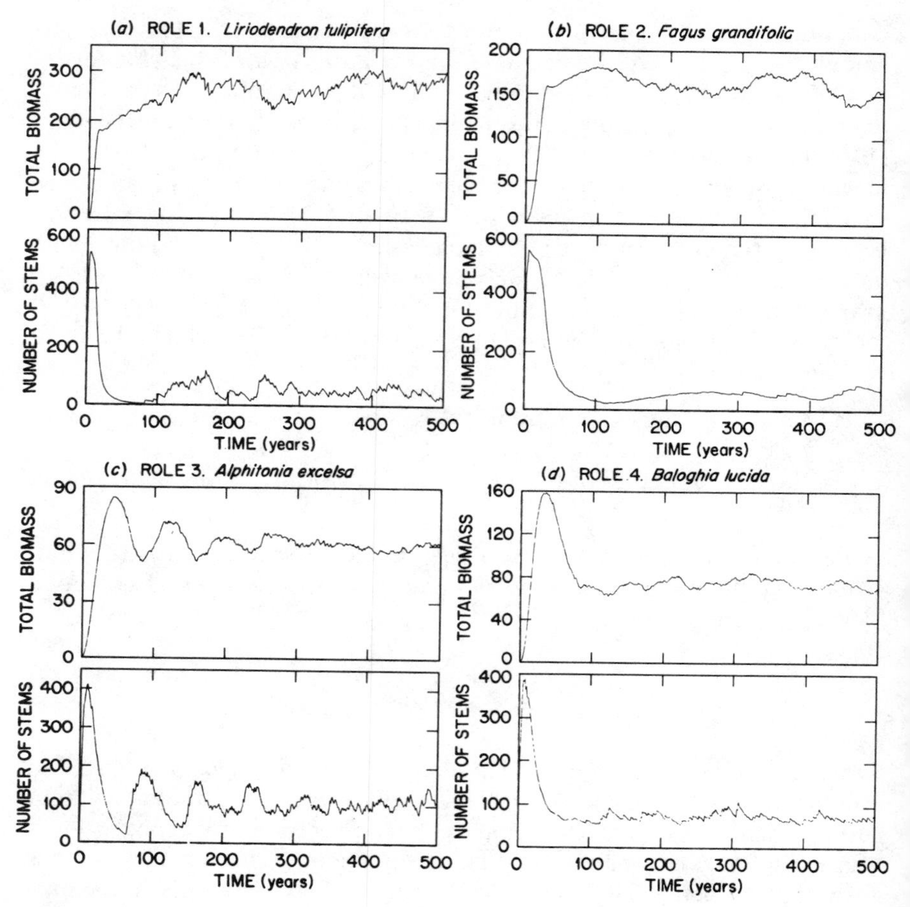

Figure 6.3. Biomass and number dynamics of monospecies-forested landscapes according to the roles of the species (*see* Figs. 5.5, 5.6). (a) Role 1: Average of 100 plots with *Liriodendron tulipifera* on 1/12-ha plots as simulated by the FORET model. (b) Role 2: Average of 100 plots with *Fagus grandifolia* on 1/12-ha plots as simulated by the FORET model. (c) Role 3: Average of 100 plots with *Alphitonia excelsa* on a 1/30-ha plot as simulated by the FORICO model. (d) Role 4: Average of 100 plots with *Baloghia lucida* on 1/20-ha plots as simulated by the KIAMBRAM model.

mass increase and having so many trees that the competition and suppression of trees greatly diminishes the overall increase of biomass at the patch. The curves shown in Figure 6.3 are not optimized with respect to spacing, and they all are at fairly high stocking densities. The notion of optimal spacing is the basis of plantation forestry, and it has been studied elaborately as both a field problem and as a forest modelling problem (see Chapter 2).

Generally, the landscape biomass dynamics of species of different roles (Fig. 6.3) are similar. The dynamics tend to differ by the growth and death rates of the species that tend to stretch or shrink the time scale of the response, rather than by species roles. This similarity in landscape pattern is in fairly sharp contrast to the difference in patterns seen at the scale of a patch (*see* Fig. 5.6).

The monospecies landscape-scale biomass dynamics show the behavior that is expected from the graphic models shown in Fig. 6.1 and is discussed above as the general model (Peet [1981]). When the detailed simulation output is inspected, there is a relation between the tendency of a species to form even-sized forests and either episodic recruitment or shade-intolerance. In simulations with an optimal stocking density, one finds an amplification of the overshoot of the biomass curve. A synchronization of mortality across landscape patches also amplifies the overshoot of biomass, because such synchrony makes the regeneration and recruitment episodic. Synchrony also increases the tendency for the biomass dynamics to oscillate following the overshoot.

The rules are complex for combining these responses in the hope of obtaining a general prediction of any landscape's biomass response. For example, a landscape that is comprised of patches in which two Role 1 species competed for dominance could be expected to have biomass dynamics the same as those shown in Fig. 6.3a, only if the growth rates, sizes, and mortality probabilities were very similar. Otherwise, large parametric differences would reduce both the synchrony of growth pattern that produces the biomass peak and the synchrony of mortality that produces the oscillatory transient. Mixtures of different species roles could produce very complex patterns. This particularly is the case if the species pairs replaced one another in the successional sequence. For example, a mixture of a Role 1 and Role 4 species on the landscape mosaic would have a transient response that would have a large biomass overshoot (due to the even-aged cohort of large Role 1 trees) with an abrupt drop to a relatively low equilibrium biomass (with a mixed-age, smaller Role 4 tree-dominated forest).

Idealized Landscape Dynamics for Multispecies Forests

Knowledge of the patterns of monospecies tree patches of differing roles is very useful in interpreting the patterns generated by gap models for more complex landscapes. The simulated response from a landscape in Arkansas that is dominated by loblolly pine (*Pinus taeda*) under the influence of fire (Fig. 6.4) is what one would expect for a monospecies-dominated landscape. This simulated response from the FORAR model has a biomass overshoot and a return to a stochastic equilibrium. A slighter over-

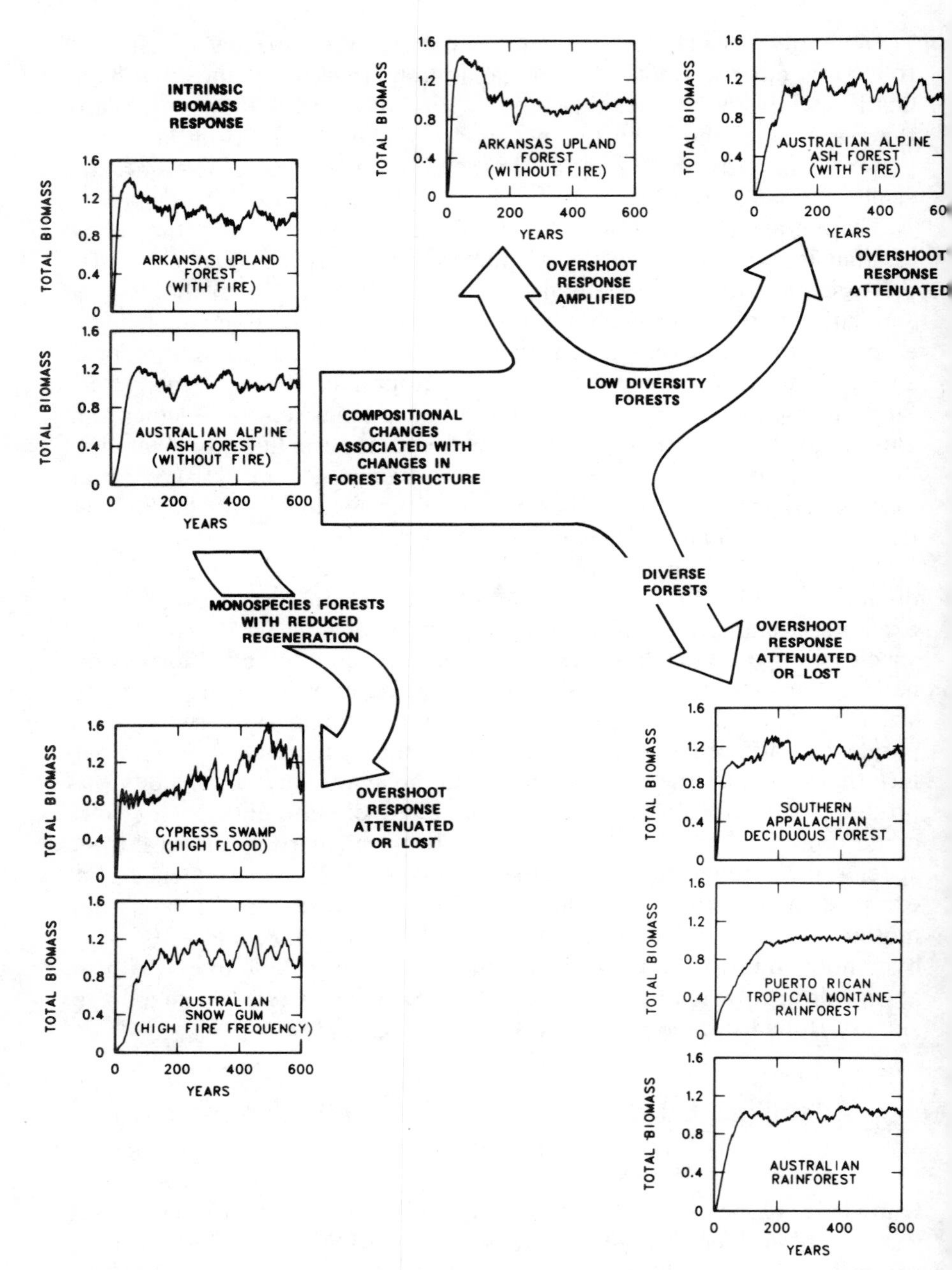

Figure 6.4. Biomass dynamics of several forest landscapes as simulated by gap models. Each simulation is the sum of 50 plots, with open plots in year 0 simulated for 600 years. For aid in comparison each graph is scaled so that the biomass at year 600 is given a value of 1.0.

shoot is in evidence for simulations of alpine ash (*Eucalyptus delegatensis*) without fire from the BRIND model. Both of these responses are logical extensions of the attributes of the two dominant species (intrinsic biomass response, *see* Fig. 6.4).

In forests of low diversity, in which there is a composition shift following the breakup of the even-aged initial cohort's canopy (Fig. 6.4), the overshoot of the landscape dynamics can be amplified if the initially dominant species are larger than the species that come to dominate the mixed-size forest that develops over long periods of time. The case shown in Figure 6.4 is for Arkansas forests without wildfire, in which oak (mostly *Quercus falcata*) replaces pine (mostly *Pinus taeda*) as the dominant species. Waring and Franklin (1979) and Franklin and Hemstrom (1981) reported a similar pattern in the extremely long (greater than 1,000-year) replacement sequences in the Pacific Northwest. In this case, Douglas-fir (*Pseudotsuga menziesii*) stands with biomass values up to 1,500 mg ha^{-1} (Fujimori, et al [1976]) are replaced in a very long successional sequence by stands of much lower biomass that are dominated by other conifers of smaller sizes. A similarly dramatic biomass overshoot, augmented by compositional change, occurs when the tall eucalypts, *Eucalyptus regnans* (mean height greater than 95 meters) and *Eucalyptus olbiqua* (mean, height greater than 90 meters) are replaced in the absence of a regenerating fire by *Nothofagus-Atherosperma* rain forests of much lower stature (less than 40 meters) (Gilbert [1958], and Jackson [1968]) in Tasmania.

In other forest systems of low diversity, a compositional shift at the time of a first canopy breakup does not amplify the biomass overshoot response, but the overshoot pattern still is quite discernible. This case has been discussed in detail by Bormann and Likens (1979a,b) for northern hardwood forests. The persistence of such dynamics in the presence of moderate diversity is a consequence of a similar size and longevity seen among the species that cause the biomass dynamics. This maintains the synchrony needed for the overshoot response to be seen. Some low diversity forests (*see* Fig. 6.4, Australian alpine ash forest with fire) show a loss of the overshoot biomass response in relation to the pattern of variation about the ecosystem's biomass asymptote.

Generally, in diverse forests with a mixture of growth rates, tree species roles, mortality rates, and tree sizes, the synchrony across several patches is lost due to the differences in the attributes of the species that are locally dominant on a given patch (Fig. 6.4). Similarly, even for monospecies forests, if the regeneration is irregular or reduced and thus is made asynchronous across the patches of the forest landscape mosaic, then the overshoot behavior also is lost.

The behavior of landscape biomass dynamics can be understood in the context of the dominant species (and their attributes) that make up the forest. This finding has been taken as an indictment of both the ecosystem concept (*see* Drury and Nisbet [1973]) and the notion that ecosystems have emergent behavior (Sousa [1980]). Edson, et al's (1981) discussion on Salt's (1979) comment on the topic of emergent properties probably is the most appropriate view of this issue: "In cases where our concern is primarily with the relationships between different levels of ecological organization, these interests are best served by explicit attention to the study of these relationships. . . claims of emergence, or of non-emergence, can contribute nothing substantial to such a study." The landscape responses of forests seem to be simple and explainable by relatively simple general models (Fig. 6.1); however, beneath this apparent simplicity, the causes are complex and difficult to observe directly over meaningful time periods. Also, the behavior of the biomass dynamics of landscape systems is considerably constrained in relation to the potential range of behaviors that even the simple general models of forest behavior might produce.

The Effects of Species on the Landscape Biomass Dynamics in the Frequency Domain

Prigogine and Nicolis (1971) and Platt and Denman (1975) have explored the consequences of using a nonlinear, statistical mechanical description of biological systems. This work has sought to investigate the origins of periodic occurrences of structures in space and time (Nicolis and Auchmuty [1974]) within ecological systems. Platt and Denman (1975) note that ecosystems should exhibit periodic behavior in time and space, and they further state that it should be a central goal in ecology to identify the characteristic frequencies of ecological systems. A method of analysis used to detect system periodicities of an ecosystem that are excited by random fluctuations is spectral analysis (Platt and Denman [1975], Shugart [1978], and Pielou [1981]). Spectral analysis, at the most elementary level, is an investigation of the periodicities and their magnitudes that make up a time (or space) series. Typically, one inspects the power spectrum (a plot of amplitude relative to power or spectral density versus frequency) that is determined by the Fourier transform of the autocorrelation function (*see* Emanuel, et al [1978a] for a mathematical development).

The immediate problem is to understand the pattern of variation in landscape biomass (e.g., as evidenced in the model-generated landscape biomass responses in Fig. 6.4) in terms of the periodicities in these biomass responses. Green (1976, 1981) used a spectral analysis of pollen abundance in sediment cores to postulate the successional responses for the Nova Scotian landscape during the Holocene climate and tree species range changes. Although the periodicities of tree-ring width variations have been computed for variations in growth for individual trees for periods up to 3,000 years (Bryson and Dutton [1961]), an actual determination of the spectral response of the biomass of landscape systems has been limited by data availability.

Spectral analyses of gap model responses, which have been averaged to produce an expected biomass response have been developed to formulate a preliminary theoretical understanding of landscape-scale biomass dynamics as simulated by models (e.g., Emanuel, et al [1978a,b], and Tharp [1978]). In developing a pattern of biomass variation over time, at the level of a landscape, by averaging the response of 100 gap model simulations and eliminating the first several years of data to eliminate the effects of transient responses, Emanuel, et al [1978a]) found that the response of the FORET model was band-limited; almost all of the variations were associated with frequencies (l) of less than 0.03 radians per year. This indicates that, according to the Nyquist sampling theorem, the appropriate frequency for sampling forested landscapes is of the order of 16 years. The shapes of the power spectra of landscape biomass dynamics (sum of 100 simulations of the FORET model, Fig. 6.5) under a variety of experimental conditions generally are similar (Emanuel et al [1978b]), with a strong cyclical component of 200–250 years per cycle. This periodicity corresponds to the intrinsic saw-tooth biomass curve (discussed in Chapter 5, Fig. 5.2) that is synchronized on the landscape and excited by the random deaths of large trees. The addition of the American chestnut (*Castanea dentata*) as a viable species results in a spectrum that clearly is richer than that for the base case, also, the dominant cycle is shifted to a higher frequency (Fig. 6.5b). Also, there is a marked decrease in spectral richness that has been observed for both a climate change perturbation (Fig. 6.5c) and a pollutant effect perturbation (Fig. 6.5d).

Each significant peak in the power spectral density corresponds to a cyclical phenomenon in the modeled forest landscape dynamics. The cyclical behavior is not necessarily the result of a system periodicity (e.g., the feedback cycles among roles that are shown in Fig. 5.8), but it is identifiable as a component of the time series of interest (i.e., total biomass across a landscape). In each of the time series, there is an important cyclical component with a period of 200–250 years. This component appears to coincide with the average longevity of a dominant canopy tree. From the biology of the tree species involved, the Role 1 species, tulip-poplar (*Liriodendron tulipifera*)—because of its rapid growth rate and its ability to grow to great size (5 meters in diameter, 70 meters in height)—tends to amplify the cycle created by the growth, establishment, and death of a canopy tree. The Role 2 species, chestnut (*Castanea dentata*), because of certain unique attributes in its biology (largely because it is an extremely successful tree that reproduces vegetatively), tends to suppress this cycle. The variations, from case to case, in the length of the dominant

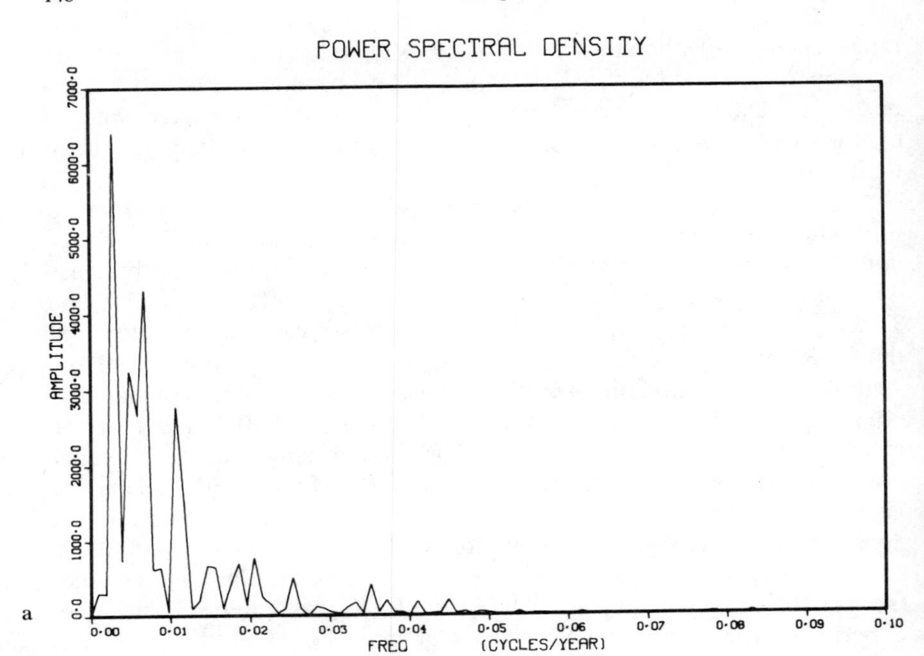

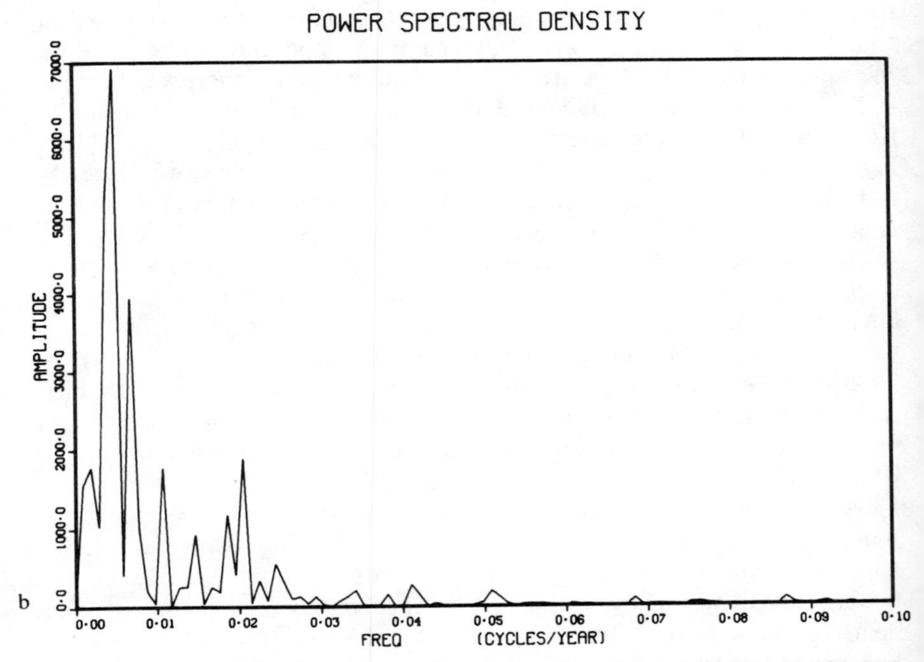

Figure 6.5. Power spectral density for the landscape biomass variable for: (a) base case; (b) American chestnut added as a viable species; (c) Increase in growing degree-days; and (d) Differential decrease in growth rate. (From Emanuel, et al [1978b].)

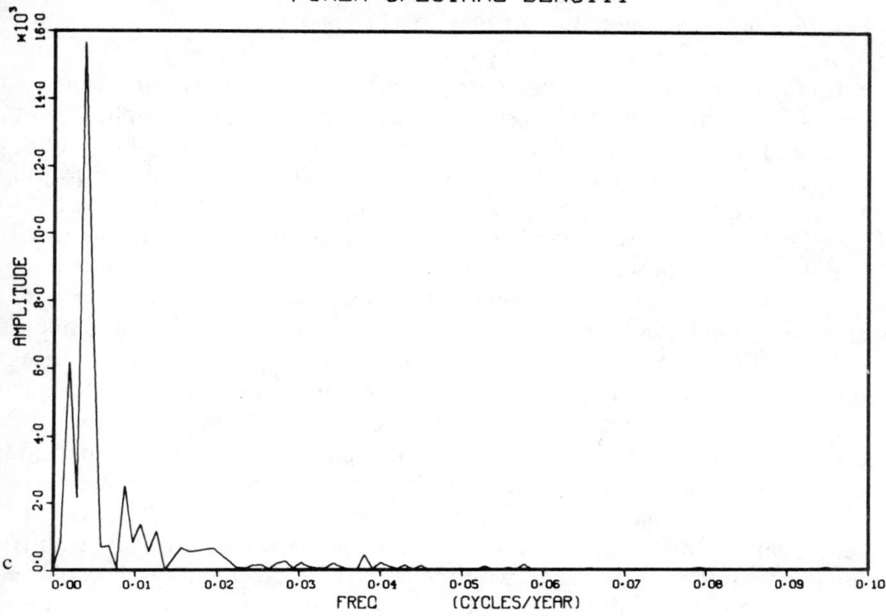

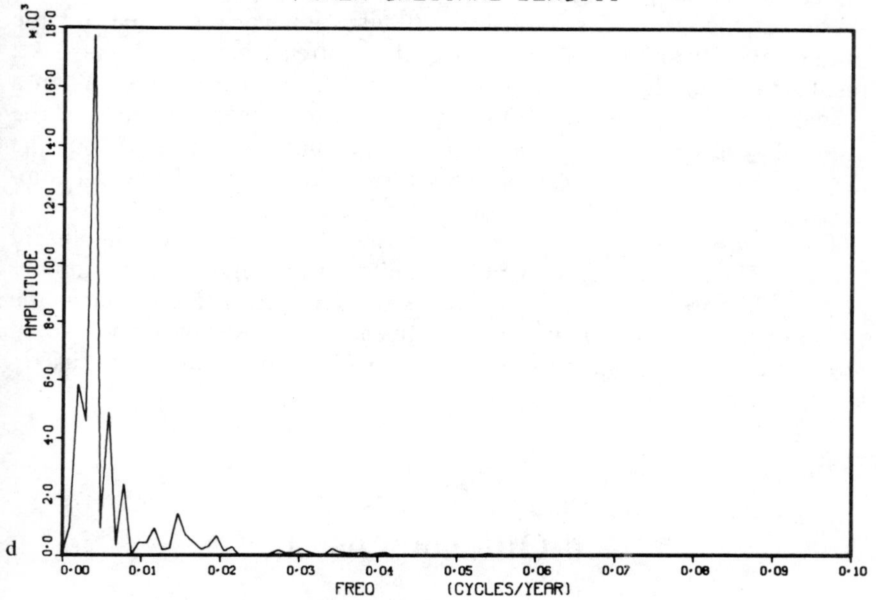

Figure 6.5. (continued)

period are attributable to differences in the biology of the important species that were abstracted as "Roles" in Chapter 5.

It is important to note that the cyclical behavior observed from examinations of the power spectral density is present in the landscape biomass time series when averaged over 100 modeled patches. This implies a fair degree of uniform phase relationship among the stands. To some extent, this results from the choice of a bare stand as an initial condition for all simulations. However, if the frequency component structures of the total biomass time series were not generally the same for each stand, any initial phase relationship would not be preserved.

Periodicities that are evident in the base case, and in the case with chestnut included as a viable species (i.e., the increased spectral richness evident in Fig. 6.5b), are attributable to the diversity of growth responses that are associated with a relatively large number of important species in the system. The tendency of the 250-year cycle to be suppressed by chestnut is apparent in Figure 6.5b, as is the tendency for longer periods to appear important in the time trace. These longer periodicities are associated with multiple-generation replacement patterns, and they are system-level responses in that they occur at periods that are significantly greater than those associated with the growth replacement process of the considered species. When the landscape system is stressed by either temperature change or pollution, the richness in the power spectra is lost and the 250-year canopy replacement cycle dominates (Fig. 6.5c,d). The response of the stressed system in both cases is clearly different (at the system level) from the rich response of the unstressed cases. This also is a result that Van Voris, et al (1980) obtained from microcosms. When a species is added or deleted from the system, the general features of the spectral responses are similar (e.g., rise in important peaks), but the magnitudes of the various peaks are altered. Considering the actual composition of the forests for each of the test cases, landscape systems with rich spectral responses are diverse in terms of the numbers of important species. The stressed systems have significant reductions in the number of truly important species that are associated with a reduction in the richness of the system power spectra. In each case, the landscapes would look different, because the species composition would vary from case to case.

Landscape Dynamics with Patch Interaction in One Dimension

Until this point in this chapter, landscapes have been simulated as the summation of the responses of independent patches that make up the landscape. This corresponds to a landscape that is sampled by inventorying the trees on a sample plot at the intersection of an inventory grid (e.g., The U.S. Forest Service Continuous Inventory Data Set); it also corresponds to many other landscape surveys using representative samples. Beyond the abstraction of the landscape as being a partially sampled system, the summation of independent patches to a idealized landscape also could be applied to actual landscapes in which patch-to-patch contagion was of reasonably little concern. If this were not the case, one would likely attempt to model the forest with a spatially explicit model (*see* Chapter 2)—an approach that seems proscribed for diverse forests. In diverse landscapes with contagion, it still seems appropriate to use an understanding that is gained from the simpler models as an interpretive tool.

However, there are some cases of actual landscape pattern and process that seem to be particularly interpretable in terms of gap model responses, even in the face of spatial interactions among patches. Sprugel (1976) studied the phenomenon of wave regeneration in the balsam fir (*Abies balsamea*) in the higher elevation forests in New England (the United States). Sprugel described that pattern as: "In certain areas of the northeast, the high-altitude fir forest exhibits a rather striking pattern which may be described as 'wave regeneration.' From a distance, such a forest is seen to be highly heterogeneous, with large areas of green canopy broken by numerous more-or-less crescent-shaped bands of dead trees whose exposed trunks show up silver in side view. These areas of dead and dying trees have frequently been thought to be small windfall areas, where particularly violent gusts of wind during a heavy storm flattened a few dozen trees and left a visible hole in the otherwise uniform canopy . . . each of the supposed windfalls is actually an area of standing dead trees, with mature forest beyond it and an area of vigorous regeneration below it." To indicate the generality of such a pattern, Sprugel (1976) also quotes the account of Oshima, et al (1958) with regard to the high altitude forests of Japan: "In the dark green of the gentle southwest slope of Mt. Shimagee covered with subalpine forest, several whitish stripes horizontally running in parallel with each other can be seen in distant view so distinctly that the mountain has been named 'mountain with dead trees strips' according to its curious physiognomy" (*see also* Iwaki and Totsura [1959]).

The regeneration wave (Fig. 6.6a) is essentially a spatial expression in one horizontal dimension of the saw-tooth biomass curve (see Fig. 5.2) that is expected for the temporal response of a single forested patch of a landscape. According to Sprugel (1976), the regeneration wave is caused by trees dying continually along the exposed front of the wave. Thus, the movement of the notch in the canopy that is diagramed in Figure 6.6b would be to the reader's left as trees at the edge died. Along with actually determining the age structure through a transect across a regeneration wave, Sprugel (1976) and Sprugel and Bormann (1981) found that older

Figure 6.6 The canopy of a montane rain forest at the Luquillo site in Puerto Rico. Because of the variation in leaf shape and canopy structure, the mosaic nature of a mixed-species forest canopy is obvious. (Photo by T.W. Doyle.)

trees near the crest of the regeneration wave did have higher mortality rates than those in the closed-canopy forest. The elevated mortality was related to prevailing winds, rime-ice, winter-desiccation, the cooling effect of summer winds, and tree age, with the expectation that the likely cause of the wave was due to the multiple actions of these factors. The wave regeneration pattern occurred almost exclusively in pure balsam fir stands. Balsam fir functions as a Role 3 species in this ecosystem (i.e., fast-growth, short-lived, small, and regeneration favored by disturbance). Sprugel (1976) noted that: "If certain areas are inherently prone to wave-type disturbances due to local topography, prevailing winds, or some other factor, then those forests must almost inevitably become nearly pure fir after a few centuries of waves passing through every 60 to 70 years. In turn, the change toward pure fir probably allows waves to move through the forest more smoothly and evenly, since in a pure stand all the trees reach the same stage of degeneration and susceptibility to environmental stress at the same time." Thus, the regeneration wave phenomenon allows a Role 3 species that one would expect, because of regeneration difficulties, to be unable to maintain pure stands over landscapes to generate favorable regeneration conditions and to hold dominance. The pattern is striking because it is linear and, thus, is obvious on the landscape.

R.O. Lawton (personal communication) reports another striking linear and apparently stable landscape configuration for the Role 3 species, *Didymopanax pittieri*; it functions as the dominant shade-intolerant, gap-colonizing species in wind-exposed montane rain forests of Costa Rica. Apparently, ridge-top populations can form bands (~50 m wide and 0.5 to 1 km long) that appear to be demographically stable, which perhaps is due to the high disturbance rate on the ridge crest locations.

Landscape Dynamics with Patch Interaction in Two Dimensions

The orderly wave-regeneration pattern of forested landscapes (Oshima [1958], Sprugel [1976], and Sprugel and Bormann [1981]) is an unusually regular pattern of organization of the birth-death pattern across a landscape mosaic. It is interesting because it allows a Role 3 species that one would expect to be a gap-colonizing component of a forest landscape to persist as a (or *the*) dominant tree species. The organizing phenomena that allow this persistence is synchronous mortality along a linear front due to environmental conditions. One could also expect synchronous mortality to induce regeneration conditions in two dimensions as well.

This type of case has been studied in detail for the Hawaiian tree, 'ohi'a (*Metrosideros polymorpha*), which is found on the island of Hawaii. Between 1954–1977, 'ohi'a canopy trees either died or were defoliated over large areas of the montane rain forest on the windward side of Hawaii. This decline or dieback was very rapid in its onset, and it caused considerable concern over the stability of forests of this species (Jacobi [1982]). The dieback initially was thought to be an introduced disease that was brought about by the fungus *Phytophthora cinnamomi*, which is associated with a similar dieback of the Jarrah (*Eucalyptus marginata*) in western Australia (Podger [1972]). Having discovered this pathogen in 'ohi'a dieback stands (Laemmlen and Bega [1972], Kliejunas and Ko [1973] and Bega [1974]), as well as abnormally high population levels of a boring cerambycid beetle (*Plagithmysus bilineatus*), it appeared that the dieback was caused by an epidemic disease. Studies (Papp, et al [1979]) indicated that neither *Phytophthora* nor *Plagithmysus* were universally associated with dieback. Dr. Mueller-Dombois and his students at the University of Hawaii noted that a similar dieoff of 'ohi'a had occurred on Maui in the early 1900s. This historical dieback had been initially suspected to be due to fungal pathogens (Lewton-Brain [1909] and Lyon [1909]), but no conclusive evidence was found to support this hypothesis (Lyon [1918, 1919]). This led to the development of a hypothesis that the dieback was a natural landscape process and it initiated several studies by

Mueller-Dombois, his students (Burton [1980]; Jacobi [in press], and Mueller-Dombois, et al [1981]), and the U.S. Forest Service (Adee and Wood [in press]).

Although the dynamics of 'ohi'a are complex on the landscape, which is partly due to the ability of this tree to function in a wide variety of soils and topographic positions while displaying a great variation in form, the general pattern of the dieback landscape process (for more detail, *see* Mueller-Dombois, et al [1981], Gerrish and Mueller-Dombois [1980], Jacobi [in press], and Adee and Wood [in press]) is described below.

'Ohi'a is a light-seeded, wind-dispersed tree that occurs as a pioneer on young volcanic substrates (Atkinson [1970]). It also is the dominant rain forest tree in most of the rain forests of Hawaii, and it forms monospecies canopies over large areas (Adee and Wood [in press]). 'Ohi'a trees are of small stature and the tree often is in stunted stands on bog soils. It requires high light levels for regeneration (Burton [1980]). Thus, 'ohi'a corresponds reasonably well to what has been abstracted as a Role 3 species. As one would expect for such a species, 'ohi'a does not regenerate well under its own canopy (Gerrish and Mueller-Dombois [1980]), which is a feature that led Clarke (1875) and (later) Hosaka (1939) and Egler (1939) to conclude that 'ohi'a forests were decadent.

The apparent mechanism by which the 'ohi'a maintains itself is outlined in a general form in Figure 6.7, but the pattern can vary greatly with conditions (Mueller-Dombois [1980], and Gerrish and Mueller-Dombois [1980]). Starting with a young 'ohi'a stand (Fig. 6.7), tree growth produces a mature 'ohi'a forest. Eventually, competition and nutrient depletion (Kliejunas and Ko [1974]) reduce tree vigor and render the stand susceptible to dieback. A triggering event (probably a climatic instability such as excessive rainfall on wet sites, drought on dry sites, or perhaps lightning, (*see* Mueller-Dombois [1980]) causes trees to begin to die. Secondary agents, such as fungus or insects (Kliejunas and Ko [1974]), attack the dying trees and help to synchronize the dieback event across the area. 'Ohi'a, which regenerates well as an epiphyte on downed tree trunks (Gerrish and Mueller-Dombois [1980]) and requires high light levels (Burton [1980]), is able to regenerate in the dieback patch; it produces a young stand to close the regeneration cycle. Because of the synchrony in the mortality event, the landscape biomass response of 'ohi'a is of the same saw-toothed shape (*see* Fig. 5.2) that is expected at smaller spatial scales for forests (*see* Chapter 5).

Apparently, the dieback mode of regeneration at the landscape scale is not restricted to 'ohi'a, but also is found to occur in several tree species—often on islands and typically conforming to the Role 3 abstraction. Paijmans (1976), in referring to forests in the lower montane zone in Papua New Guinea, notes: "A common feature of mature *Nothofagus* stands are patches of dead or dying trees, for which no obvious cause has been found. It has been suggested (Robbins and Pullen [1965, p. 105]) that

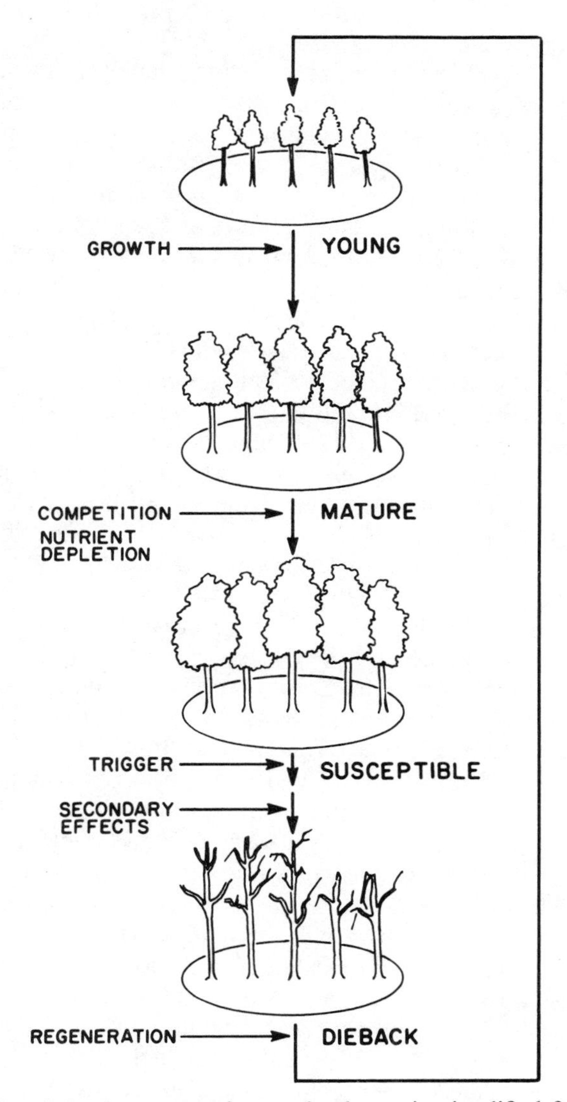

Figure 6.7. The dieback regeneration cycle shown in simplified form for 'ohi'a (*Metrosideros polymorpha*) on the Hawaiian Islands.

groups of even-aged trees die off together on reaching overmaturity." Mueller-Dombois, in visiting these forests, felt that the dieback strongly resembled the dieback phenomenon seen in the Hawaiian Islands (Mueller-Dombois [1982]). Ash (1981), in discussing the *Nothofagus* dieback in Papua New Guinea, attributed the dieback to the combination of weather, substrate, pathogens, and successional development that are implicated in the well-studied Hawaiian landscapes. Mueller-Dombois (1982), in reporting an extensive survey of large Pacific islands, found dieback patterns occurring in several forests; all were dominated by canopy species that are both shade-intolerant species and whose regeneration pattern was the two-dimensional patch-analog to the one-dimensional wave regeneration. These forests included *Eucalyptus deglupta* forests in New Britain, *Acacia* species near Rabaul in New Britain, and *Nothofagus* (*N. solaudri*) and *Metrosideros* (*M. umbellata*, *M. robusta*, and *M. excelsa*) forests in New Zealand. Mueller-Dombois (1982) also reported the phenomena in Sri Lanka and New Caledonia. The critical consideration is whether or not the species affected by the dieback event is able to successfully regenerate following the synchronized mortality event.

There are any numbers of factors that can produce synchronous mortality on an even-aged forested landscape. For example, tree-killing bark beetles (*Coleoptera*, *Scolytidae*) frequently select weakened trees as hosts and emit pheromonal cues to concentrate dispersing beetles to these hosts (Wood [1982]). These beetles are capable of killing living trees over large areas and, thus, of reinitializing landscapes with regenerating intolerant-tree seedlings.

This type of reinitialization over large areas, even if it is a natural landscape process, makes the landscape subject to radical widespread change. Such vulnerability can occur if the time period for regeneration is short; particularly if some event should alter the ability of the tree to regenerate at this critical time. This appears to be the case in the Maui 'ohi'a dieback (Lewton-Brain 1909, Lyon [1909]), in which the areas of dieback were deliberately planted with introduced tree species. As another example, the Australian landscape is undergoing continent-wide episodes of dieback among several species of *Eucalyptus* (Old, et al [1981]). In some of these cases, *Eucalyptus* regeneration has been altered by human land use (e.g., grazing—particularly sheep, or on altered fire regime) which causes regional-scale deforestation. The etiology of eucalypt diebacks shows considerable geographic variations and it is a complex consequence of natural processes and human alterations in land use (Old, et al [1981]). Nonetheless, eucalypt dieback illustrates the potential vulnerability of dieback-regeneration landscapes under land use changes.

Conclusions

The biomass response of a continuous landscape that is composed of a mosaic of forested patches can be explained by a general model of growth, regeneration, and death on the mosaic elements, if delay effects are included. The mathematical forms of biomass responses across a landscape are variable, but not nearly as variable as the general model indicates they could be. Theoretical landscapes that are composed of single-species forests of selected species can be used to interpret the behavior of landscape biomass dynamics on real landscapes. Also, analyses on landscape biomass dynamics that use spectral analysis techniques show processes from patch dynamics to be evident in the landscape dynamics; also, they provide an idea of the sampling periodicity that is appropriate for estimating biomass change across a landscape. Two rather special cases of landscape patterns (wave-regeneration and dieback-regeneration) are cases of the landscape process that generate a clear pattern of biomass response, which is typically seen at a smaller scale.

The next chapter (Chapter 7) investigates the landscape as a dynamically changing mosaic and discusses the equilibrium concept of landscape systems.

Chapter 7

Categories of Dynamic Landscapes

The dynamics of small patches of a forested landscape are driven as nonequilibrium systems via the deaths of large trees (*see* Chapter 5). This periodic disequilibrium response (*see* Fig. 5.2) also is evident in the landscape biomass time series (*see* Fig. 6.5; *see also* Emanuel, et al [1978a,b]) when landscapes are fabricated by summing the responses of patches that are simulated by gap models. In nature, the pattern of wave-regenerated forests (Oshima, et al [1958], and Sprugel [1976]) and regenerative dieback (Gerrish and Mueller-Dombois [1980]) can be described as a consequence of the nonequilibrium nature of forest mosaic patches. There also is an abundance of observations that are purported to show such small-scale nonequilibrium responses in other forests (*see* White [1979] for a review).

The saw-toothed curve that produced for small-scale biomass dynamics (*see* Figs. 5.2, 5.6) is a direct consequence of the dynamics of growth, death, and regeneration within the small patch, but this pattern also can be augmented by external factors. White (1979, p. 230) summarized this as:

> Natural disturbances have been traditionally defined in terms of major catastrophic events originating in the physical environment and, hence, have been regarded as exogenous agents of vegetation change. Problems with this view are: (1) there is a gradient from minor to major events rather than a uniquely definable set of major catastrophes for each kind of disturbance, and (2) some disturbances are initiated or promoted by the biotic component of the system.

Tansley (1935, pp. 286-287) stated:

> In 1926 (p. 680) I proposed to distinguish between *autogenic succession*, in which the successive changes are brought about by the action of the plants themselves on the habitat, and *allogenic succession* in which the changes are

brought about by external factors. "It is true of course (I wrote) and must never be forgotten, that actual successions commonly show a mixture of these two classes of factors—the external and the internal" (p. 678). I think now that I should have gone farther than this and applied my suggested new terms in the first place to the factors rather than to the successions. It is the fact, I think, that autogenic and allogenic factors are present in all successions; but there is often a clear preponderance of one or the other, and where this is so we may fairly apply the terms, with any necessary qualifications, to the successions themselves.

Odum, et al (1979) also noted:

> One of the difficulties of all practical ecological research is that of distinguishing allogenic from autogenic causes.

At a small spatial scale, almost regardless of the generating event (either endogenous or exogenous), the effect of a disturbance is manifested in a more or less similar way—a saw-toothed biomass dynamic curve with the amplitude of the curve determined by the size of the largest trees on a plot and (thus), to a degree, both endogenous and autogenic. For example, compare the basal area dynamics of Oliver and Stephens' (1977) reconstruction of a small plot in New England (Fig. 7.1) to the single-plot biomass dynamics in Figure 5.2. Clearly, the forms of the curves are similar, but the causes are endogenous in one case (Fig. 5.2) and largely exogenous in the other (Fig. 7.1). Before discussing landscapes, it is appropriate to elaborate on the dynamics of patches in the face of exogenous disturbance.

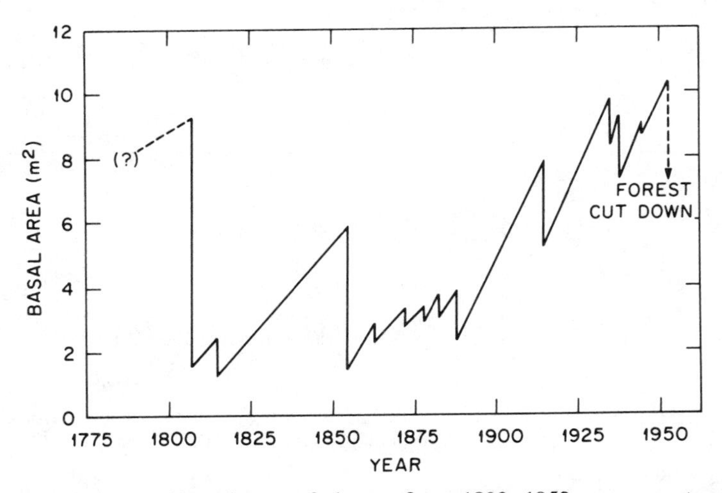

Figure 7.1. History of basal area of change from 1803–1952, as reconstructed by Oliver and Stephens (1977). The drops in basal area are due to a variety of events including logging, firewood harvest, the chestnut blight, hurricanes, and gypsy moth damage. (From Table 2, Oliver and Stephens [1977].)

Exogenous Disturbances and Patch Dynamics

At the scale of a small patch (about 1/10 ha) exogenous disturbance is manifested as an increased mortality rate of trees (Grime [1974], and Oliver [1981]). The KIAMBRAM model contains a subroutine that attempts to mimic the effects of the "*chablis*," which was described by Oldeman (1978) as the destruction associated with the death of a large tree. By altering the frequency and magnitude of the occurrence of the chablis and its associated mortality, one can use this subroutine to obtain a theoretical idea of the effect of a small scale disturbance at varying frequency and intensity (Fig. 7.2).

If disturbances are infrequent (Fig. 7.2a,b,c), both the basal area and the biomass dynamics at the scale of a patch are driven mostly by the internal birth-growth-death cycle. However, the mortality rates are ele-

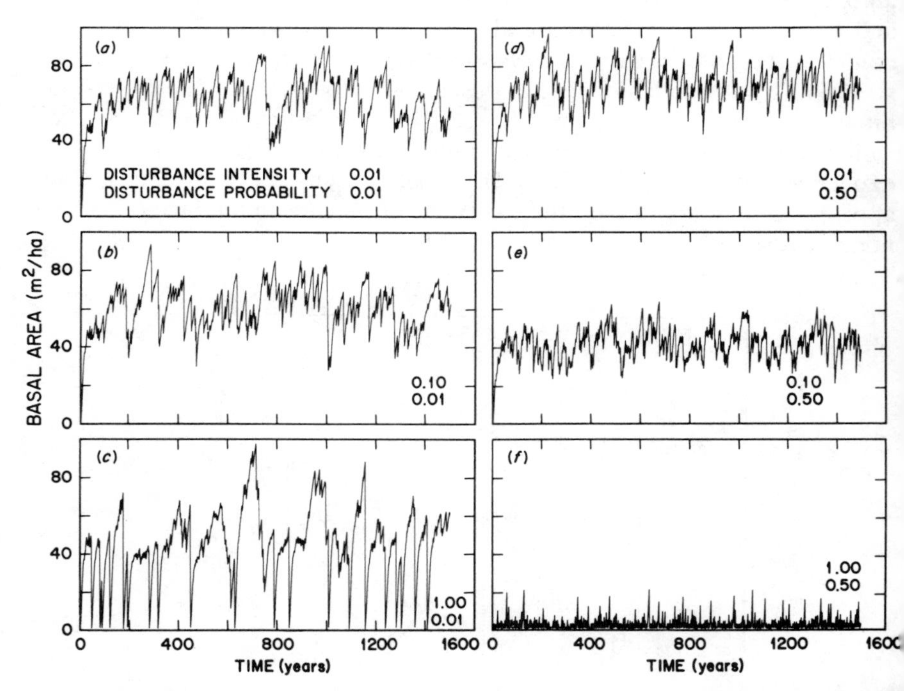

Figure 7.2. Response of the KIAMBRAM Model to changes in the probability of a disturbance and the intensity of the disturbance (likelihood of mortality). Each graph is a basal area for 1,500 simulated years. (a) Disturbance intensity (likelihood of an individual tree's death in the year of a disturbance event) = 0.01; disturbance probability (likelihood of a disturbance event in a year = 0.01. (b) Disturbance intensity = 0.10; Disturbance probability = 0.01. (c) 1.00; 0.01. (d) 0.01; 0.50. (e) 0.10; 0.50 (f) 1.00; 0.50.

vated and the number of extremely large trees are reduced. If the disturbances are frequent, but mild (Fig. 7.2d), the reduction of large gap-forming trees favors the small, numerous Role 4 type undercanopy tree species. Infrequent, but more severe, mortality events (Fig. 7.2e) are favorable to the Role 3 (nomadic or pioneer) species that use the more frequent occurrence of large gaps. A frequent, severe disturbance (Fig. 7.2f) causes the formation of a short forest that is comprised of small trees and dominated by rapidly growing species (often Role 1 or 3 species) with exceptional regeneration attributes. Harper (1977, p. 710) developed a theoretical diagram for populations (or communities) that are subjected to "disasters"; he used this diagram to introduce *r* and *k* selection in the context of plant communities. Harper (1977), in speaking of populations, noted that when disturbances are so frequent and severe (that they constantly place the populations in a growth and recovery phase (Fig. 7.3a);

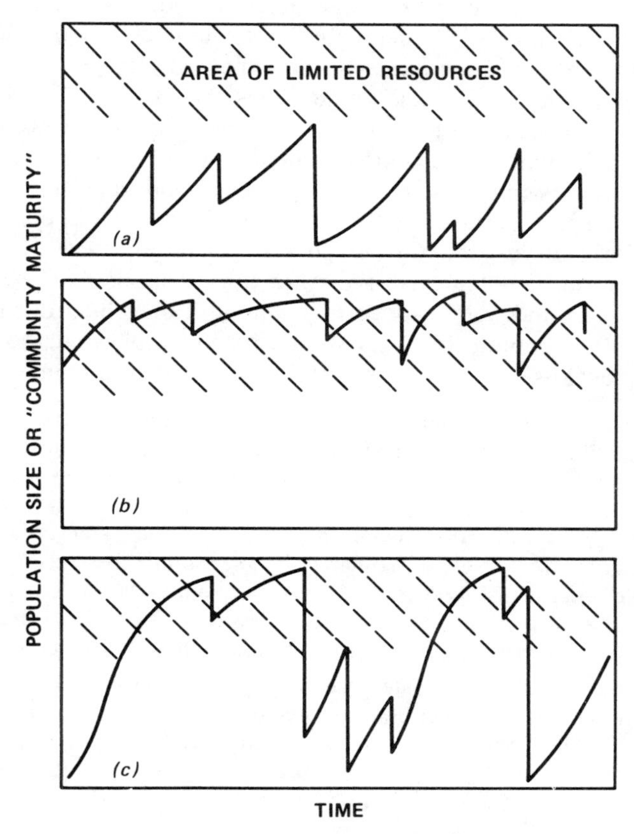

Figure 7.3. Diagrammatic representation of the growth of populations (or the maturation of communities) in environments with disasters and resource limits: (a) Always in *r* phase; (b) Always in *k* phase, and (c) Mixed. (From Harper [1977].)

then one would expect species with colonizing ability, high dispersibility, and high reproductive effort (*r*-species). Grime (1974) also felt that regularly disturbed habitats should contain species that followed a "ruderal strategy" of life history (i.e., rapid seedling establishment and rapid growth). When disturbances are less frequent, then competitive interactions among species become more important (Fig. 7.3b). It seems that the dynamics of most forests at small spatial scales (particularly if they contain species with different roles) could be expected to contain a mixture of *r* and *k* species (Fig. 7.3c).

Statistical Interpretations of Landscape Systems

In Chapter 6, landscapes were constructed as mosaics of small patches and a gap model was used to simulate the fate of each tree on each patch. By tabulating 100 of these patches, the expected biomass response was produced first for monospecies landscapes (*see* Fig. 6.3) and then for mixed-species landscapes (*see* Fig. 6.4). These responses were produced with each of the mosaic elements initially being open (no trees), and they could be taken as the expected response of a landscape to an arbitrary disturbance as large in area as the landscape itself. The important single aspect of these curves (*see* Figs. 6.3, 6.4) is that (after a period of time) the synchronization of patch birth-growth-death cycles is lost and the biomass varies around some apparently equilibrium value. Bormann and Likens (1979a,b) noted this larger scale response in what they call the "shifting-mosaic steady state" concept of ecosystem dynamics:

> . . . the Shifting-Mosaic Steady State may be visualized as an array of irregular patches composed of vegetation of different ages. In some patches, particularly those where there has been a recent fall of a large tree, total respiration would exceed GPP [gross primary production], while in other patches the reverse would be true. For the ecosystem as a whole, the forces of aggradation and of decomposition would be approximately balanced, and gross primary production would about equal total ecosystem respiration. Over the long term, nutrients temporarily concentrated in small areas in fallen trunks would be made available to large areas by root absorption and redistribution by litterfall.
>
> The structure of the ecosystem would range from openings to all degrees of stratification, with dead trees concentrated on the forest floor in areas of recent disturbance. The forest stand would be considered all-aged and would contain a representation of most species, including some early-successional species, on a continuing basis.

As was indicated in Chapter 6, this is a venerable concept in ecology (Aubreville [1933, 1938], Watt [1925], Whittaker [1953], and Whittaker and Levin [1977]) that also can be attributed to Clements:

Nevertheless, all areas within the sweep of climate and climax, whether bare or denuded by man, are marked by more or less evident successional movement of communities and, hence, belong to the climax in terms of its development. In consequence, each climax consists not merely of the stable portions that represent its original mass but also of all successional areas, regardless of the kind or stage of development." (Weaver and Clements [1938, p. 479], cited in McIntosh [1981].)

In coupling the idea that a landscape should be viewed as the statistical average of the mosaic elements which comprise that landscape with the idea that the mosaic elements are themselves nonequilibrium systems (*see* Chapter 5, Fig. 5.2), the resultant theoretical landscape will necessarily contain a mixture of patches of varied successional ages (Forman and Godron [1982]). If the number of patches is very large, the landscape dynamics (a statistical average) become extremely predictable. The proportion of patches in various successional (or vegetative structural) classes also becomes regular as the variance of the landscape is stablized by large numbers of patches. If the landscape is sufficiently large, then random disturbances that simultaneously affect several patches of the landscape mosaic also can be averaged into the total landscape response (Whittaker and Levin [1977]). If the disturbance spatial scale is large in relation to the size of the landscape, then the statistical description of the landscape begins to break down due to small sample-size effects. In the most extreme case, if a disturbance is larger than the landscape area, the disturbance would act as a synchronizing event. Then, one would obtain the landscape responses, such as those shown in Chapter 6 (*see* Fig. 6.4), in the case of a single disturbance event. If such a large disturbance were sufficiently frequent, the resultant landscape response (for biomass) would resemble the nonequilibrium saw-toothed curve that is seen at the scale of a gap model.

Thus, for relatively frequent exogenous disturbances, one can describe two types of responses, which depends on the size of the disturbance and the size of the landscape. Large disturbances on small landscapes produce responses that are effectively nonequilibrium*; small disturbances on large landscapes produce responses that are statistically regular or quasi-equilibrium (Shugart and West [1981]). The division between these two extremes, of course, is arbitrary. Using gap models both to simulate the biomass variance versus the sample size curve and to simulate the

---

* The terms equilibrium and nonequilibrium are used to describe, respectively, the case in which the number of patches entering a given condition (e.g., a biomass range, a forest type, etc.) on a landscape over a time interval is balanced by the number changed from that condition to some other conditions and the case in which this balance does not occur. The number of patches in each condition at equilibrium is the landscape steady-state. If the landscape system is altered from the steady state and the processes in the system eventually cause a return to the steady state, then the system can equilibrate and the steady-state is stable.

Table 7.1. Some properties of effectively nonequilibrium and quasi-equilibrium landscapes in the extreme cases

Property	Effectively Nonequilibrium Landscape	Quasi-equilibrium Landscape
Disturbance size	Large	Small
Landscape size	Small	Large
Forest age structure	Even-aged for frequent disturbances	All-aged
Total landscape biomass	Unpredictable	Regular
Age distributions of populations	Unstable for long-lived organisms	Stable

tendency for the landscape biomass spectral response to converge as a function of sample size (W. R. Emanuel, pers. comm.), it appears that the biomass quasi-equilibrium for a landscape requires 50 or more patches to be averaged. This number would be larger in diverse forests, particularly if the vegetative composition were of interest. Also the minimum number of patches of a landscape that must be averaged is a direct consequence of the amount of variation in the quasi-equilibrium that is deemed acceptable. The properties of effectively nonequilibrium landscapes and quasi-equilibrium landscapes in the extreme cases are reasonably clear (Table 7.1) even in the face of a relatively arbitrary boundary; they are discussed in the next section.

Examples of Effectively Nonequilibrium and Quasi-Equilibrium Landscapes

If one needs to average the behavior of, for example, 50 patches to obtain a quasi-equilibrium at the landscape level, then there should be effectively nonequilibrium landscapes that simply are not large enough to reach a state of equilibrium. One can use the 1/50-disturbance/landscape size ratio to identify some example cases (Fig. 7.4). Because the patch disturbed by the fall of a large canopy tree is of the order of 100 to 1000 m^2, the minimal landscape that is large enough to absorb the effects of tree falls using the 1/50 ratio is of the order of 10^4–10^5 m^2. Most small watersheds that feed small streams in the southeastern United States are large enough to remain in quasi-equilibrium when a tree falls or some other small-scale disturbance occurs (Fig. 7.4a). However, a wildfire of the size that is commonly found in the southern Appalachian region would disequilibrate

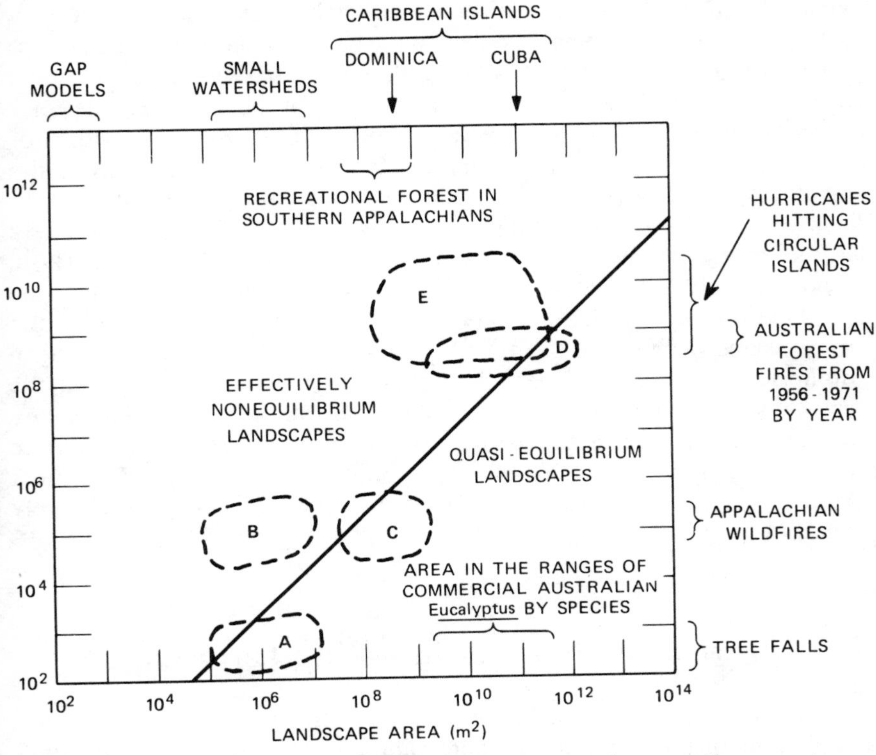

Figure 7.4. Scale of disturbance and scale of landscapes for example ecosystems. The line between the effectively nonequilibrium and the quasi-equilibrium landscapes is based on a 50:1 ratio of landscape area to disturbance area. Combinations of disturbance and landscape scales illustrated are: (a) Treefalls on small watersheds; (b) Wildfires on small watersheds; (c) Wildfires on recreational forests; (d) Australian forest fires on the range of Australian *Eucalyptus* species; and (e) Hurricanes on Caribbean islands. (From Shugart and West [1981].)

these same watersheds (Fig. 7.4b). Since large proportions of the small watersheds synchronously respond to each wildfire, the amount and type of vegetative cover would differ over time. It appears that only the larger national forests and recreational parks in the southern Appalachians may be of sufficient size to absorb the random wildfires and remain quasi-equilibrium landscapes.

It is difficult for a land manager to provide constant levels of habitat for some plant or animal species in nonequilibrium landscapes. This habitat constancy could be obtained by altering the scale of disturbance or by increasing the area under management; however, neither of these strategies often is an available option. In the case of the Appalachian forest landscape, sufficient area is involved so that a large national park might be managed with the diversity of landscape patches remaining in relatively constant proportions, even when given the scale of wildfires in the region.

The amount of forest that is burned annually by wildfire in Australia approaches the size of the entire range of many of the *Eucalyptus* species that make up these forests (Luke and McArthur [1978]) (Fig. 7.4d). For example, the effects of a holocaust in 1939 are manifested today in the unstable age and size structure of the total population of the alpine ash, *Eucalyptus delegatensis* (Hillis and Brown [1978]). Disproportionately large numbers of the alpine ash population were established in 1939 in response to favorable conditions for germination that was caused by the fire. Trees established before 1939 are relatively rare. Similarly, only the largest island of the Caribbean, Cuba, might possibly have a landscape in equilibrium despite hurricane disturbance (Fig. 7.4e). When given the scale and regularity of hurricanes in the Caribbean, the vegetation on any smaller island could be expected to be in an effectively nonequilibrium state (Doyle [1981]). A case in point is Dominica, where some of the most extensive stands of mature rain forest in the Caribbean were severely damaged by hurricane David in 1979.

Intrinsically Nonequilibrating Landscapes

All the landscapes discussed above have a common feature in computer simulations of their dynamics when using a gap model—if enough landscape patches can be averaged, the vegetative composition of these landscapes tends, over time, to approach an equilibrium. However, there also is the possibility that some landscapes in certain environmental conditions are intrinsically nonequilibrating systems regardless of the landscape size.

For example, in the model experiment that was discussed earlier and shown in Figure 5.10, the percentage of *Liriodendron* in a two-species *Liriodendron-Fagus* hypothetical forest shows apparent hysteretic behavior in response to a slowly varying climate variable. In the region of the hysteresis, they are dominated by either *Liriodendron* or *Fagus*—each of which has a strong tendency for self-replacement. Thus, with respect to composition, the hysteresis region has two stable states. Similar, multiple stable states also occur in the dynamics of the BRIND model for *Eucalyptus* (*see* Fig. 4.15) (Shugart and Noble [1981]). In general, if the presence of a parent tree at a site greatly enhances the likelihood of the replacement being the same species as the parent, then such multiple compositional stable states also tend to occur. For ecosystems that are dominated by species with quite different typical life histories, it also is possible to have two stable vegetation types at a given location (Walker [1981]). The results from the gap models and from other, somewhat less detailed, models (Cattelino, et al [1979], Noy-Meir [1982], and Noble [in press]) indicate that these situations would most likely occur in transition zones between different types of vegetation.

In nature, Griggs (1946) and Leak and Graber (1974) have found evidence on mountains of the presence of more than one potential, persistent vegetation type; Marie-Victorin (1929) and Polunin (1937) found similar evidence for larger areas. The simplest case involves two different types of vegetation that—once established in a landscape—would each be self-sustaining in relatively constant proportions.

Landscapes that are made up of mosaics of these multiple stable state patches may not equilibrate around a single expected landscape stability point. These landscapes would be the average of patches in which the composition of each patch will tend to remain the same over time; but, if changed, it will stay in the new configuration until changed again. When pushed, (e.g., by a management decision) to a new configuration and then released to natural processes, these systems would remain in the new configuration. These landscapes could be most unforgiving to a land manager who made an incorrect decision regarding the stewardship of the land. An undesirable landscape configuration, once produced, would not revert to its former state when left alone; it would remain in the undesirable state until it was actively transformed to some other state.

The Severity and Frequency of Disturbance

White (1979) categorizes disturbance by likelihood (both disturbance frequency and predictability) and magnitude. Disturbance magnitude, according to White (1979), is more difficult to define. For example, windstorms vary in physical size, in the size of area they disturb, and in the amount of damage they do to vegetation. The spatial aspect of disturbance magnitude has been discussed in developing an equilibrium/nonequilibrium categorization of landscapes. It is appropriate to inspect changes in landscapes in response to disturbance severity and frequency.

Comparative investigations of equivalent landscapes of sufficient size to be quasi-equilibrium systems but under different natural disturbance regimes would require large amounts of land. Such investigations are reasonable theoretical modelling problems but are logistically difficult to actually attempt. For this reason, discussions about the ecological consequences of disturbance severity and frequency often have focused on the adaptive traits of species in response to a disturbance. Studies of adaptations of species to disturbance regimes are numerous, and fire adaptions (Table 7.2) provide some good examples.

One problem with inspecting the adaptive traits of species in relation to a given environment is that (as Grime [1979b] has pointed out) successful

Table 7.2. Some species adaptations to variations in wildfire frequency and/or severity (see also Kozlowski and Ahlgren 1974, Noble and Slatyer 1977, Gill 1981)

General Adaptation	Example(s)	Comment
Soil protection of buried buds (Gill, 1981)	1. Rhizomes in Eucalyptus porrecta (Lacey 1974) 2. Root buds in Acacia dealbata (Gill 1981 and Sequoia sempervirens (Weaver 1974) 3. Lignotubers in most of genus Eucalyptus (Pryor and Johnson 1971) and in other taxa (Gardner 1957)	Vegetative regeneration allows established individuals to occupy sites after fire. Since very young or very old individuals may have reduced ability, trait may be best at intermediate fire frequencies.
Bark protection of buds (Gill 1981)	1. Stem buds in Eucalyptus regnans (Cremer 1972) 2. Bark insulation in Pinus (Reifsnyder et al. 1967) 3. Bark thickness in Pinus (Reifsnyder et al. 1967) and Eucalyptus (Vines 1968).	Allows individuals to survive all but the more severe fires.
Post-fire flowering	1. Xanthorrhoea australis (Gill 1981)	Increased seed production to colonize site after fire. Since individuals must survive fire, fires should not be severe.
Storing seeds in canopy (Gill 1981, Noble and Slatyer 1977)	1. Heat required for dehiscence in Banksia ornata (Gill 1981); Pinus banksiana and P. contorta (Heinselman 1981)	Trait often present or absent as varieties within a species. Often fire frequency and severity effects proportions of these types in a population.
Seed storage in the soil (Gill 1981, Noble and Slatyer 1977)	1. Long seed dormancy in Ceanothus (Zavitkovski and Newton 1968); Prunus pensylvania (Heinselman 1981); several other taxa (Ballard 1963) 2. Heat-resistant seeds in several chaparral species (Biswell 1974)	Allows fire-adapted species to withstand long inter-fire frequencies or very severe fires.
High dispersability (Noble and Slatyer 1977)	1. Several Pinus, Populus, and Picea species in Boreal ecosystems (Heinselman 1981)	Allows species to invade burned sites from other sites. Provides some independence from both frequency and severity.

species often have more than one adaptive trait that is operative under a particular environmental regime. Thus, a species might have traits that allow it to survive severe (but very infrequent) fires (e.g., long-lived buried seeds), and also to resprout in the face of less severe (but more frequent fires). Another problem is that a particular trait may be adaptive to more than one environmental factor (Gill [1981]). For example, the ability to resprout from epicormic buds allows a species to develop a canopy both after a fire and after insect defoliation. Hard seed-coats confer the ability to survive fires as buried seeds, but they also reduce the loss of seeds to seed predators. Thus, identifying the "cause" of species adaptations to environmental conditions can be problematic, particularly since much of the evidence is in the form of correlations (e.g., a given morphological character often is found in species that are typical of a given environment). Harper (1982) noted:

> "If all the plants in a waterlogged habitat contain aerenchyma this can be seen (and taught!) as a splendid example of convergent evolution; if some have aerenchyma, some have superficial roots, others have mechanisms that prevent the formation of toxic anaerobic by-products and yet others are able to metabolize such products, we have a splendid example of evolutionary divergence, a variety of 'solutions' to a single environmental 'problem.' Thus, if the biologist finds similarities between organisms in a habitat, he can feel satisfied that he has an ecological convergence and he can find equal satisfaction in demonstrating differences that illustrate necessary ecological divergence."

A disturbance (frequency and magnitude) can sort the abundances of species according to their attributes. This ordering or sorting is something of an ecosystem analog to natural selection that is working on the individuals of a population; however, it obviously is not an identical process to natural selection. For example, natural selection can be thought of as eliciting the goal-oriented response of maximizing fitness (*see* Roughgarden [1979]). The "goal" of ecosystem self-ordering is less easily abstracted as a maximization procedure, although Odum (1971) proposed that maximization of power is a reasonable analog. There are fairly remarkable parallels in the structure and function of ecosystems in different parts of the world, which are subjected to similar climatic and disturbance regimes (Box [1981], *see* Chapter 1), just as there are similarities among unrelated species evolved under similar conditions. The action of natural selection in the context of community interactions is a current topic of interest to theoretical ecologists (e.g., Wilson 1976); an understanding of how species roles are shaped in an ecosystem context eventually may be produced from this work.

Oliver (1981) lists five modes of regeneration in trees that he ordered in terms of competitive advantage (with respect to a species being able to occupy or reoccupy a site) following disturbances of decreasing intensity. The order was:

1. Germinating from seeds that enter the area after a disturbance (advantaged in the most severe disturbances such as soil erosion, mudflows, and alluvial deposition, or glacial advancement or retreat).
2. Germinating from dormant seeds.
3. Sprouting of new stems by several different mechanisms.
4. Layering (branches touching soil establish roots and initiate the growth of a new tree).
5. Accelerated growth of trees that survive the disturbance and are still on the site (in the least severe disturbances). These attributes also are found in the fire adaptation examples in Table 7.2.

In the same manner that disturbance severity can differentially favor species in an ecosystem, the frequency of disturbance also can have an effect on the species composition. This is evident in experiments with the BRIND model (Shugart and Noble [1981]) shown in Figure 4.11, in which—for simulated forests occurring at different altitudes, the compositional mixture is altered as a function of the likelihood of a wildfire. Several authors have discussed the effect of varying the frequency of disturbance on ecosystems (Odum [1969], Loucks [1970], Whittaker [1970], Heinselman [1973], Wright [1974], Connell [1978], and Oliver [1981]), with the consistent view that variations in disturbance frequency are expected to alter the mixture of species that dominate a landscape. This view also is supported by an abundance of theoretical studies (*see* Levin [1976], and Huston [1979] for review).

Doyle (1981) used the FORICO model to explore the interaction of disturbance frequency and severity on the diversity of rain forests in the mountains of Puerto Rico. Doyle tested his model by ascertaining the agreement of the simulated dominance-diversity curves with the actual curves for Puerto Rico—given the current hurricane frequency and apparent severity (Fig. 4.13). By varying the simulated hurricane frequency (probability of a hurricane) and magnitude (average percentage of individual trees killed by a hurricane), Doyle developed a response surface of diversity as a function of severity and frequency of disturbance (Fig. 7.5). The diversity index used in this case was a measure of evenness (Pielou's 1969 'J' Statistic) that takes on a value of 1.0 when all the species are represented by an equal number of individuals. From this case, it appears that the pattern of dominance in the natural Puerto Rican forest is not at a maximum diversity value; but, it is a part of the response surface that is

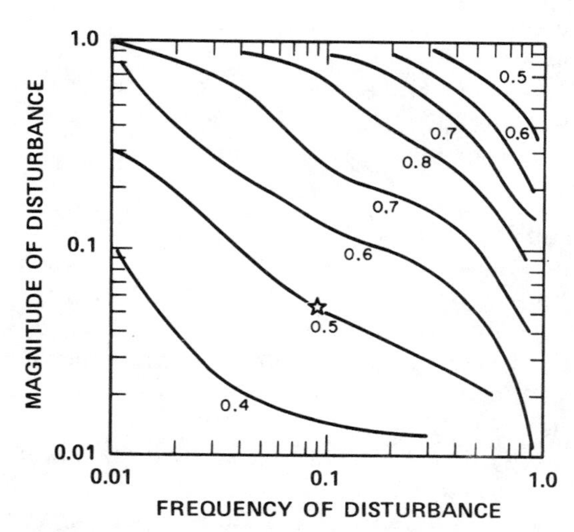

Figure 7.5. Disturbance-diversity relationship for a simulated Puerto Rican rain forest under varied disturbance regimes, as determined by hurricane disturbance frequency and magnitude. The diversity is measured as 'J' evenness (Pielou 1969). Approximate position of Puerto Rico is indicated by a star. (From Doyle [1981].)

relatively flat with respect to changes in disturbance severity or frequency. An increase in either disturbance frequency or severity causes an initial increase in shared dominance among the species; but, it is followed by an eventual decline in shared dominance under the most severe cases. The pattern of response is in the shape of a wave that has its crest at fairly high disturbance frequencies and magnitudes. The pattern of response for other forests that are simulated by gap models is not now known, but it is an area of current research. Response surfaces for other factors could also be generated, such as the growth/disturbance response surface for species diversity produced by Huston (1979).

Computer Models of Quasi-Equilibrium Landscapes

Quasi-equilibrium forested landscapes are composed of a sufficient number of mosaic elements so that the average behavior of the mosaic elements is predictable. It is the averaging of stochastic processes, such as disturbances, that produces a regular average and variance ("pattern"), which is seen as the proper interpretation of Clement's climax concept by several authors (Whittaker [1953], Whittaker and Levin [1977], and Bormann and Likens [1979a,b]). Considerations of landscape size certainly can be used to determine appropriate mathematical approaches for designing landscape-scale models.

Considering that a landscape is a number of mosaic elements, the landscape dynamics are obtained from the change of these elements over time (Shugart, et al [1973], and Weinstein and Shugart [1983]). Considering that every landscape element is a point in an abstract space whose dimensions are attributes (e.g., biomass of Species 1, numbers of trees, leaf area, and so on), the dynamics in each of the mosaic elements is represented as a vector that indicates change in this abstract hyperspace (Fig. 7.6a). The hyperspace also could be divided into volumes that are defined along the attribute axes. Such volumes are denoted T1, T2, and T3 in the two-dimensional case shown in Figure 7.6b. These volumes correspond to forest type. The geometry that delineates them is the stand description. In the example (Fig. 7.6b), a mosaic element that (at a given point in time) was of type T1 forest would have a considerable amount of attribute X_1 and a lesser amount of attribute X_2. Allen, et al (1977) investigated the dynamics of phytoplankton communities by using ordination techniques to form the hyperspace axes and to provide an example of this conceptualization of system dynamics (*see also* Swaine and Greig-Smith [1980]).

Landscape dynamics are the movements of many mosaic points through the attribute space (Fig. 7.6c). Waggoner and Stephens' (1970)

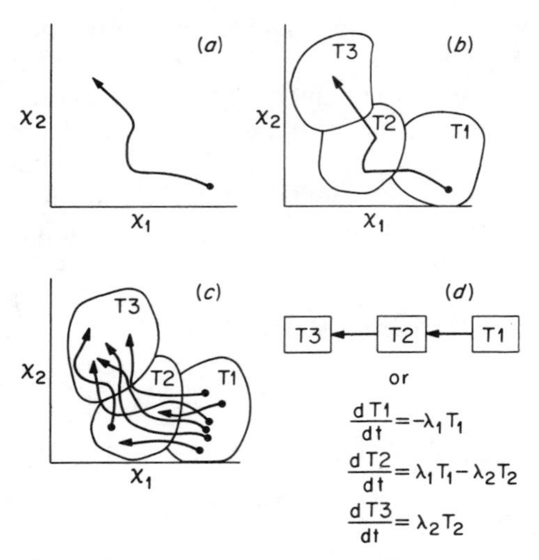

Figure 7.6. An abstract representation of change in quasi-equilibrium landscapes: (a) Change of a single mosaic element's attributes (X_1 and X_2 are quantified attribute axes) indicated by an arrow; (b) Forest types T1, T2, and T3 delineated in the spaces as in (a); (c) Landscape dynamics shown as the motion of many mosaic elements in the hyperspace; and (d) Flow of elements through forest types conceptualized as a compartment model or as a system of differential equations.

Markov model represented the change of a forest as a realization of the likelihood that—given a mosaic patch was in a forest type at one time—that patch would be in another (or the same) forest type at some subsequent time. If the number of mosaic elements is large (as is the case in a quasi-equilibrium landscape), then the motion of landscape elements can be viewed as a continuous flow of elements from one forest type to another (Weinstein and Shugart [1983]). In this case, the change can be represented in a landscape as a system of differential equations (Fig. 7.6d).

This approach to modelling landscape dynamics has been used in several large-scale succession models (Olson and Christofolini [1966], Shugart et al. 1973, Johnson and Sharpe 1976, Loucks et al. 1981). For example, Shugart, et al (1973) used the patterns described by Curtis (1959) for forest succession within the lands of the western Great Lakes region (Fig. 7.7). On xeric sites the succession progresses from intolerant oak-dominated sites to sites that are dominated by mixed oak species (red oak-white oak, *see* Fig. 7.7). Initially, stands on the wettest sites are characterized by tamarack, followed by black spruce or stands dominated by northern white cedar, respectively. Tamarack generally is the first tree species to become established on sphagnum-sedge mats; also, spruce has a higher occurrence on firmer organic substrates. Northern white cedar,

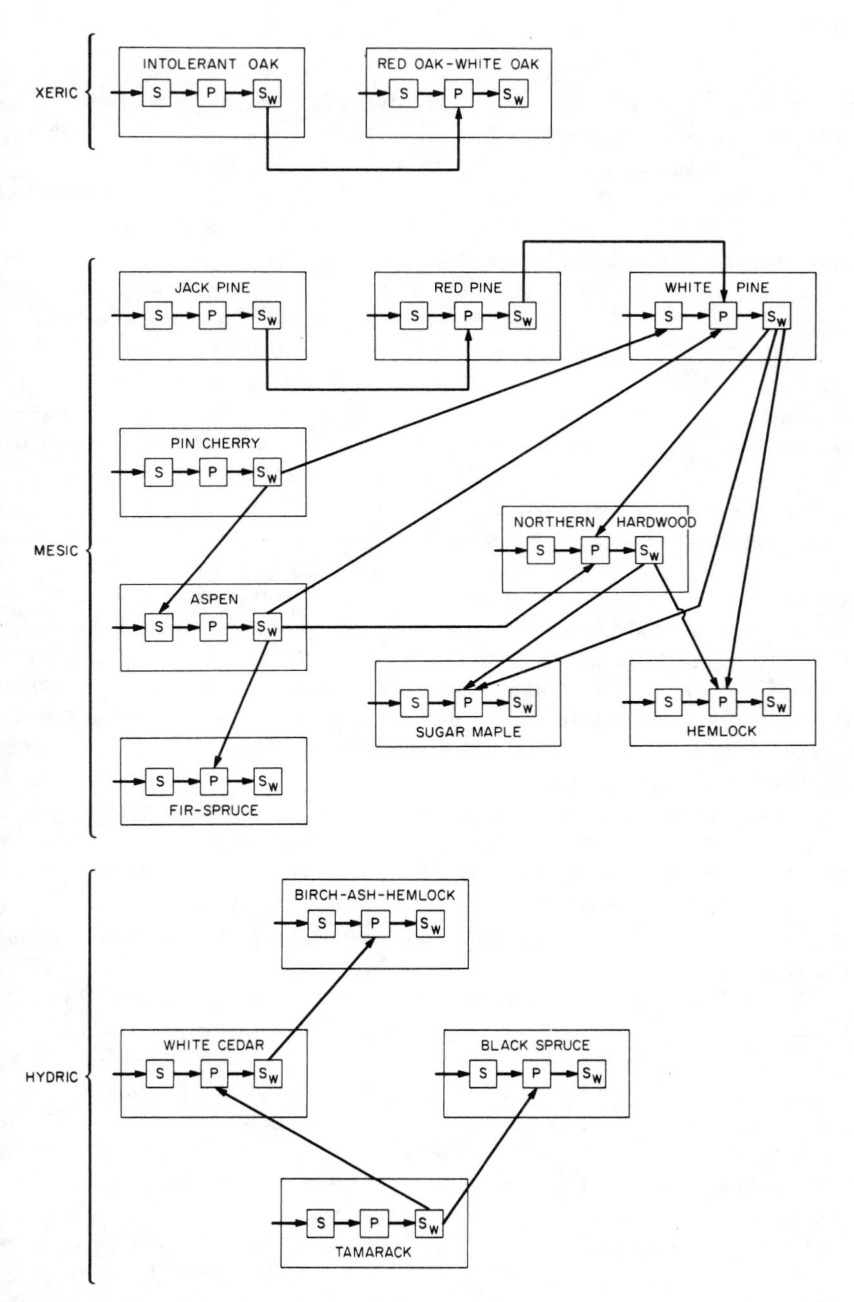

Figure 7.7. Model topology of western Great Lakes regional succession model. The labeled blocks (modules) indicate forest types that are identified by dominant tree species. The three blocks within each module indicate the dominant size category of trees (S = seedling and saplings, P = pole timber, and S_w = saw timber) within the given forest types. Arrows represent transfers of acreages of land from one block to another. (From Shugart, et al [1973].)

which is more shade-tolerant, occurs more frequently on partly to well-decomposed woody peats. Hemlock, yellow birch, and black ash are common components of the cedar type and eventually predominate.

On mesic sites (Fig. 7.7); 1) Aspen, pin cherry, and jack pine are initial dominants; 2) White pine, red pine, mixed northern hardwood are intermediate; and 3) Hemlock, sugar maple, and fir-spruce are dominant on the least recently disturbed sites. On drier mesic sites, the jack pine is a precursor for forest that are typified by the slightly more shade-tolerant and longer-lived red pine, which (in turn) is followed by white pine. Balsam fir-white spruce forests form a single link from some sites that are dominated by aspen state. Within the Great Lakes region, an aspen association (i.e., *Populus tremuloides*, *P. grandidentata*, *Betula papyrifera*, *Prunus pensylvanica*) is by far the most predominant forest type on disturbed sites. On some sites, especially on clayey soils (Braun [1950]) or following a severe burn, pin cherry can be the dominant species for a short time. The acreage of the white pine type, with its subdominant associates of hemlock and hardwoods, has been severely reduced by logging. The mixed northern hardwood type can be considered an admixture of maple, beech, basswood, aspen, birch, and oaks, with no one species significantly dominant over the others. Beech, hemlock, sugar maple, and (to some extent) yellow birch are characteristic species in the mature forest of the Great Lakes region. Yellow birch is the least shade-tolerant of this group; but, once established, is potentially long-lived. It reaches its greatest prominence in the hardwood forests of northern Wisconsin (Curtis [1959]). Hemlock is an extremely shade-tolerant species that can survive long periods of suppression and it grows best on moist sites such as valleys or slope bottoms. Sugar maple, which is the most shade-tolerant species, is a ubiquitous member of the mature forest within the region.

The pattern of landscape dynamics that is shown in Figure 7.7 can be represented as a system of ordinary linear differential equations (*see* Table 7.3; the parameters were developed by Shugart, et al [1973] as an example case). When these equations are initialized with values that are appropriate (for example) for northern lower Michigan, it is possible to simulate the change in the area of landscape that is associated with the different forest types (Fig. 7.8). This particular example simulation shows the dynamic change of the northern lower Michigan landscape when it is initialized (year 0) with the 1966 cover types for the area (from Chase, et al [1970]; see Shugart, et al [1973] for details) and projected for 250 years in the absence of timber harvest and fire. This simulation features the rapid diminution of aspen- and jack pine-dominated stands (due to the hypothetical elimination of large disturbances that favor these dominant species) and an increase of terminal cover types on the landscape.

For quasi-equilibrium landscapes, either deterministic differential equation models or stochastic Markov models seem to be capable of

Table 7.3. Differential equations with rates for western Great Lakes regional successional model

Cover state	Site category[a]	Differential equation
Tamarack	1	$\dot{x}_1 = f_1 - 0.01x_1$
	2	$\dot{x}_2 = 0.01x_1 - 0.01x_2$
	3	$\dot{x}_3 = 0.01x_2 - 0.0111x_3$
White cedar	1	$\dot{x}_4 = f_2 - 0.0176x_4$
	2	$\dot{x}_5 = 0.0176x_4 + 0.00222x_3 - 0.00869x_5$
	3	$\dot{x}_6 = 0.00869x_5 - 0.001x_6$
Black spruce	1	$\dot{x}_7 = f_3 - 0.0125x_7$
	2	$\dot{x}_8 = 0.0125x_7 + 0.00889x_3 - 0.0125x_8$
	3	$\dot{x}_9 = 0.0125x_8$
Birch-ash-hemlock	1	$\dot{x}_{10} = f_4 - 0.025x_{10}$
	2	$\dot{x}_{11} = 0.025x_{10} + 0.001x_6 - 0.017x_{11}$
	3	$\dot{x}_{12} = 0.017x_{11}$
Fir-spruce	1	$\dot{x}_{13} = f_5 - 0.033x_{13}$
	2	$\dot{x}_{14} = 0.033x_{13} + 0.04x_{21} - 0.0222x_{14}$
	3	$\dot{x}_{15} = 0.0222x_{14}$
Pin cherry	1	$\dot{x}_{16} = f_6 - 0.10x_{16}$
	2	$\dot{x}_{17} = 0.10x_{16} - 0.0667x_{17}$
	3	$\dot{x}_{18} = 0.0667x_{17} - 0.20x_{18}$
Aspen	1	$\dot{x}_{19} = f_7 - 0.05x_{19}$
	2	$\dot{x}_{20} = 0.05x_{19} - 0.02857x_{20}$
	3	$\dot{x}_{21} = 0.02857x_{20} - 0.1x_{21}$
Jack pine	1	$\dot{x}_{22} = f_8 - 0.0333x_{22}$
	2	$\dot{x}_{23} = 0.0333x_{22} - 0.0333x_{23}$
	3	$\dot{x}_{24} = 0.0333x_{23} - 0.10x_{24}$
Red pine	1	$\dot{x}_{25} = f_9 - 0.01x_{25}$
	2	$\dot{x}_{26} = 0.01x_{25} + 0.10x_{24} - 0.01x_{26}$
	3	$\dot{x}_{27} = 0.01x_{26} - 0.01x_{27}$
White pine	1	$\dot{x}_{28} = f_{10} + 0.04x_{18} - 0.01x_{28}$
	2	$\dot{x}_{29} = 0.01x_{28} + 0.005x_{21} + 0.01x_{27} - 0.00667x_{29}$
	3	$\dot{x}_{30} = 0.00667x_{29} - 0.005x_{30}$
Hemlock	1	$\dot{x}_{31} = f_{11} - 0.01x_{31}$
	2	$\dot{x}_{32} = 0.01x_{31} + 0.001x_{30} + 0.003x_{36} - 0.0067x_{32}$
	3	$\dot{x}_{33} = 0.00667x_{32}$
Northern hardwood	1	$\dot{x}_{34} = f_{12} - 0.01x_{24}$
	2	$\dot{x}_{35} = 0.01x_{34} + 0.056x_{21} + 0.0035x_{30} - 0.01x_{35}$
	3	$\dot{x}_{36} = 0.01x_{35} - 0.02x_{36}$
Sugar maple	1	$\dot{x}_{37} = f_{13} - 0.02x_{37}$
	2	$\dot{x}_{38} = 0.02x_{37} + 0.017x_{36} + 0.0005x_{30} - 0.01x_{38}$
	3	$\dot{x}_{39} = 0.01x_{38}$
Intolerant oak	1	$\dot{x}_{40} = f_{14} - 0.01x_{40}$
	2	$\dot{x}_{41} = 0.01x_{40} - 0.02x_{41}$
	3	$\dot{x}_{42} = 0.02x_{41} - 0.02x_{42}$
Red oak-white oak	1	$\dot{x}_{43} = f_{15} - 0.00667x_{43} + 0.02x_{42}$
	2	$\dot{x}_{44} = 0.00667x_{43} + 0.02x_{42} - 0.00667x_{44}$
	3	$\dot{x}_{45} = 0.00667x_{44}$

[a] 1 = seedling-dominated stands, 2 = pole timber, and 3 = saw timber.
From Shugart, et al. (1973).

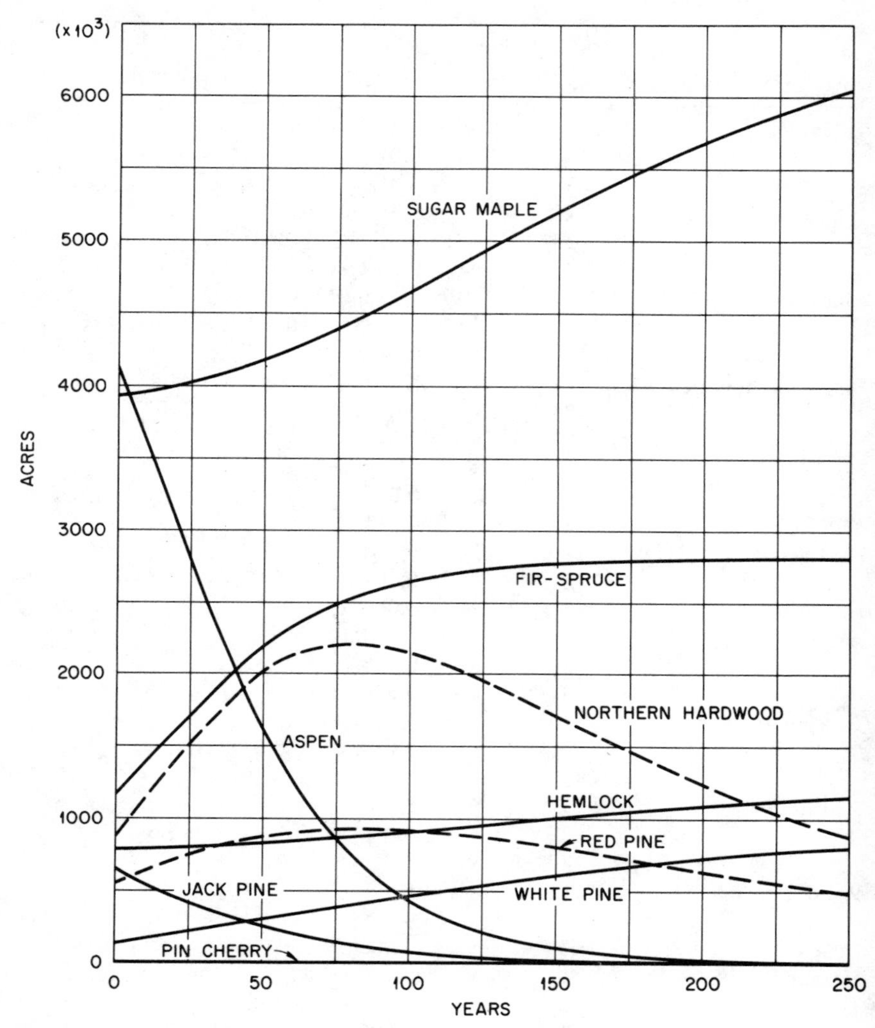

Figure 7.8. A 250-year simulation of the mesic portion of the western Great Lakes regional succession model. (1 acre = 0.4047 ha). (From Shugart, et al [1973].)

capturing general dynamics of the landscapes insofar as they are (or can be) known. The two modelling approaches are not competitive; also, for any Markov model of succession, one can develop a differential equation model that has the same average behavior as a large number of Markov simulations. Indeed, Horn (1976) considers them to be essentially the same sort of models with respect to the important underlying ecological assumptions that are associated with the models. There are several basic scientific considerations that are attendant with using these approaches to simulate landscape patterns. Usher (1979), after inspecting several successional studies expressed as Markovian processes, found that the majority of successional processes can be considered as nonindependent sequences, (by implication of Markovian processes)—and not merely as random sequences. He also found that in the two cases where there was sufficient data to test for stationarity, neither process was stationary. There has been an apparent success with Markov approaches both in theoretical (e.g., Horn [1975a,b]) and basic (e.g., Wilkins [1977], Henderson and Wilkins [1975], and Brown and Podger [1982]) studies. The possibility that landscape systems may lack stationarity (the probabilities of change from one state to another themselves change over time), in a differential equation context, would be manifested as model parameters that changed as a function of time (Shugart, et al [1973]). Problems of parameter estimation in such systems are difficult and—given the problems in sampling forest dynamics over time and space—are virtually insurmountable. In general, the best manner in which to proceed may be to accept these model approaches as approximations of a more complex nature and to use these models with caution. Usher (1979) proposes the interesting approach of using methods of successive approximation to the modelling of field successional processes: 1) Multivariate statistical procedures to define the stages and to indicate the data to include in a Markovian model, and 2) A Markovian model to approximate the successional process and to refine the stages.

Applications of Landscape Models

One of the first Markov models used for forest dynamics was developed for use in optimizing forest product harvests in a commercial forestry operation (Hool [1966]). Johnson and Sharpe (1976) also used their differential equation model to inspect the effect of fire and harvesting policy on a regional landscape. The most natural applications for models such as these are at the scale of the quasi-equilibrium landscape, and they are involved in assessing the regional effects of land-use policy or practice.

More detailed applications of Markov succession models for landscape management are in the recent work of Kessell and Potter and their colleagues (Cattelino et al. [1979], Potter, et al [1979], Kessell and Potter [1980], Kessell [1979a,b, 1981a,b, and Kessell, et al [1982]). These authors developed several models for landscapes in the United States, Canada, and Australia; they used the models in the context of natural landscape management. The models typically consist of an underlying Markov simulation model that may have parameters that vary according to environmental gradients or other conditions (Fig. 7.9; Table 7.4). This simulation model is used to project the temporal change of each element

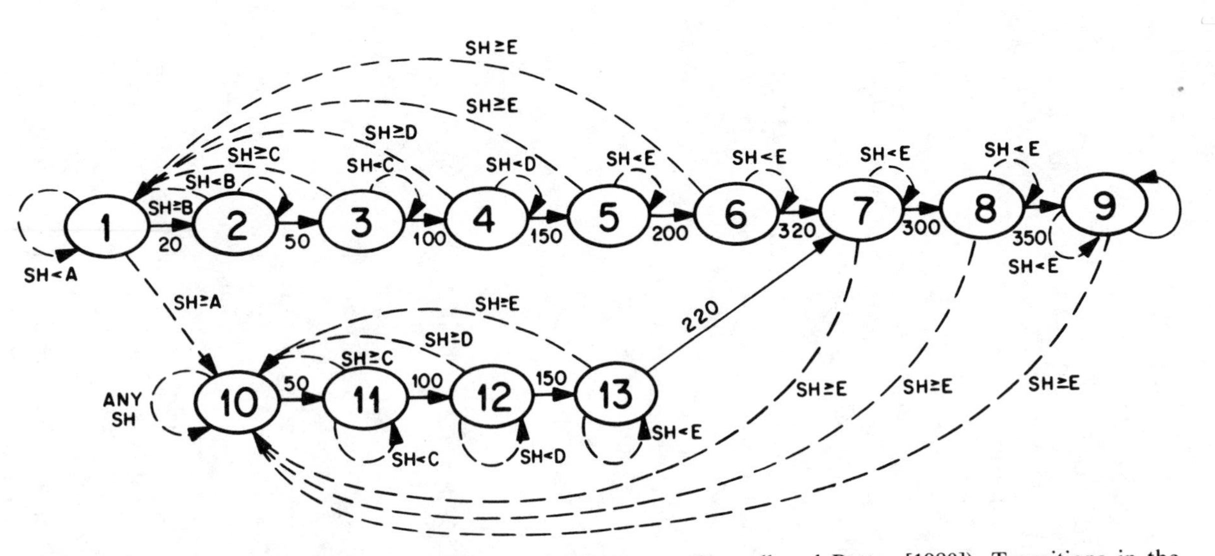

Figure 7.9. Succession model for nine Montana habitat types (Kessell and Potter [1980]). Transitions in the absence of disturbances are determined by stand ages (in years) shown by the solid arrows. Transitions in response to fire are a function of stand age and the fire scorch height (Van Wagner [1973]). Scorch height (SH) is compared to parameters A,B,C,D,E according to habitat type. States 1–6 that are stands typified by lodgepole pine (*Pinus contorta*) of increasing age. States 7–9 are for forest development beyond the longevity of lodgepole pine. States 10–13 are successional stands, but without lodgepole pine. Species composition for each state for one habitat type is shown in Table 7.4. (From Kessell and Potter [1980].)

Table 7.4. Matrix relating species composition to successional states for the *Pseudotsuga menziesii*/*Symphoricarpos albus* habitat type.[a] (From Kessel and Potter 1980.)

	Succession model states (see Fig. 7.9)												
	1	2	3	4	5	6	7	8	9	10	11	12	13
						Overstory Species							
Trees > 1.4 m tall													
Pinus contorta (Lodgepole pine)			3	2	1								
Pinus ponderosa (Ponderosa pine)			2	2	2	2	1	1			3	3	2
Pseudotsuga menziesii (Douglas-fir)			4	4	5	5	5	6	6		4	4	5
Tree seedlings (< 1.4 m tall)													
Pinus contorta (Lodgepole pine)	1	1											
Pinus ponderosa (Ponderosa pine)	1	1								1			
Pseudotsuga menziesii (Douglas-fir)	1	1	1	1						1	1	1	
Total canopy cover			4	4	4	4	4	4	4		4	4	4
						Understory Species							
Shrubs and subshrubs													
Juniperis communis			2	1							2	1	
Potentilla fruticosa				1	1	1	1	1	1			1	1
Ribes spp.	2	2											
Rosa spp.			1	2	2	2	2	2	2		1	2	2
Shepherdia canadensis			2	1	1	1	1	1	1		2	1	1
Spiraea betulifolia			2	2	2	2	2	2	2		2	2	2
Symphoricarpos albus	1	1	2	2	2	2	2	2	2	1	2	2	2
Berberis repens			1	1	1	1	1	1	1		1	1	1
Total shrub cover	2	2	4	4	4	4	4	4	4	2	4	4	4
Forb and grasses													
Calamagrostis rubescens	1	1	3	1	1	1	1	1	1	1	3	1	1
Carex geyeri	2	2	3							2	3		
Festuca scabrella	1	1	1		1					1	1		1
Gramineae	1	1		1	1					1		1	1
Arica spp.			2		1	1	1	1	1		2		1
Astragalus spp.			1	1	1	1	1	1	1		1	1	1
Fragaria spp.	2	2	1		1	1	1	1	1	2	1		1
Total forb and grass cover	5	5	3	2	1	1	1	1	1	5	3	2	1

[a]The successional state numbers correspond to the model shown in Fig. 7.9. Importance values are expressed on a seven point scale where: blank = <1%; 1 = 1-5%; 2 = 6-25%; 3 = 26-50%; 4 = 51-75%; 5 = 76-95%. Importance values for trees at least 1.4 m tall are relative density; all other importance values are absolute cover. The "Gramineae" entry includes all grasses not previously listed by species.

of a map of vegetation types for a landscape by using small computers. These computer programs can be accessed by using a conversational computer language that allows a land manager to sit at a computer terminal, to make decisions about the landscape (e.g., should these blocks of land be cleared, burned, and so on), and to project the consequences of these decisions as a dynamically changing map. The potential of these models as planning tools is considerable (Franklin [1979]), and it may be imperative for proper long-term land use.

Modelling effectively nonequilibrium landscapes is a considerably more difficult exercise, because the ecosystem that one would like to predict has, as one of its attributes, great variability. One approach is to simulate patches of the landscape by using a gap model and to attempt to develop the landscape's response as a distribution of likelihoods (e.g., the BRIND simulations in Fig. 4.11). If the nonequilibrating disturbances are controllable, it also is possible both to determine the consequences of a particular sequence of disturbance events and to then manage the disturbance regime toward some goal. Some logical examples of natural nonequilibrium landscapes are islands, which is a topic that will be discussed in part of the following chapter.

Conclusions

Landscapes can be thought of as being quasi-equilibrium or nonequilibrium ecosystems, which depends on the comparative spatial scales of the disturbances and the landscape. This is true regardless of whether the disturbances are endogenous or exogenous. Disturbance frequency and severity can have an ordering effect on landscape systems due to either selection or the favoring of one species or functional group over another. Fire adaptations are a reasonable example of the range of ways in which a disturbance can sort the pattern of species adaptations at the scale of a landscape. Considering that a landscape is a mosaic of forested patches (e.g., as may be simulated by a gap model), one obtains a quasi-equilibrium ecosystem for large landscapes that has been proposed by several authors (Watt [1947], Whittaker [1953], Bormann and Likens [1979a&b]) as a statistical alternative to Clement's climax concept. Some natural landscapes (e.g., islands) appear to be effectively nonequilibrium systems.

In the next chapter, the response of animals to a dynamic landscape mosaic will be discussed as both a theoretical and a management problem. The relationship between island biogeography and the dynamics of effectively nonequilibrium landscapes also will be discussed.

Chapter 8
Animals and Mosaic Landscapes

This chapter is primarily focused on the response of animals to a landscape that changes as a dynamic mosaic. However, it is appropriate to initially consider the ways in which a forested landscape acts and reacts with animals in a more general context. In many discussions (e.g., Mattson and Addy [1975], Lohm and Persson [1976], and Zoltin and Khodashova [1980]), the term, "role of animals," has been used to connote the functional aspects of animals in relative to the larger system—especially those functions that can manifest changes in the behavior of the total ecosystem. Some may find the word, "role," to be teleological; however, it also brings to mind the rich analogy that is implied in the title of G. E. Hutchinson's (1965) book, The Ecological Theater and the Evolutionary Play. The use of the word also forms a bridge to the concept of the ecological niche, particularly as "niche" is used by some ecologists (e.g., "The niche is an animal's job; its habitat is its address." Odum [1971]).

Roles of Animals in Ecological Systems

The roles of animals in the mosaic landscape probably are as numerous as one could wish to elaborate. The present discussion will be restricted to a brief review of some of the major roles of animals in forested ecosystems. These include:

(1) The regulation of important ecological processes, such as the breakdown of leaf litter, the release of nutrients, or the formation of the soil: In forests, the balance between the production of organic matter and

the return of this organic matter to the atmosphere by various respiratory losses is dominated by plants (Olson [1963]). Trees obviously dominate the production of fixed carbon. Fungi and bacteria are the predominant agents of the breakdown of this carbon. The role of animals in the ecosystem cycling and release of essential elements seems to be as regulators of ecosystem processes (Chew [1974], Lee and Inman [1975], Mattson and Addy [1975], Weiner [1975], Lohm and Persson [1976], O'Neill [1976], Springett [1978], Kitchell, et al [1979], and Schowater [1981]). Kurcheva (1960) demonstrated an approximately 50% reduction in the decomposition of *Quercus robor* leaf litter when the soil arthropods were excluded by treatment with napthalene. Witkamp and Crossley (1966) repeated this experiment, with similar results, using *Quercus alba* leaves . Reichle and Crossley (1967) found that although the energetic requirements of the forest floor arthropods in a *Liriodendron* forest in east Tennessee only amounted to about 3% of the carbon dioxide (CO_2) efflux from the forest floor ("soil respiration"), when the forest floor arthropods were removed from the system (by selective poisoning) the soil respiration dropped by more than 30%. These responses were attributed to the action that increased the surface-to-volume ratio of the leaf litter (van der Drift [1958], and Crossley [1976]). Also, the recycling of coarser material through coprophagous arthropods, (e.g., millipeds) can amplify the rate of carbon dioxide efflux quite beyond the rate expected from the relatively small energetic requirements of these species (McBrayer [1973, 1977]). Along a similar vein, Swank, et al (1981) found that the total ecosystem export of nitrate was enhanced in a forested watershed that was undergoing insect defoliation. In this case, the effect of animals was to alter the rate and timing of the input of leaf material to the litter decomposition subsystem. Zoltin and Khodashova (1980) found that litterfall decomposition rates doubled during an outbreak of oak leaf roller.

There appear to be a reasonable number of examples of animals acting in efficient ways to alter the movement of material in ecosystems far beyond the degree of influence that one would ascribe when given the relatively small biomass and miniscule energy demands of animals. This rather surprising influence of animals on the larger ecosystem frequently has been documented in forested systems (*see* Mattson [1977], and Zoltin and Khodashova [1980] for several examples).

(2) The reduction of the efficiency of the photosynthetic surface: Changes in the efficiency of photosynthate production, when directed against a subset of the tree species in a competitive web, can have an effect that is amplified by this competition. Morrow (1976, 1977) found that these effects of canopy grazing by phytophagous insects could influence the composition of plant associations in Australian *Eucalyptus* forests. In general, the reduction in available photosynthate can be a consequence of partial, as well as, total defoliation. It also can be the consequence of tapping the photosynthate without the obvious destruction of leaves, as is done by some sucking insects. Assuming that plants

would evolve to allocate photosynthate reserves for maximizing fitness, Rhoades (1979) hypothesized that herbivory would cause a plant to develop chemical defenses against herbivory in proportion to the risk associated with it. Along with developing several corollaries to this central hypothesis, Rhoades predicted that environmentally stressed individual plants should be less well defended against herbivores (*see* the paragraph 3 below); also, because chemical defenses are a cost to the photosynthetic budget of the tree, an effect of herbivory is to reduce the amount of photosynthate that is allocatable to other uses. The positive feedback loop of reduced photosynthetic surfaces from herbivory which causes a reduction in the amount of photosynthate available for chemical defense against herbivory allowing greater herbivory, invokes the potential for canopy grazing (often a seemingly small part of the total energy budget of a forest) as being a very sensitive parameter in the total ecosystem's functioning.

(3) The weakening of damaged trees and the killing of weak trees: The synchronization of mortality that is brought about by such action has the landscape-level consequence of allowing the widespread regeneration of Role 1 and (in particular) Role 3 tree species. Some insects demonstrate an ability, and even an inclination, to produce a widespread synchronous mortality that can be directed to trees that are quite healthy, but that are near dying trees. For example, Raffa and Berryman (1983) discuss the case of the bark beetle, *Dendroctonus ponderosae*, which infests the lodgepole pine, *Pinus contorta*, in northeastern Oregon. These species exhibit a positive feedback between the increased density of the beetles (as the number of trees in an area are successfully attacked) and the enhanced ability of the insects to invade healthy trees with increased numbers. The same positive feedback relation is suggested in Waring and Pitman's (1980) analysis of bark beetle attacks; success is arranged as a response surface with respect to a general index of tree vigor (Fig. 8.1). In Chapter 6, the stem-boring beetles were mentioned as one of the mortality synchronizing agents in the 'ohi'a regenerative dieback phenomenon in Hawaiian landscapes. It is conceivable that the negative short-term effect of insect predation on trees can, in the case of Role 1 or Role 3 trees, can have the long-term benefit of providing a synchronized regeneration event for the trees. Noteworthy in this regard is that many of the native insects of North America that are capable of mounting large-scale epidemics have, as their hosts, trees that conform reasonably well to what have been called Role 3 and Role 1 trees (see Chapter 5). It is possible that these apparent predator/prey relationships could be thought of as including a measure of mutualism when tree regeneration is considered. However, the direct testing of this hypothesis is elusive and largely based on the interpretation of present patterns.

(4) The alteration of the likelihood of a given tree species' successful regeneration at a site: These effects of animals already have been considered, to a degree, in discussions of regeneration in Chapter 3. The role of

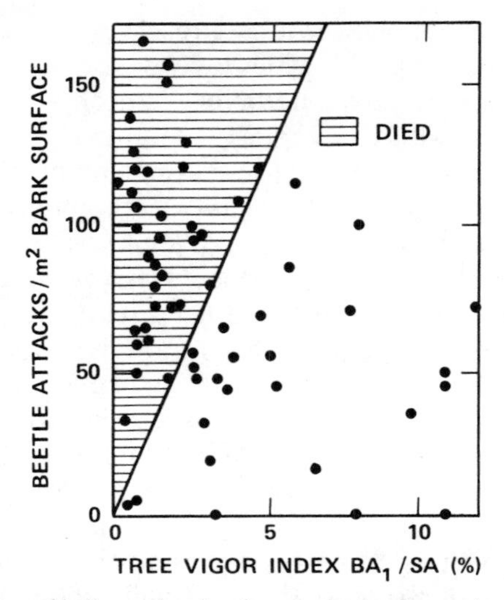

Figure 8.1. Mortality of lodgepole pine in relation to intensity of beetle attack and tree vigor index. (From Waring and Pitman [1980].)

animals in this regard can be either positive (e.g., the burying of seeds or the pollination of plants by animals) or negative (e.g., the eating of seeds or the excessive grazing of sprouts). One consequence of these actions is found in the masting patterns of trees. Masting is the tendency of certain trees to produce crops of seeds with a large year-to-year variation. One explanation for masting (Silvertown [1982]) is that the production of "bumper crops" of seeds makes it impossible for animals feeding on seeds to consume a tree's entire crop within a mast year. Silvertown (1980a) found that, in an array of species, those with the greatest amount of seed predation also had the greatest tendency to have the masting habit. Silvertown (1980b) also found that seeds from fleshy-fruited species (which generally were dispersed by passing through the gut of an animal) did not mast, while nonfleshy-seeded species typically did have masting as part of their life history. Janzen (1971) provides a detailed review of the importance of seed feeding by animals in a variety of ecosystems. When the trees are strongly competitive with one another, the role of animals influencing the pattern of regeneration can affect the long-term pattern of forest composition.

The roles of animals in forests are manifested in the way that they can work to alter or control the functioning of the system. The examples that have been provided here are intended as a sampler of the wide array of ways in which animals can affect the whole system behavior of forests. The study of the niches of animals attempts to add a theoretical understanding of the causes of patterns in these animal roles in ecosystems. In addition, some of the methods used to quantify the niches of animals can be used to predict the possible occurrence of a given animal species as a response to a dynamically changing vegetative structure.

Niche Theory: A Brief Review

Hutchinson (as early as 1944 and in 1957 and 1965) presented the n-dimensional hyperspace concept to unify the so-called Eltonian and Grinnellian concepts of the ecological niche. Grinnell (1917) often is credited with coining the term "niche" in an ecological context; however, the word was used earlier by Johnson (1910) in an ecological context (*see* Gaffney [1975]). Cox (1980) traces the term to Bradley's (1725) The Family Dictionary and provides the quotation: "The way to destroy the Niches of Spiders in our Gardens, is to pour some Turpentine Oil upon them." Regardless of these lexicographic considerations, Grinnell developed a niche concept (1928). He defined the niche as, ". . . the ultimate distributional unit within which each animal is held by its structural and instinctive limitations." Elton (1927) also developed a concept of the ecological niche, but he chose to strongly emphasize the functional aspects of the animal. For example, the niche of an animal in the Eltonian context could be a "large active carnivore" or "small grass-eating herbivore." In time, the Eltonian niche became strongly associated with the ecological energetics of the animal (e.g., Weatherly [1963]), and the Grinnellian niche was associated with the animal's habitat. Animals in different places could have the same or similar niches, as evidenced in the phenomenon called "ecological convergence."

Hutchinson's (1957, 1965) niche hyperspace concept was an attempt to unify the differences in the then current use of the term, "niche." The n-dimensional hyperspace model represented the niche of an animal as a volume in a space whose axes were relevant environmental variables. The space had n-dimensions in which n was the number of environmental variables. The fundamental niche of an animal was the volume in the n-dimensional space that described the points where the animal could potentially live. The set of points that was associated with a particular place was called the biotope. The volume in the space that was the intersection between the fundamental niche and the biotope was the realized niche. The realized niche was the consequence of the potential of the species that was manifested with respect to the possibilities in a given environment; thus, it was a quantification of what Grinnell (1928) might have intended in his term, "ultimate distributional unit." The overlap between the realized niches of two species was presumed to be an indication of the competition between the species. Both this concept and the possibility of two species having the same niche in different environments appealed to important aspects of the Eltonian niche concept. Hutchinson's idea of expressing the niche with an abstract geometrical model captured the attention of theoretical ecologists for a decade (Horn [1966], Maguire [1967], McNaughton and Wolf [1970], Colwell and Futuyma [1971], Pielou [1972], and May [1973, 1975]); however, the strictly geometrical interpretation has been lost in the more recent applications.

MacArthur's (1958) classic work on the fine-scale habitat use of five forest warblers in New England strongly emphasized the relation between individual microhabitat use and the hyperspace niche theory model of Hutchinson (1944, 1957, 1965). Fujii (1969) realized that the n-dimensional hyperspace was clearly analogous to the n-dimensional sample space that was used in multivariate statistics. Cody (1968) and Fujii (1969) both developed the idea of using multivariate statistics in studies of animal niches. In 1971 and 1972, a number of papers by individual independent workers attempted to use multivariate statistics to quantify the niches of a variety of animals (Green [1971], Hespenheide [1971], James [1971], Martinka [1972], and Shugart and Patten [1972]). The fundamental idea in this work was that by understanding the pattern of association between the utilization of a small area by a given species and the set of environmental variables typifying the small area, one could use ecological inference to interpret these associations as quantifications of the animal's niche.

The Mosaic Element as a Habitat Element

The use of multivariate statistics as a quantified analog to the niche has several important theoretical considerations for modern ecologists. Among these topics is the degree and pattern of similarity that found in assemblages of animals or plants (McNaughton and Wolf [1970], and May [1973]). Another question concerns the degree to which abundance of animals is related to niche metrics (McNaughton and Wolf [1970], and Dueser and Shugart [1979]). Several of the methods initially used to quantify the n-dimensional niche for animals in field conditions also are useful for associating animals with vegetative structures. It is this strongly practical spinoff from an initially theoretical topic that will be used to explore the expected animal habitat dynamics in mosaic landscapes.

There are several different multivariate statistical techniques that actually have been used to quantify the niches of animals. The choice of technique often was the actual topic of interest in the earlier studies. For example, Fujii (1969) explored principal components analysis as a statistical analog to the Hutchinson (1965) n-dimensional niche model. Green (1971) chose to use discriminant function analysis; and Shugart and Patten (1972) used the Mahalanobis statistic for the same purpose. Williams (1981), in a recent review of the use of discriminant function analysis in wildlife research, notes that, "Discriminant analysis is a technique which has come to be much used in ecological investigations. It is applicable to the study of niche breadth, niche overlap, resource partitioning, habitat selection, community structure, and many other topics. In fact, the methodology is potentially useful for any ecological situation in which an association is desired between well-defined groups and a set of ecologically meaningful measurements." Williams, in this review, continues to caution against the use of discriminant function as an inferential procedure (this case is referred to as "canonical analysis"); later (on page 195, Capen [1981]), Williams endorses an emphasis of the classification ability of the discriminant function methodology. This also is a type of multivariate analysis that may prove to be useful in using forest dynamics models as simulators of animal habitat.

Chapters 4–7 have attempted to develop an understanding of forest dynamics at increasingly larger spatial scales. The classic concept of animal habitat also can be inspected along this same gradient of spatial resolution. The most general definitions of animal habitat consider the habitat to be the place (size of the place is not often specified) where an animal is typically found. In the cases of many of the forest birds, the apparent scale at which one conveniently views habitat agrees reasonably well with hierarchical scales that have been used in this volume to consider forest dynamics. Thus, along with the existence of a fairly large number of bird habitat studies to draw from as examples, there also is a degree of heuristic value in avian examples as well.

The approximate spatial scale of the forest gap [approximately 0.10 ha [0.25 acre]) also is the scale that is frequently referred to as the avian microhabitat. Interest in this smaller scale use of landscapes is related to the classical development of animal habitat selection (von Uexkull [1909], Kohler [1947], Tinbergen [1951], and Harris [1952]). In the mid-1960s, there was an intensified interest in habitat selection as an ecological phenomenon (e.g., Wecker [1963], Klopfer [1965], and MacArthur and Pianka [1966]), with a strong emphasis placed on the responses of organisms to a multiple set of environmental variables—often variables that, in some sense, related directly to small-scale vegetative structure.

For example, Smith, et al (1981a,b) used the FORET model in conjunction with detailed data on the microhabitat associations of several bird species. A subroutine, DISCRIM, was developed to tabulate the dynamics of a species habitat through succession and in response to forest harvest procedures. The DISCRIM subroutine sorted the suitability for birds of the vegetative structure that was simulated each year by the FORET model by using a two-group discriminant function analysis (Morrison [1967]) to separate potential from non-potential bird habitat.

Figure 8.2 is a hypothetical example of the actual classification procedure that was used in DISCRIM. For each bird species, each of 298

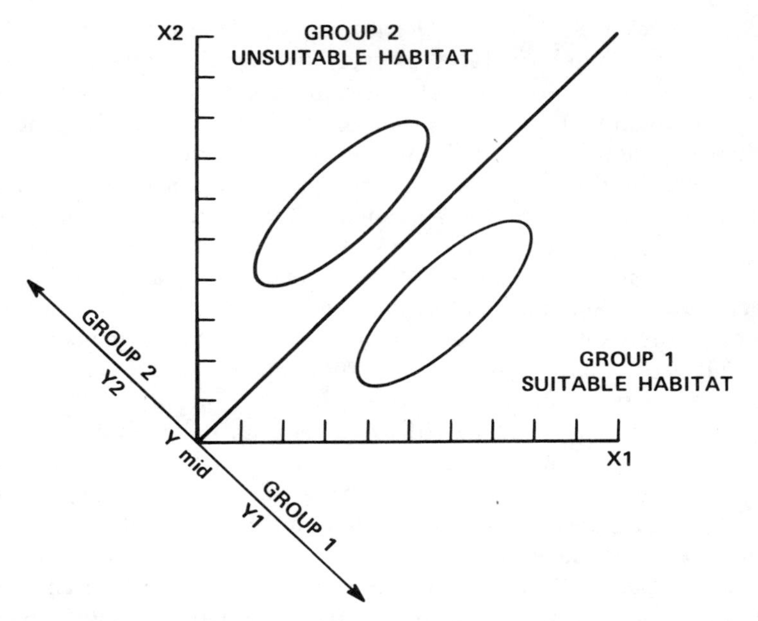

Figure 8.2. Linear decision scale for subroutine DISCRIM. (From Smith, et al [1981a].)

census plots was placed into one of two groups, suitable habitat or unsuitable habitat. The census plots were forest inventory plots located on Walker Branch Watershed, which was a site of intensive ecological investigation (during the U.S. International Biological Program) located on the U.S. Department of Energy Reservation in Oak Ridge, Tennessee (*see* Grigal and Goldstein [1971] for a detailed description of the site). A suitable habitat was represented by survey plots that were contained in the breeding territories of the population of the species of interest. An unsuitable habitat was similarly represented by survey plots that were outside these territories.

The two elipses in Figure 8.2 represent the two populations of survey plots in two-dimensional space, with x_1 and x_2 being two variables that describe the habitat structure and that are derived from the survey plot data. Discriminant function analysis is a statistical procedure for finding a linear combination of the original predictor variables (x_1 and x_2), which results in the largest difference between the two mean vectors (i.e., maximizes the ratio of the between-group to the within-group sums of squares). If $\bar{X}_1$ and $\bar{X}_2$ are the sample mean vectors for the two groups and S is the pooled estimate of the variance-covariance matrix, the determination of the discriminant function amounts to finding the vector (a) of the linear compound ($a\bar{X}$) of the responses that will maximize the distance between the two groups in p-dimensional space. It can be shown that:

$$a = S^{-1}(\bar{X}_1 - \bar{X}_2)$$

and that the general form of the resulting discriminant function is:

$$Y = (\bar{X}_1 - \bar{X}_2)^T S^{-1}\bar{X}$$

with

$$y_1 = a_1x_1 + a_2x_2 + \cdots + a_px_p$$

where p is the number of variables in the classification (*see* Morrison [1967], p. 130).

In the two-group case, one is interested in classifying by using what is called the "linear decision scale" (line y in Fig. 8.2). Observations are classified as most likely being members of either Group 1 (e.g., suitable habitat) or Group 2. These evaluations of the discriminant function scores are used in the Smith, et al (1981a,b) DISCRIM subroutine to decide whether a given simulated habitat provides for a potential breeding habitat of each species of interest.

Figure 8.3 shows results that were simulated under two conditions: with and without timber harvest. In the without-harvest case, a random subsample of the 298 0.08-ha (0.20-acre) survey plots that characterize the vegetation of the Walker Branch Watershed is loaded into the modified

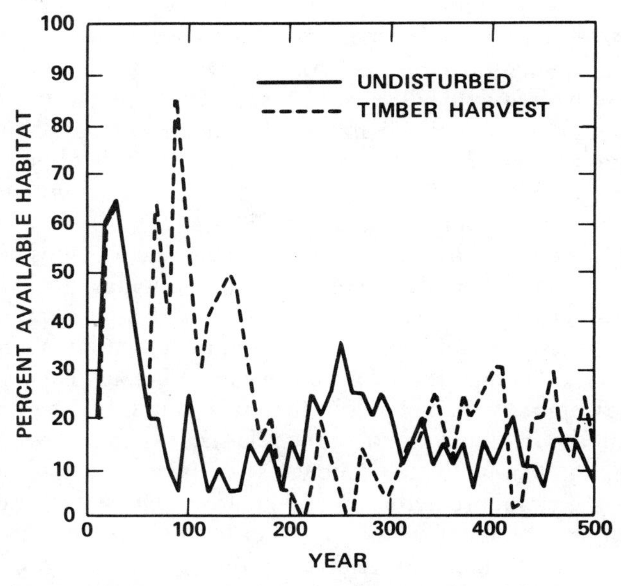

Figure 8.3. Ovenbird habitat on the Walker Branch Watershed from two 500-year simulations with and without timber harvest. (From Smith, et al [1981a].)

FORET model and projected in time for 500 years. The linear decision scale for the ovenbird (*Seiurus aurocapillus*), in the case of Figure 8.3, is used to determine habitat suitability for each plot for each of the 500 simulated years. The with-harvest case is analogous, except that on year 60 (and subsequently at 60-year intervals) a diameter-limit cut that removes all commercially valuable timber above 23 cm dbh is applied to all of the simulated stands. This form of timber management is not unlike what was actually practiced on the watershed before 1940. Simulations of undisturbed forest dynamics show an initial increase (of available habitat for the ovenbird) ranging from about 20% of the watershed to about a 65% general decline in potential habitats until there is an oscillation in the range of 10–20% from about year 250 on (Fig. 8.3).

With timber management, there is considerable divergence from the simulations of the undisturbed forest. The dynamics are identical for both cases before the first cut at year 60. Then, the managed stands show a sharp increase in available habitats, reaching a maximum of 85% in year 90; Then there is a continuous decline until the forest is cut for a second time in year 120. This second cut also produces an increase in the habitat of the ovenbird, but not to the degree of the earlier cut. For the remainder of the simulation, the pattern shortly after a timber harvest is to a small increase in ovenbird habitat followed by a decline in habitat availability.

Along with showing the importance of historical considerations in determining the effects of a particular timber harvest procedure on a given forest, these results demonstrate the possibility of simulating the dynamics of potential animal habitats for either a species or an entire community. The simulation does not consider the ability of the ovenbird population to demographically track these changes in habitat. Similarly, the model does not simulate dynamics in habitat preference. Nonetheless, the application of this model to predict potential effects of proposed and untried management schemes on specific forest areas is a potentially useful tool for both forest and wildlife managers.

The prediction of animal occurrence from the relationships between an animal's use or nonuse of a small area and the ecological attributes of the small area is based on a parsimonious theory of animal presence and absence. The theory is minimal in the sense that a great number of potential factors (e.g., competition for space with other animals, predation, interference competition, and so on) that might influence the distribution are not included. In actual practice, the relatively simple view that microhabitat selection can be taken as the cause of animal pattern seems quite powerful as a general predictor of actual pattern. Of course, the underlying causes of microhabitat selection [either for a single microhabitat type as in James' (1971) "niche gestalt" or in the case of many animals that utilize the edge habitat which has become increasing common on many man-altered landscapes—a set of microhabitats] are themselves potentially very complex. It is appropriate to consider how the microhabitat theory relates to other basic concepts in community ecology.

Consequences of Habitat Selection for Mosaic Elements on Animal Communities

The response of an animal population to a dynamically changing mosaic is strongly related to the rate of increase of the animal population with respect to both the rate of change and the frequency of disturbance of the landscape mosaic. If these landscape dynamics are slow with respect to the dynamics of the population, then the animal population response is manifested mostly in the birth and death rates of the population. In this case, factors such as the availability of resources or space in suitable habitat are relatively constant factors; the traditional modeling approaches of population dynamics (i.e., autonomous nonlinear differential equations) can be used to project the expected population response. When the landscape processes are fast, in relation to the population dynamics, one would analogously expect the population dynamics to be dominated by the dynamics of the landscape processes. This latter case will be treated here in some detail. When the time constants of the population and landscape are similar, it is difficult to generalize about the specific nature of the population response. The potential for population oscillations, instabilities, and extinctions could be greatly increased, if the general tendency for instability in systems that are driven with periodic inputs of a frequency near that of the system's natural frequency prevailed (*see* May [1973]).

The example that was developed above for the ovenbird is a reasonable approximation of a population response, under the condition that the landscape processes (manifested as suitable habitat availability over time) dominate the population process. In the logistic equation, the k term would fluctuate as a function of the dynamics of the landscape. In a landscape that behaves as a shifting mosaic of habitats, it appears (Shugart and Seagle [in press]) that several important relationships in community ecology (particularly the species-area curve that has been an important concept in the development of the theory of island biogeography) can arise as a consequence of such a seemingly simple model (Fig. 8.4).

Shugart and Seagle (in press) used a model of the vegetation dynamics of the island of Tasmania (Noble and Slatyer [1980]) as the basis model for simulating the patch dynamics of islands of different sizes. Ten habitat types were recognized, based on Noble and Slatyer's analysis of the successional pattern in Tasmanian wet-sclerophyll and temperate rain forest landscapes. The transition probabilities were calculated as the re-

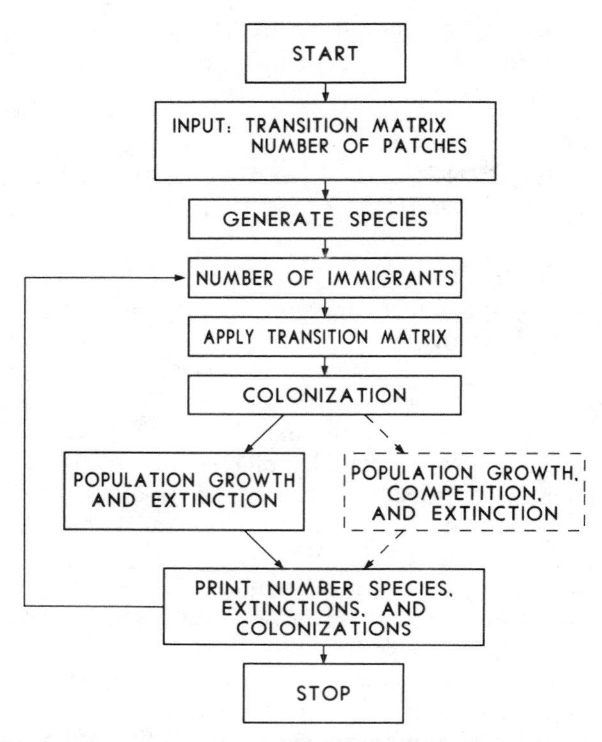

Figure 8.4. Flow chart of a model of animal abundance on landscapes of different sizes. (From Shugart and Seagle [in press].)

ciprocal of the successional time between these habitat types. The resultant 10×10 matrix was implemented as a first-order Markov process. The age of each patch was also computed as the length of time that the patch had been classified as a given type. The landscape area was presumed to be the number of patches in a particular simulation. For each simulation, the starting conditions were patches that were allocated in proportion to the equilibrium implied by the underlying Markov matrix. The successional pattern in Tasmania was pictured by Noble and Slatyer as including the effects of wildlife of different intensities, as well as landslips that caused later successional habitat types to be reset to earlier types. At equilibrium, the landscape simulated by the model represents all habitat types to some degree on the landscape due to the inclusion of these disturbances.

To simulate the colonization of these model island landscapes by animals, a pool of 150 hypothetical species was generated from uniform statistical distributions of different species-level attributes. These characteristics included:

1. The range of habitats used by each species;
2. The range of habitat ages used by each species;
3. The intrinsic rate of increase of each species;
4. The number of habitat patches needed to support an individual of a given species; and
5. For cases in which one species competed with another, the species-specific efficiency (used to determine the competition coefficient) by using patches that overlapped in their potential occupancy (due to points 1 and 2, above) with other species.

Points 1 and 2 are consistent with the idea of letting the carrying capacity term in the logistic equation vary with the habitat availability. Points 1, 2, and 5 are directly related to the main ideas that were discussed earlier, in terms of the niche-hyperspace concept.

Each model run was initialized with no species occupying the simulated landscapes. Species were selected at random from the pool of available species (mean = 12 invasions; standard deviation = 2.0). During each year of a model run all populations on the simulated island were allowed to grow or decline according to the logistic equation, in which the carrying capacity was the sum of the patches that were suitable to be occupied by a given species divided by the number of patches required to support an individual of the species. When the species were allowed to compete (*see* Volterra equation, Equation 5.1), the competition coefficient was the proportion of patches that a species shared with the competitor multiplied by the ratio of their use efficiencies. The probability of a population (of size N) becoming extinct was the inverse of 2 raised to the $N - 1$ power (the probability of all individuals in a given generation being of the same sex).

The effect of increasing landscape area on the number of established species (Fig. 8.5) displays a rapid increase in the number of species, with a slowing of this increase as the area becomes large. The shape of the curves are not different, but there are fewer species with competition. In both cases, the smaller landscapes support small populations that are sensitive to extinction due to fluctuations in suitable habitat. These fluctuations are a product of statistical variation in successional processes on small landscapes. The small landscapes with the high extinction rates (and consequently low richness of animal species) are what were called "effectively nonequilibrium landscapes" in Chapter 7.

The model:

$$S = CA^z,$$

where: S is the number of species, A is the area of an island, and C and z are model parameters, has been used to describe the relationship between the size and the species richness of an island (Preston [1962], and MacArthur and Wilson [1967]). There is considerable debate on the interpretabil-

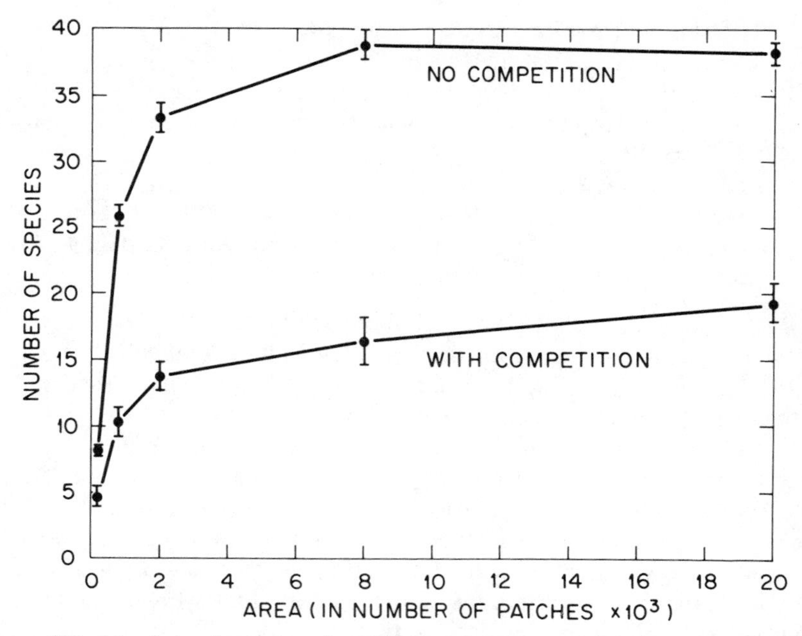

Figure 8.5. Number of species after 300 simulated years on landscapes of different size and with or without interspecific competition. (From Shugart and Seagle [in press].)

ity of the parameter z (Connor and McCoy [1979], and Sugihara [1981]), but its value has consistently been reported from empirical studies to be in the range of 0.20 and 0.35. Using a log-log transformation to estimate z for the curves shown in Figure 8.5, Shugart and Seagle (1983) obtained values of 0.29 for z in the no-competition case and a value of 0.31 in the case with competition. While these results are suggestive rather than conclusive, they indicate the possibility that patterns of animal richness on islands could be (at least in part) a consequence of the size-related statistical variations in landscape dynamics.

Conclusions

This chapter has treated in a simple way some of the complexities of animal populations that interact with vegetation on mosaic landscapes. The discussion was in two parts: 1) The role of animals in altering the functioning of forest ecosystems; and 2) The response of animals to changes in the forest ecosystems. These two facets need to be unified, but this is a tremendously difficult problem that involves considerations of widely different space and time scales. The results shown in this chapter are hopeful starts in this direction. That the task is worthy of undertaking should be considered in light of the following statement of Sukachev (1964):

> To understand the life of the entire biogeocoenose, to discover the nature of the process of regeneration of the tree stand, it is necessary to know more precisely how the atmosphere, soil, and organic life are differentiated in the mosaic elements of such forest biogeocoenoses. Although the differences may be comparatively small they still may play a definite and sometimes a considerable role in the life of the stand and of other parts of the biogeocoenose. Therefore there is great theoretical and practical value in the organization of studies of these mosaic elements from all aspects'' (p. 560).

The interaction of mosaic heterogeneity with animal-mediated ecosystem processes is, indeed, an important area for investigation in our understanding of the dynamics of ecosystems. The inclusion of decomposition and element cycling in models of forest dynamics (Aber, et al [1978], and Weinstein, et al [1982]) probably is a step in the right direction, but the problem of interaction among the mosaic elements (i.e., the simulated gaps in a gap model) has not yet been attacked in this way.

The use of simulation models as habitat simulators has potential applied value—particularly in the management of landscapes for the preservation of particular populations or for the maintenance of some desired level of species diversity. The apparent relationship between the larger scale behavior of such models and the theory of island biogeography provides a tie to the considerable body of theoretical work that has developed for island systems.

Chapter 9

Predicting Large-scale Consequences of Small-scale Changes

Several problems in forest ecology are concerned with the extrapolation of changes. Typically, these problems involve predicting larger scale consequences from typical smaller scale observations. Extrapolation is an intrinsically difficult procedure almost regardless of how one goes about it. Nonetheless, it seems particularly important in the context of current problems in the management of natural forests. In Chapter 2, the desire to predict the longer term response of managed forests to change (e.g., from improved genetic stock or with forest fertilization) was noted as a major impetus in the development of forestry simulation models. This situation also is found in the management of natural forests. Chapter 8 treated one such problem, the management of forests for the maintenance of animal habitats. In this chapter, other cases will be presented in which gap models have been applied to extrapolate the expected dynamics of forests. These examples include: 1) Projecting the larger scale regional consequences of air pollutant effects on forests; 2) Predicting the long-term response of forests to changing climate conditions; and 3) Evaluating the response of a forest to increased ambient levels of carbon dioxide (CO_2).

Possible Effects of Air Pollutants on Forest

The Problem

In the United States, the current pattern of generation of electricity has about 80% of the total oil- or coal-fired power plants distributed in the eastern United States. Most of this area either is currently in forests or was once forested. At present, along with increased urbanization and industrialization in general, large fossil-fueled electrical generating plants with tall stacks have created an increased potential for regional-scale changea in the air quality. The major focus of predicting the effects of pollutants has shifted from site-specific studies of the acute effects of gaseous pollutants near a source to understanding the effects of chronic exposure to multiple pollutants from multiple sources in a regional airshed. The recent and widely publicized recognition that more strongly acidic rainfall (an apparent product of the long-distance transport and transformation of sulfur and nitrogen oxides from fossil fuel combustion) now falls over an increasingly larger area of the eastern United States (Cogbill and Likens [1974], and Likens and Butler [1981]) has served to further increase the interest in regional-scale effects on forests.

Despite this interest, much of the information on the impacts of air pollution on forests is from observations made near large smelting operations during the first half of this century. Forests underwent decline and sometimes demonstrated increased mortality over thousands of hectares in the immediate vicinity of then unregulated smelting operations (*see* Scurfield [1960], Hepting [1968], Miller and McBride [1975], and Smith [1981] for reviews). How can one use these local studies (which demonstrate that pollutants such as sulfur oxide, heavy metals, and hydrogen fluoride—often at some unknown level—can cause severe localized effects) to predict regional-scale effects (at presumably lower levels)? The problem is formidable. It involves including the potential effects of anthropogenic stress on the growth of trees along with the complexity of tree growth processes and their interactions with "natural" environmental factors (Kozlowski [1981]). This problem cannot be solved simply by the applying a model, but models can be used to project the possible consequences (over long time or space scales) of inferences about the possible effects of pollutant events. A logical initial question is; "If the pattern of response of plants under acute pollutant stress is an index of the magnitude of their response to chronic stress, then how large a systematic reduction in the growth rate of trees is needed to cause a significant effect over a region?"

The effects of sulfur dioxide on forest tree growth recently have been addressed in reviews by Miller and McBride (1975), Kozlowski (1981), and Smith (1981). An overlay of forested regions in the United States, with isopleths of air stagnation and acidic deposition (Fig. 9.1) clearly demonstrates the potential scale of the impingement of sulfur dioxide (and other pollutants—principally ozone and acidic rain) on forests. Several problems limit our ability to project the response of forests to low levels of sulfur dioxide at a satisfactory scale of resolution. They include:

(1) The need for a better characterization of atmospheric chemistry in nonurban areas and within forest stands in these areas. This lack of information, which characterizes pollutants in rural areas stems from an allo-

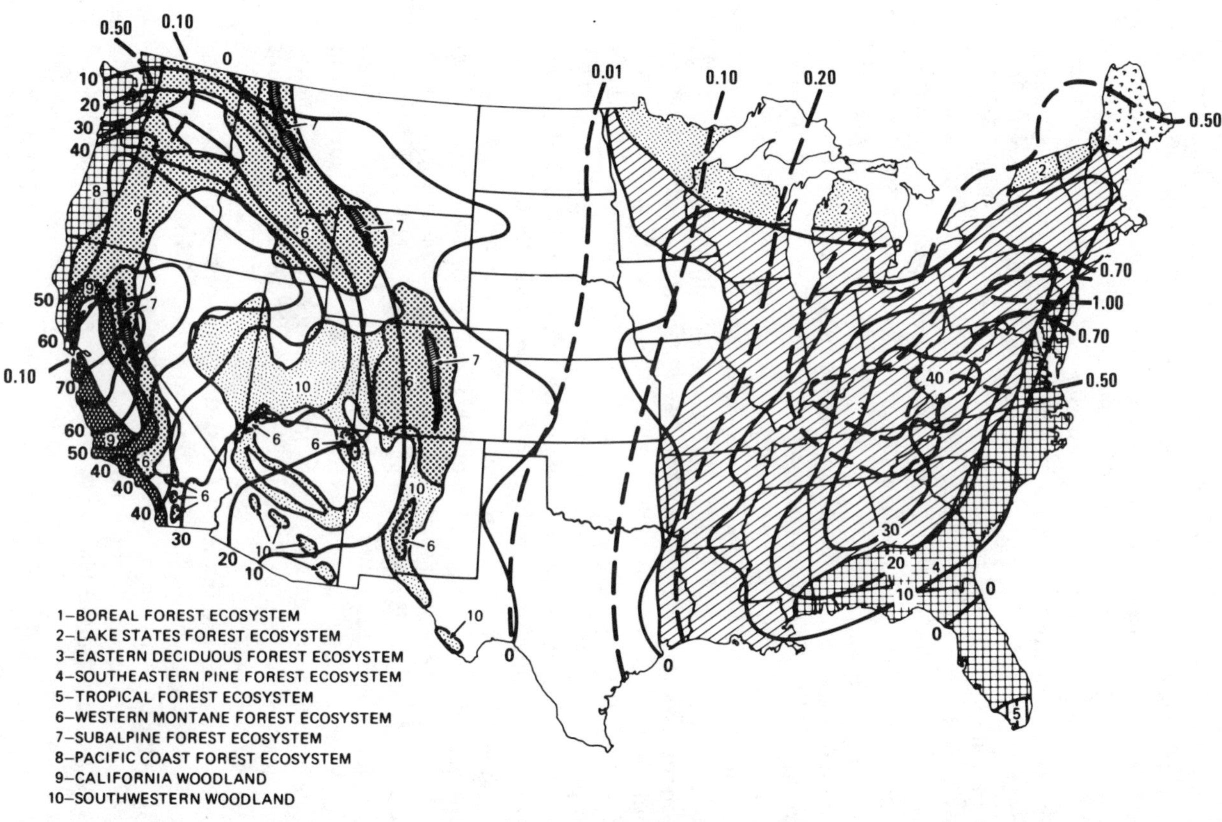

Figure 9.1. Solid lines are frequency isopleths for the total number of forecast days with high meteorological potential for air pollution over a 5-year period. These isopleths are shown in relation to major forest types in the contiguous United States. (After Miller and McBride [1975].) Dashed lines are isopleths of mean hydrogen ion deposition (kg ha^{-1}

cation of air quality monitoring to almost exclusively determine exposures to urban populations.

(2) The lack of documentation of growth the effects of chronic exposure of forest trees to sulfur dioxide and other associated regional-scale pollutants under field conditions and over a range of pollutant concentrations.

(3) The need to characterize both the short- and long-term effects of sulfur dioxide (and associated pollutants) on forest stand growth, competition, and successional dynamics. It is to shed light on this part of the problem that gap models have been used.

Model Application and Results

Many of the verifications and validations that have been used on gap models (*see* Chapter 4) involve testing a model's ability to simulate conditions along gradients and under other conditions that reflect the reliability of the model to extrapolate. Of course, the fact that a given model has been successfully used to extrapolate one condition (e.g., the pattern of vegetation along a complex altitudinal gradient) provides no guarantee that the same model will successfully extrapolate some new condition (e.g., the response of a regional forest to chronic pollution). For this reason, any model extrapolations of this type should be regarded with caution and should be augmented with primary observations as much and as soon as possible. The FORET model has been used to perform model experiments on the potential effects of a chronic air pollution stress, which is expressed as a systematic change in the growth rates of pollutant-sensitive trees (McLaughlin, et al [1978], West, et al [1980], Shugart and McLaughlin [1984]; *see also* Kercher, et al [1980] for an application that uses a different gap model). The results of these studies are:

(1) There can be profound effects on tree populations in response to relatively small (approximately 5–10%) changes in annual growth rates. In some cases, tree species that are subjected to small decreases in their rate of growth can be totally eliminated from the simulated forest. In other cases, species that are subjected to equally small effects can actually increase as their competitors are diminished by growth reductions.

(2) The response of the ecosystem *in toto*, under these relatively small changes in tree-growth rates, can be greater than inferred from simply averaging the reduction in growth across all the species that comprise the stand. This same result also holds if the average that is computed is a weighted average based on the composition of the stand.

(3) The age (or stage of development) of a forest stand can have a radical effect on the nature of the pollutant response. Using the FORET model, West, et al (1980) found that when pollutant-sensitive trees (trees that showed extreme responses in the vicinity of point sources such as

smelters) were given an annual growth-rate reduction of 20%, moderately sensitive species were given a reduction of 10%, also, tolerant species were given no reduction in growth rate; species showed differing responses depending on forest development at the time of addition of the stress. Trees in younger forests, or in forests at a stage of development that had many suppressed trees, tended to be more responsive to pollutant stress. Furthermore, species with equivalent sensitivity (e.g., black oak [*Quercus velutina*] and yellow-poplar, [*Liriodendron tulipifera*]—both are given growth reductions of 10% in Fig. 9.2) can show different responses at different stages of forest development.

(4) Using an extended version of the FORET model, which included soil moisture effects, Shugart and McLaughlin (1984) found that the effect of stand age in altering a given species pollutant response also had a strong interaction with stand site conditions. For example (Table 9.1), when black oak was subjected to a 5% growth reduction in a simulation of a mixed forest stand in which sensitive species had a growth reduction of 10%, intermediate tolerance species had a reduction of 5% and insensitive species were given no reduction in growth; the resultant response varied from as much as a 22% reduction in biomass (the biomass averaged 78% of control over 50 years) for young stands on dry sites to a 20% increase in some of the older stands. There are two implications of this work for evaluating of pollutant effects on forest systems. First, if these effects actually are found in natural ecosystems, it may be very difficult to project the consequences of pollutant stress from the laboratory or green-

Table 9.1. Black oak under intermediate stress (5% growth reduction year^{-1})

Stand age at stress initiation (years)	Biomass Change (% of control)				Duration of stress
0	0.85	0.87	0.85	0.78	50 years
100	0.99	0.98	0.96	0.96	
200	1.02	0.93	0.89	1.00	
300	0.98	1.01	0.99	0.95	
0	0.81	0.81	0.82	0.74	100 years
100	0.99	0.94	0.91	0.92	
200	0.88	0.89	0.84	0.91	
300	0.94	1.13	1.03	0.99	
0	0.85	0.83	0.84	0.73	200 years
100	0.95	0.96	0.89	1.04	
200	0.88	0.83	0.89	0.89	
300	0.88	1.20	0.93	0.95	
	Wet—			—Dry	
	Site moisture status				

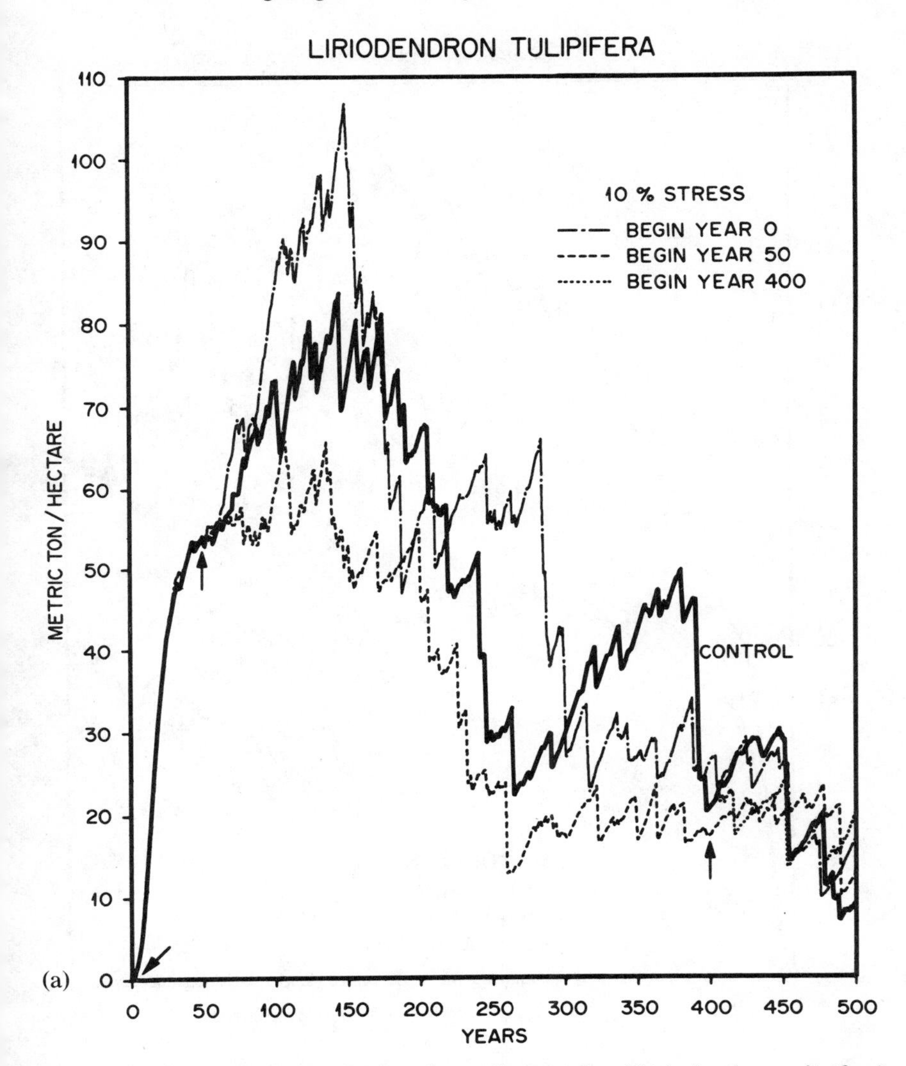

Figure 9.2. (a,b) Biomass changes for yellow-poplar (*Liriodendron tulipifera*) growing in a mixed forest in which species deemed pollution-sensitive were reduced in growth by 20%, species of intermediate tolerance by 10%, and tolerant species were unchanged. This growth stress was introduced at years 0, 50, and 400 during a 500-year simulation (100 plots), starting with an open plot as the initial condition. Control is the same design with no stress applied. Both yellow-poplar and black oak were reduced by 10% in growth (intermediate tolerance). (From Journal of Environmental Quality 9:43–49, 1980, with permission of the American Society of Agronomy, Crop Science Society of America, and Soil Science Society of America.) (See Figure 9.2(b) on page 202.)

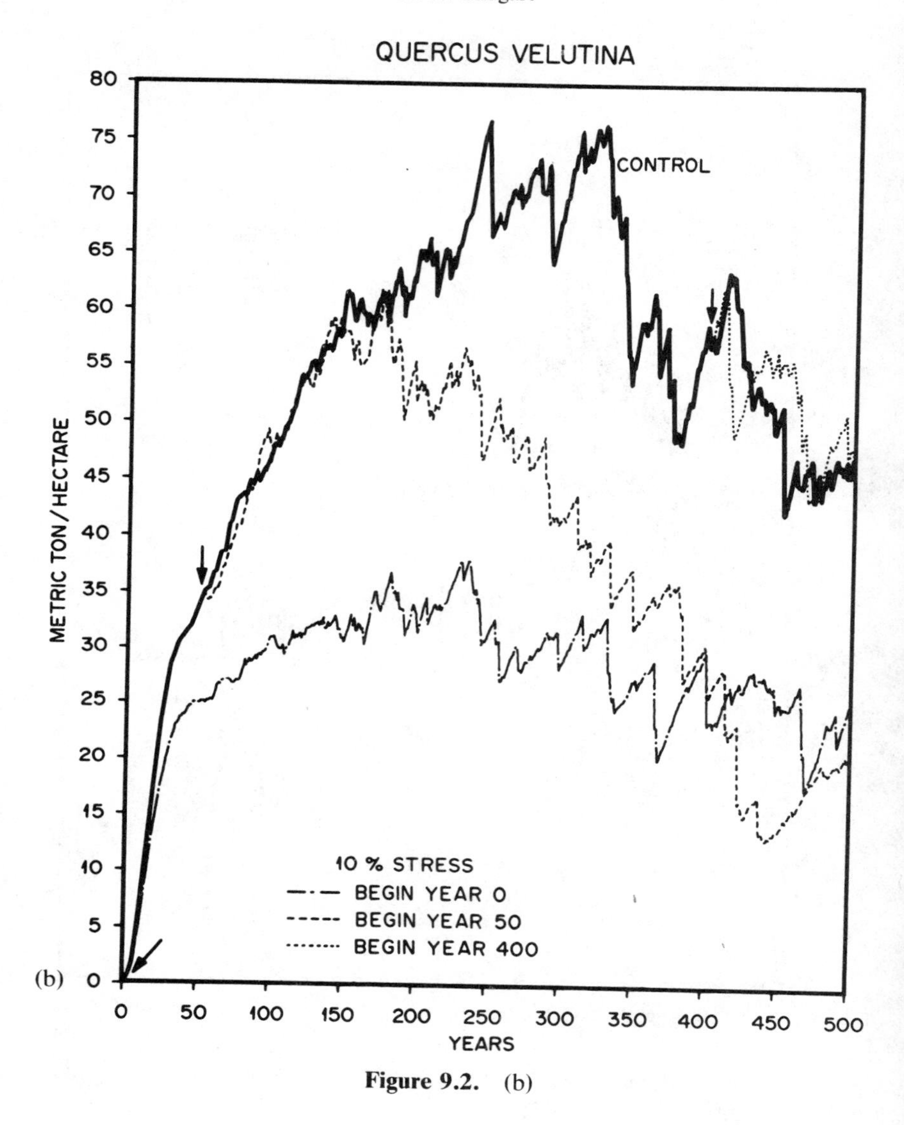

Figure 9.2. (b)

house to field conditions. This potential difficulty can compound the already difficult problem of determining the whole plant response to pollutants of organisms as large as trees. There is an obvious need for an understanding of forest ecosystems as an adjunct to more physiological studies of pollutant stress. Second, there is a need to exercise care in the design of field studies on the ecosystem effects of pollutants. The results that are indicated in paragraphs 3 and 4 (above) indicate a:

complex web of interactions among site factors, age and species that could be diabolically difficult to unravel in studies that did not use considerable effort to control these interactions. In particular, indiscriminate comparisons among forests without consideration of stand age, stand structure, and site factors could well produce a tangle of seemingly conflicting results. (Shugart and McLaughlin [1984].)

These difficulties probably will mean that models will play a role in the evaluation of the longer term and larger scale projection of the consequences of pollutants for some time to come.

Assessing the Potential Effects of Carbon Dioxide Fertilization

The Problem

Carbon dioxide (CO_2) in the earth's atmosphere is relatively transparent to incoming solar radiation; however it absorbs energy in the infrared portion of the spectrum that is radiated back into space by the earth. This creates the potential for what has been called a "greenhouse" effect, which could significantly alter the earth's climate if a substantial increase in CO_2 occurs (e.g., Manabe and Stouffer [1980]). Carbon dioxide as measured by Keeling, et al (1976) since 1958 at Mauna Loa Observatory in Hawaii, has a systematic increase (Fig. 9.3) that generally is attributed to the input of fossil fuel to the atmosphere (Bacastow and Keeling [1981]). The complex exchanges among the oceans, land, and atmosphere

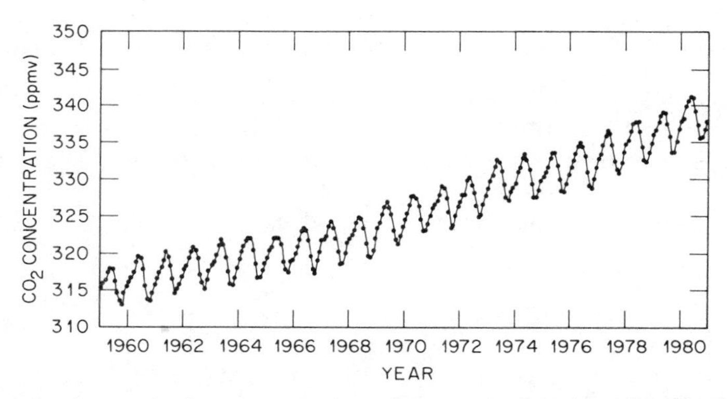

Figure 9.3. Atmospheric concentrations (in ppmv) of carbon dioxide at Mauna Loa Observatory, Hawaii. (From Keeling [1982].)

and the potential of the increased CO_2 to change the climate have produced a significant and difficult scientific problem. Bolin (1981) notes:

> Man's interference with the natural global carbon cycle by fossil fuel burning and changing land use has been receiving increased attention in recent years. Increasing amounts of carbon dioxide in the atmosphere may cause a detectable change of the global climate sometime early in the next century . . . it therefore becomes urgent to develop models of the global carbon cycle to predict the likely future changes as dependent on the use of fossil fuels, deforestation and expansion of agricultural land.

One small, but nonetheless important, part of this larger problem is the question of whether or not the increased level of CO_2 will stimulate plant growth by increasing the supply of CO_2 for photosynthesis (carbon fertilization). The increased plant growth could, in turn, cause more carbon to be stored in the terrestrial systems, thus reducing the rate of increase in the atmosphere. Kramer (1981) discusses this problem with a view towards the plant physiological aspects of the problem,

> Many scientists seem to assume that the increasing carbon dioxide concentration of the atmosphere will automatically bring about an increase in global photosynthesis and dry matter (biomass) production. . . . Some scientists also assert that the increase in carbon fixation by photosynthesis will be large enough to slow down the increase in atmospheric CO_2 produced by burning fossil fuels (Bacastow and Keeling [1973]). The validity of these assertions is of great importance to policy makers who need to know the global effects of the increasing use of fossils fuels. However, they are based on the assumption that the potential rate of photosynthesis is limited chiefly by CO_2 concentration and that the rate of photosynthesis is limited chiefly by a low potential rate of photosynthesis. But are these assumptions valid?

Kramer continues, in his review, to point out that there are potential difficulties with both of these assumptions and that

> we cannot make reliable predictions . . . until we have information based on long-term measurements.

One question that can be approached by using models is determining the maximum effects, if both of the assumptions that Kramer identifies were met and the response of each tree in the stand was similar. The question is; "What is the biomass response expected of a forest if the growth rates of all trees are uniformly increased in a systematic fashion?"

Model Application and Results

The models that are used in this investigation are the FORET, FORAR, FORICO, KIAMBRAM, and BRIND models. A brief description of the models and a tabulation of the model tests are in Chapter 4. All of these models successfully simulate a considerable variety of the longer term responses of forested landscapes, and they are reasonable tools for assessing the potential consequences of perturbations on forests. The models are particularly useful for inspecting the potential consequences of carbon dioxide fertilization because the fundamental equation (*see* Equation 3.1, Chapter 3) that is used in the models to calculate the annual increase in the diameter of a tree in the simulated stand is derived from a simple growth model that balances photosynthate production against the respiratory cost of maintaining living tissue. A change in the rate of photosynthate production can be simulated by adjusting the "G" parameter directly.

To inspect the potential magnitude of a fertilization of photosynthate production, the intrinsic growth rate (G) of each simulated tree was systematically increased by 100%, 50%, 25%, 12.5%, 6.25%, and 3.125% (six levels of photosynthate change) in 25 replicate computer runs of 500 simulated years each for seven different gap model cases (Fig. 9.4). The percentage change in the amount of organic matter that is found in living plants (phytomass) was determined by averaging the phytomass on each of 25 simulated plots over the time period of 300-500 years and by comparing these results to a control with no fertilization of the photosynthesis production process. These responses are complex at low levels of growth enhancement, but they generally are linear at higher levels of fertilization. One case (the FORET model; *see* Fig. 9.4 G) actually demonstrates a diminution of phytomass at low fertilization levels. Under inspection, this is caused by the uniform increase in the growth rate, which allows an increased survival of suppressed understory trees. These trees are species that tend to grow more slowly and attain a smaller mature size; hence, the competition-induced change in total phytomass. The increase in phytomass ranges from 20–60%. The increase is not due to an elevated maximum phytomass at any point in the landscape. It is caused by the increased rate of filling gaps made in the canopy by the mortality of large trees. For this reason forests with a high rate of disturbance and more growth (Fig. 9.4A,B) show a larger landscape-scale fertilization response. In the present simulations, the amount of phytomass on a single simulated plot is limited either by shading (for the smaller trees) or by the maximum size (for the largest trees). In actual forest systems, other limiting factors (such as nutrient limitation) could likely work to reduce any CO_2-induced growth enhancement. These results should be regarded as the maximum that is expected for a uniform increase in growth rate across natural forested landscapes. If the changes in growth rate were not uniform across all of the species involved, then the potential expression of CO_2-induced fertilization effects could involve changes in the community composition and could be very complex. Such responses were noted above in the case of air pollutant stresses.

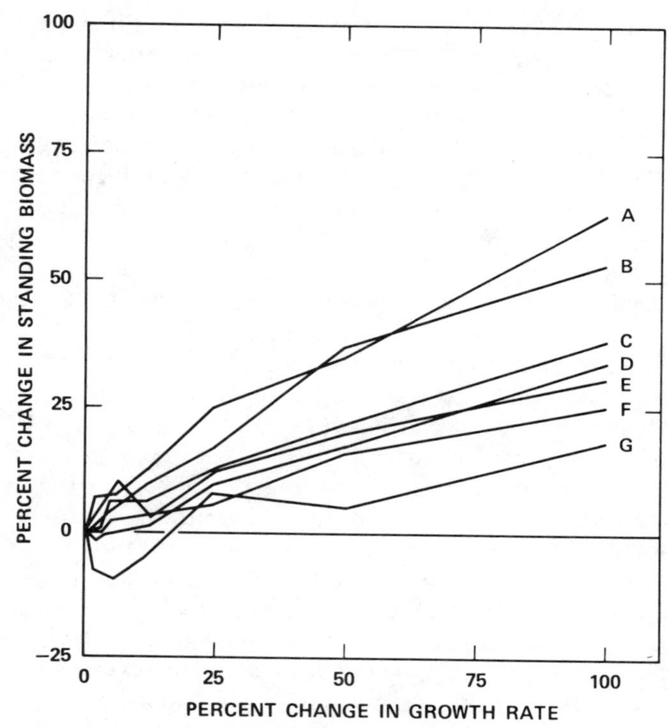

Figure 9.4. Response of seven different simulated cases of increased growth rate mimicking a maximum, nonspecies-specific response to carbon dioxide fertilization. Lines are the percentage of standing crop over years 300–500 of a 500-year simulation for systematic increases in all growth rates of 3.125%, 6.25%, 12.5%, 25%, 50%, and 100% in relation to a control case with no alteration in the growth rate. The cases are: (A) FORAR model with wildfire; (B) BRIND model with wildfire; (C) FORAR model without wildfire; (D) FORICO model; (E) KIAMBRAM model; (F) BRIND model without wildfire; and (G) FORET model.

Reconstructing Prehistoric Vegetation

Background

Pollen analysis techniques often are used to reconstruct the composition of prehistoric plant communities. This also is an area in which forest dynamics models are beginning to be tested as tools for aiding in the inferential testing of theories. Reconstruction of the changes in forest communities over the past 16,000 years at Anderson Pond in Tennessee already has been mentioned as a test on the FORET model (*see* Chapter 4). Before further discussing an example application of simulation models to reconstruct a prehistoric forest, both the weaknesses and strengths of fossil pollen data should be understood. The principal advantages of using pollen grains for paleological reconstruction are:

1. Pollen grains are ubiquitous; they are produced in about the same amounts every year.
2. Species composition of the pollen grains is related to the species composition of the surrounding forests.
3. Most pollen grains can be identified at the genus level and some can be identified to the species level.
4. Pollen that is deposited under anaerobic conditions in lakes and bogs is preserved almost indefinitely in the contemporaneous sediments.

For all these positive aspects, sedimentary pollen composition does not directly reflect the numerical proportions of the species, because the pollen release varies from species to species (e.g., Solomon and Harrington [1979]). Since insect-pollinated species produce and release little pollen, these species are under-represented or even absent from sedimentary pollen records. Pollen types also are differentially transportable in the atmosphere, and they sink in the atmosphere at different rates according to their volume and density (Gregory [1973]). Additional complexities in relating fossil pollen to the forests that generated this pollen is derived from the irregularities in soils and topography in the area surrounding the lake or bog. Soil differences could have mediated species competition and induced large local differences in the prehistoric forests. Topographical relief could have induced similar variations (due to slope and exposure differences) and also could have introduced irregularities in pollen transport (Solomon and Harrington [1979]).

An important consequence of these considerations is that while fossil pollen is an extremely valuable object of study in understanding the patterns of forests in the past, each taxon in a sedimentary pollen sample did not necessarily occupy the same portion of the landscape or the same total area of landscape as other equivalently abundant taxa in the sediment sample. Webb, et al (1978) found that sedimentary pollen generally records the variation in vegetation at the scale of 100 km^2 or larger. *Acer* pollen, however, greatly underestimates the presence of maple trees on the landscape and reflects only trees in an area of about 3 km^2 around the sample site (Bradshaw and Webb [1983]). The deposition pattern of maple can be contrasted to that of pine (*Pinus*), which is deposited in sediments to reflect pine trees over a landscape that is well beyond 3,000 km^2 in size (Bradshaw and Webb [1983]). Pollen records from very small lakes (about 1 ha) and bogs tend to have less pollen from distant sources (Tauber [1965, 1977], and Heide and Bradshaw [1982]), while larger lakes reflect less of the extremely local vegetation (Maher [1977], and McAndrews and Power [1973]). Lake basins also can control the temporal resolution of fossil pollen records. For example, burrowing invertebrates (Davis [1967, 1974]) and seasonal water currents (Davis [1968, 1973]) can vertically mix the sediments. Deep lakes in temperate and boreal regions can contain much less stirred, laminated sediments with layers (varves) that are deposited annually. These varved sediments clearly have a much finer time resolution (Craig [1972], and Swain [1973, 1978]) than the more stirred samples.

It is fair to say that with respect to interpreting fossil pollen data, the reconstruction of the prehistoric forest pattern is an art as well as a science. In Europe, these reconstructions have been accomplished by relating fossil pollen percentages (proportions) to tree frequencies (e.g., Anderson [1970, 1973], and Faegri and Iverson [1975]) and then sorting out separate influences from the pollen diagrams. Using this sort of approach, Iverson (1941, 1949) was able to detect prehistoric land-use patterns in Denmark. Iverson (1964) also was able to detect the influence of soil development over time in other pollen reconstructions, as Tauber (1965) has been able to do for climate change.

Due, at least in part, to the increased complexity of the North American forest flora, the reconstruction of forest communities by applying a pollen/tree-taxon representation value that has been successful in Europe has not worked nearly as well in North America (Davis and Goodlett [1960], Davis [1963], and Janssen [1967]), although there is promise in recent work in this area (Webb, et al [1981], and Heide and Bradshaw [1982]). The more successful approach has been to match the multiple proportions of fossil pollen to a modern pollen pattern that is collected from the surface samples of a lake or bog (McAndrews [1966], and Davis [1969]). If the fossil pollen is sufficiently similar (Ogden 1969, 1977) to the modern pollen, then one assumes that they were derived from similar forests. The validity of this assumption is improved when lake basins that are used to build the modern "reference" data base are of similar size and morphometry to the lake that is producing the fossil pollen.

In the same way, one can use the weather station data from near the sites that are involved in the calibration of the modern pollen analog to provide a tool for paleoclimate reconstruction (e.g., Delcourt [1979]). The more direct method of accomplishing this is to relate modern climate to pollen without reconstructing the forests as an intermediate step. This approach uses multivariate statistics ("transfer function"), and it was initially used by Cole (1969) and Webb and Bryson (1972).

The Problem

Bryson and Wendland (1967) suggested that in North America (but not in Europe) the colder summers during the full-glacial period may have been accompanied by winters with temperatures that were hardly different from present winters. This concept was based on relating features of the Laurentide ice sheet to biotic anomalies, in particular, to explain the presence of thermophilous deciduous trees that grow within boreal conifer forests over large geographical areas (Wright [1968], Birks [1976], King [1981], Delcourt [1979], Ogden [1966], and Watts [1975]). Such mixtures have little areal extent at present. Traditional approaches to analyzing prehistoric plant communities (Davis [1969], Webb and Bryson [1972], and Webb, et al [1981]) are precluded when no modern analogs exist. Since the occurrence of this vegetation has been thought of as a response to climate, the problem is to determine if any of several plausible hypotheses on the climate during full-glaciation are simultaneously consistent with fossil pollen data and with simulated responses from the FORET model to these climates.

Model Application and Results

To explore this problem as an application of the FORET model, Solomon and Shugart (1984) developed a series of model simulations that explored the vegetation pattern expected for four different scenarios of the climate at a point during the last full-glacial period. The forest that was reconstructed back to 16,000 years ago (Delcourt 1979) at Anderson Pond in White County, Tennessee, has the mixed boreal and deciduous forest character required. It was used as the test data to judge the four simulations. The four cases were:

(1) The Periglacial Climate Scenario: July temperatures were lower by 11.7C. This case is developed from Moran's (1972) reconstruction of full-glacial climatic conditions based on periglacial indicators.

(2) The Climate Model Scenario: This case is based on Adem's (1981) reconstruction of a full glacial climate by using an energy balance climate model with enhanced albedo to map January, April, July, and October temperatures. This case featured a late cold summer (12C colder than present) and a less strongly altered winter temperature regime (5.6C less than present).

(3) The Traditional Pollen Analog Scenario: This climate was produced by matching the pollen pattern for Anderson Pond in 16,000 BP (before the present) with the most similar present-day analog (Davis and Webb [1975], and Webb and McAndrews [1976]). Because of the mixed nature of the prehistoric forest, there are no suitable modern analogs; the best of the imperfect matches was used. The matching to a traditional pollen analog indicates that the climate at Anderson Pond 16,000 years ago was most like that of Kenora, Ontario, at present.

(4) The Composite Pollen Analog Scenario: Following the logic of Bryson and Wendland (1967) and Bryson and Kutzbach (1974), a composite was developed by using only the boreal taxa to compute the summer

climate and by using the deciduous taxa to determine the winter climate. Under this reconstruction, the climate at Anderson Pond (16,000 BP) had summers like those of Kenora, Ontario, but the winters were like Duluth, Minnesota.

For each of the four climate scenarios, forest succession was simulated under three different soil and slope conditions: 1) Rolling uplands adjacent to Anderson Pond (and the likely source of much of the pollen in the pond) with well-drained loams; 2) Steep south-facing slopes (temperatures 2C warmer) with a thin loamy surface mantle and a plastic clayey subsoil; and 3) Steep north-facing slopes (temperatures 2C cooler), also with a thin loamy surface mantle and a plastic clayey subsoil. One would expect the fossil pollen in Anderson Pond to contain pollen from forests growing on all three of these soil and slope conditions. Most of the pollen could be expected to come from the uplands (*see* paragraph 1, above).

Results of the simulation are shown in Figure 9.5. The pollen spectrum for 16,500-15,500 ^{14}C-years ago from Delcourt (1979) appears along the left margin. During that time, spruce pollen ranged from 15–25%, pine averaged from 50–80%, and deciduous trees varied from 10–20% of the total pollen. A summary of the response of the simulated forest under the different scenarios follows:

(1) The Traditional Pollen Analog Scenario: The simulated winter conditions were too cold to allow oak and other thermophilous trees to grow and survive in any of the simulation cases, except for a small proportion on the south-facing slopes. The model produced a boreal forest that is much like the one found growing near Kenora, Ontario, at present, but it is not particularly like the one found 16,000 BP at Anderson Pond.

(2) The Climate Model Scenario: Under the climate model scenario, the growth of thermophilous trees also was greatly reduced, but (in this case) by the coldness of the summers. This is logical, because the growing conditions simulated by the Adem (1981) climate model correspond to what is now the northern limit of the closed boreal forest rather than to the southern limit (Hare and Thomas [1979], and Rowe [1977]). This scenario provides the forest simulation model with little power to reproduce the Anderson Pond 16,000 BP conditions.

(3) The Periglacial Climate and The Composite Pollen Analog Scenarios: Both of these scenarios produced responses from the forest simulation model that are consistent with pollen spectra from Anderson Pond. The two scenarios produced forest responses that were similar, except for the conditions on the north-facing slopes. These differences are due to the differences in moisture conditions in the summer in the two scenarios.

The overall indication from these results is that a decrease in seasonality that is associated with full-glacial climate could indeed produce the stable mixed boreal and deciduous forest ecosystems that are not currently extant in North America, but recorded in the fossil pollen at Anderson Pond, Tennessee, 16,000 years ago.

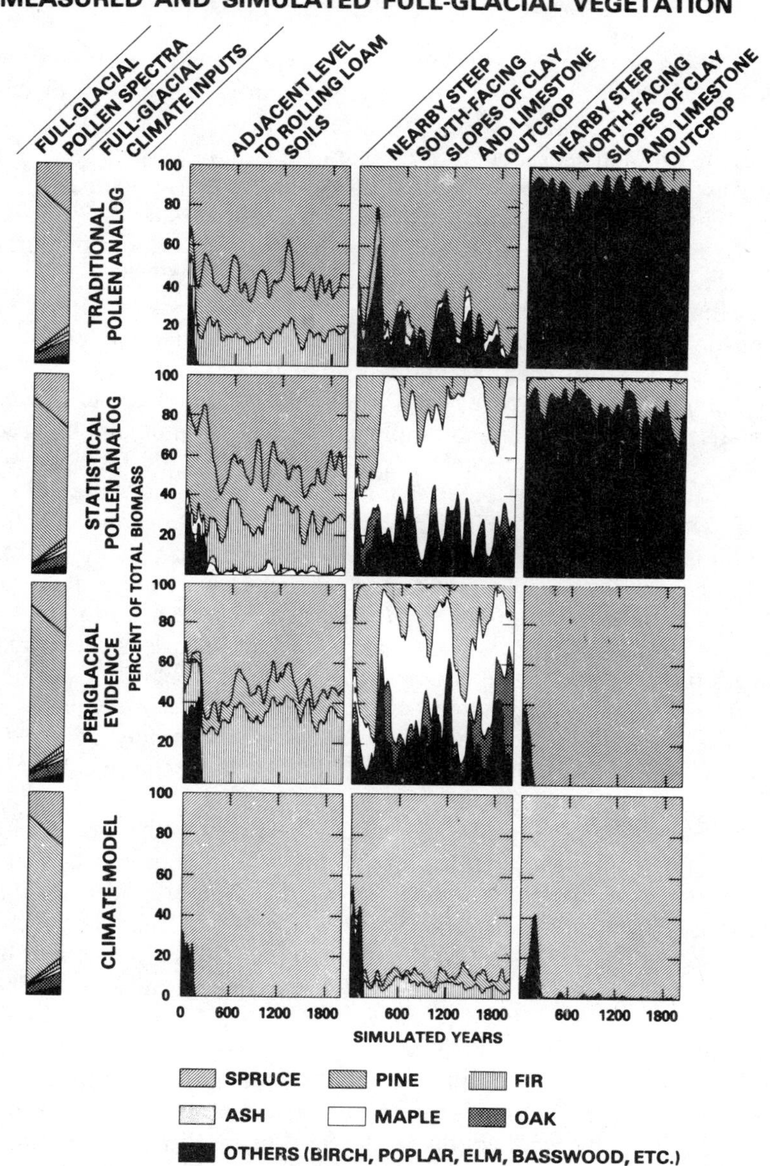

Figure 9.5. Measured and simulated full-glacial vegetation at Anderson Pond, White County, Tennessee. Full-glacial pollen spectra at the left consist of the maximum and minimum pollen percentages from 16,500–15,500 ^{14}C-years ago (Delcourt [1979]) and are repeated in each row for ease of comparison. Each row is a simulation under one of four different climate scenarios for three different types of sites.

Conclusions

The three examples in this chapter provide a sample of the potential range of the ways that forest simulation models can be used to project the consequences of change in natural forests. Probably, the best use of models in problem solving derives from their ability to project forest dynamics to answer questions such as "What might happen if . . .?," "How long might it take for . . .," or "Is it possible that . . .?" It is the desire to extend what we know (or think that we know) about the internal dynamics of forested systems to the consequences of change in these forests that probably will dominate the applied interests in natural landscapes for several years to come. This is, in part, a direct result of the fact that humanity is directly affecting natural ecosystems in ways that often were not possible even a few years ago. For example, the change in regional-scale, ambient air-pollution conditions is enough of a recent problem that it has many potential ramifications that are far beyond the scope of the data that we have available (or that we have the ability to collect in the near future). Similarly, the measured change in the ambient CO_2 atmospheric concentration changes the base state under which the forests that provide our current historical record developed. Regardless of the effect of the change in CO_2, the climate has changed in just a few tree generations and perhaps in even less time. There is an increased realization on the part of scientists and decision-makers of the possibility that today's forests (which almost without exception have been altered in some way by either a human or natural disturbance) may be seeking a quasi-equilibrium state that is different from anything that has been seen in history.

The desire to reliably project forest change (even in the face of altered internal dynamics and given a recognition of the omnipresence of disequilibrating disturbance) has served to make computer models a valuable tool for applied ecological problems. The models of forests currently are developing a reasonable record for these applications. It is important to stress, in the light of the fairly recently developed ability of models to predict the expected pattern of forests, that these models are profiting largely by the foresight of field-oriented ecologists in viewing the forest system as a dynamic entity and in collecting data on the dynamics of these systems in an appropriate manner. The success of models does not proscribe additional data collection on the workings of forests. On the contrary, they campaign for an increased intensity in field ecology. For example, the body of this book has centered on the theoretical and practical implications of a small, but representative, subset of the large array of forest dynamics models. Three ingredients are important in forming the next step in this work. First, there is a need to have more data that record the response of forests that have power to test the models. In reconstructing fossil pollen chronologies, for example, a model is presented a set of changing forest patterns and changing ambient conditions that tend to force the model to do more than simply reproduce the data that was fed into it. Such data sets can be used to design good model tests for this reason. Second, there is a need to gain additional experience in simulating forests of different essential character. There are hundreds of forest simulation models (*see* Chapter 2); however, if one eliminates models of monospecies plantations and models of North American temperate forests, then the number of models is much smaller. Third, there is a clear need for a model of the growth of a single tree that is computationally fast at annual time steps, and that includes greater detail of tree physiology—including, in particular, the allocation of photosynthate to various organs. The development of such a model (and the data needed both to create and test this model) could require a large and extensive investigation of the physiology of whole plants. Such an effort also should be focused on the understanding of trees over their full range of sizes.

Chapter 10

A Theory of Forest Dynamics

There is a theory regarding the dynamic nature of forests that probably can be best attributed to the now classic discussions of A. S. Watt (1925, 1947) on pattern and process in forests. The central idea behind this theory (that may apply to all sessile organisms; (*see* Paine and Levin [1981]) is that a forest's dynamic response can be characterized as a cycle:

> Following the death of a large tree and its fall, a canopy gap forms. The area below this gap becomes the site of increased regeneration and survival of trees. Trees grow, the forest builds, the canopy closes, and the gap disappears. Eventually, the now mature forest in the vicinity of the former gap suffers the mortality of a large tree and a new gap is formed and the cycle is repeated.

This view of the interaction of spatial pattern and temporal dynamics is found in the works of Raup (1957), Whittaker and Levin (1977), Connell and Slatyer (1977), Bormann and Likens (1978a,b), and others. The generality of this theory recently has been endorsed by Whitmore (1982) as:

> . . . The forests throughout the world are fundamentally similar, despite great differences in structural complexity and floristic richness, because processes of forest succession and many of the autecological properties of tree species, worked out long ago in the north temperate region, are cosmopolitan.

While some might use the "new" mosaic-dynamic view of ecosystems as an alternative to the holistic older view of ecological systems that often is attributed to Clements (1916, 1928, 1936), McIntosh (1982) has reviewed this controversy and has pointed to some of the difficulties in such arguments (*see also* Chapter 1). Probably, a most revealing insight in this regard is the comment of Watt (1961, in Greig-Smith [1982]):

> The quantitative assessment of the components of the plant community leaves you with a pile of bricks, timber and chimney pots without showing how they are arranged—and even when this is done it is a static picture that is presented. Now, whatever criticism one may have of Clement's notion of the climax, his injection of the dynamic principle into a static discipline did for ecology what Lyell did for geology, Darwin for biology and Dokuchaiv for soil science—he brought new life into it.

Recognition of the spatial heterogeneity of the forest ecosystem does not negate the view that the system can be considered as a dynamic whole. Nor does the development of a model that allows the inclusion of the autecological details of individual trees necessarily imply the exclusion of models that use more aggregated ecosystem variables.

The models presented in this volume are not revolutionary alternatives to our understanding of ecological succession. To the degree that they are successful, they are extensions of the ability of ecologists to weave more details of nature into a dynamic concept of successional processes in ecosystems. They are a product of the computer age. Without the modern high-speed digital computer (in particular the greatly increased speed and reduced cost of computation) the detail and stochasticity (of both parameters and structure) in the models would make them unsolvable.

Allen and Starr (1982) discussed the behavior of the FORET model in terms of its model structure and viewed the model as an implicitly hierarchical model. They note that in the FORET model:

> The individual trees and individual species assert themselves as quasi-autonomous wholes, while at the same time contributing to the vegetation-mediated environment within the stand. It is this dual existence of simultaneous wholeness and partness which is the cornerstone of hierarchy theory as developed by Koestler (1967). This explicit separation of wholeness and partness of trees contributes to the hierarchical architecture of the model.

One of the more initially surprising aspects of gap models from the point of view of a systems ecologist is the relative insensitivity of the model to virtually all of its parameters taken one at a time. The only parameter that seems to have a great effect on the model's performance, compared to other ecological models with which I have worked, is the size of the simulated plot (*see* Shugart and West [1979]). This would seem to be a most unlikely candidate for a dominant parameter until one realizes that the size of the simulated gap is, in fact, a major structural change in the competition equations that serve to create most of the tree-to-tree competition. It is ecologically appropriate that the gap models display sensitivity to gap size. Gap sizes were initially determined by verifying abilities of the models to simulate the structural patterns of the forests in question. The result of these model explorations of the appropriate plot sizes to simulate forests (assuming that competition in the horizontal dimensions was homogeneous) was to use gaps of about 0.10 ha in temperate forests and of a somewhat smaller size in tropical forests. This model-derived space scale is identical to the empirical value of 1000 m^2 or 0.10 ha reported by Whitmore (1982) as being the minimum size of a gap that is typically needed for the successful regeneration of pioneer species (*see* Role 1 and Role 3 species in Chapter 5).

Gap models form a special theory of the expected dynamics of forested ecosystems that has its antecedents in Watt's (1947) concepts of pattern and process in ecosystems. This theory is special because its applicability is limited by the underlying model assumptions. For example, since the models are constructed with the assumption that competition has horizontal homogeneity within the simulated quadrat, phenomena that are explicit functions of the spatial location of individual trees are clearly outside the gap-model special theory (although they are in the domain of some of the spatially explicit models discussed in Chapter 2). There seems to be a natural tendency for scientists to suspect computer simulation models that have been established as theoretical paradigms. Perhaps this suspicion has its origin in comparisons to the mathematical elegance and historical importance of the development of theory in physics. But, should one necessarily expect one science to mimic another science in all its aspects? Perhaps computer simulation models appear to lack the parsimony that has characterized the development of western science since William of Ockham. In fact, this appearance is not true of gap models on inspection. The functions in gap models are relatively simple (*see* Chapter 3) and are related one to another by equally simple functions. The complexity of the models is more apparent than real and its origin is in the multilevel nature of the models mentioned by Allen and Starr (1982 and above).

If gap models can be used as a special theory of forest dynamics, it is appropriate to inspect this theory in terms of its range of applicability and in terms of the general predictions that spring from this theory. These are the tasks of this concluding chapter.

The Domain of Applicability of Gap Models

The theoretical implications of gap models are constrained by the underlying assumptions in these models. In general, the set of successful applications of models or theories to predict new observations of nature is the best indication of the domain of appropriateness of these models. In this sense, the examples provided in Chapter 4 are samplers of the types of events that the gap-model theory can be expected to predict. These include a wide array of structural and compositional responses of forests over periods of time and in response to gradients.

One of the principal problems in testing models is procuring data sets that are not so confused by historical (and often not well-known) events as to make the test difficult or impossible. For example, while the FORMIS model of the forest dynamics of the Mississippi River floodplain was being developed, Dr. J. T. Tanner (Professor Emeritus, University of Tennessee) made data available on the detailed structure of "virgin" forests on the Singer Tract near Tallalah, Louisiana. These data were collected in the late 1930s as part of Tanner's (1942) doctoral work on determining the status of the ivory-billed woodpecker (*Campephilus principalis*) in North America. To avoid biasing the model development in favor of reproducing this unusual data set, thus reserving the data as a test of the model, the data were not examined until after the model construction and parameterization. When the model/data comparison eventually took place, it was apparent that the sweetgum (*Liquidambar styraciflua*) comprised as much as 50% of the canopy trees in Tanner's data and displayed a size structure that was heavily skewed toward very large trees. The biology and growth rate of sweetgum, as encoded in the FORMIS model, would require that most of these canopy sweetgum trees had been established during a disturbance event(s) some 150–200 years earlier than Tanner's samples. This would have been in the period 1740–1790 and in country so remote that (even in 1930) Tanner was obliged to ride overland on horseback to reach his study plots. Historical records of the hypothetical 1740–1790 disturbance have not been obtained, but they are a necessity if the model is expected to reproduce the Tanner data.

This type of problem is essentially a data collection problem and a direct consequence of the slow dynamics and long-time scales that are associated with forests. Apart from such data collection problems, which are apt to plague any theory of forest dynamics, what are the intrinsic limitations of theories about forests based on gap models? Delcourt, et al (1982) have produced a diagram that relates disturbances and spatio-temporal scales (Fig. 10.1a). This comparison provides a reasonable heuristic picture of some of the boundaries of a special theory of forest dynamics based on gap models. Gap models have not been extensively applied to what Delcourt, et al called the "microscale." However, the most numerous applications of the models have been at the microscale/macroscale boundary in the vicinity of "2" on Figure 10.1b. These applications generally have involved the simulation of longer term forest dynamics (500–1000 years, 100 replicate 0.10-ha plots—the typical simulation cases found in much of Chapter 4). The longest temporal scales simulated to date (*see* discussions of Solomon's model applications in Chapters 4 and 9) are slightly below the temporal scale in which Delcourt, et al would expect soil development to be observable. The gap models presently do not at include pedogenesis, although the work of Aber, et al (1978) and Weinstein, et al (1982) consider many of the elemental and organic matter balances that probably would go into such an extension. The upper space-scale limitations of the gap model applications (4 and 5 in Fig. 10.1b) are near the macroscale/megascale boundary of Delcourt, et al (Fig. 10.1a).

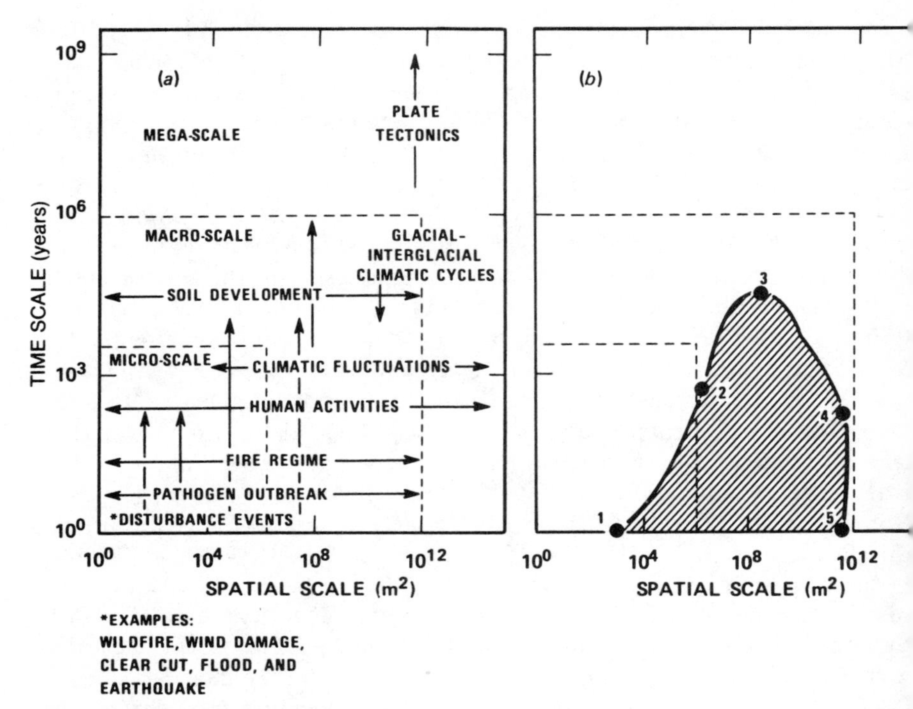

Figure 10.1. (a) Biotic disturbance regimes viewed in the context of the space-time domain in which the scale for each process reflects the sample intervals typically required to observe it. (From Delcourt, et al [1982].) (b) The same space-time diagram with the domain of various applications of gap models shaded. Typical points determining the shaded area are: (1) Space-time scale of a typical gap model (about 1 ha, 1 year); (2) most simulations of forest succession dynamics (*see* Chapter 4); (3) Landscape reconstructions around Anderson Pond, White County, Tennessee (*see* Chapter 9); (4) Long-term effects of air pollutants on regional landscapes (see Chapter 9); and (5) short-term effects of air pollutants on regional landscapes (*see* Chapter 9).

Several of the disturbances that were considered by Delcourt, et al to be in the domain of the gap models actually have been treated in gap model applications (e.g., pathogen outbreak, fire regime, human activities, and climatic fluctuations).

The area in Figure 10.1b can be taken as a reasonable indication of the domain of gap models, but there is one important consideration; the models do not treat phenomena that operate below annual time scales with any degree of mechanistic detail. In some cases (*see* Chapter 3) there are subannual time-step computations made in a model; however, these computations are usually related to indices that alter the annual growth rates of trees. As a direct consequence of this internal time boundary in the model, any subannual forest responses are clearly out of the domain of the model. Furthermore, any longer than annual time-scale phenomena, in which a detailed understanding of subannual forest dynamicsis absolutely essential for predicting the phenomena also are proscribed because of this lower time-scale boundary (unless these subannual dynamics can be included as, for example, an annual average).

If given an indication of the models' spatio-temporal domain (Fig. 10.1b) from the model applications, one can consider the abiotic variables that might drive an ecosystem that displayed responses like that of a gap model in this domain. One important abiotic-driving variable that (according to Fig. 10.1a) is important in the domain of applicability of gap models is listed by Delcourt, et al (1982) under the category of climatic fluctuations. Mitchell (1976) has discussed the overall pattern of variation in the earth's climate and has synthesized this discussion in a diagram within the frequency domain. Using Mitchell's somewhat nonstandard graphic display of the pattern of climate variation (Fig. 10.2a) that spans 14 orders of magnitude, one can inspect the range of gap models in Figure 10.1b. The upper temporal scale on gap model applications is of the order of 10^4 for some of the paleo-landscape reconstructions. The lower limit is the annual computation step. By transforming the spectral analysis of Emanuel, et al (1978b) of the frequency response on a gap model (the FORET model) to the scales used by Mitchell (1976) for his graph, one also can determine the expected performance of a gap model in the range of gap model applications (Fig. 10.2b). The part of Mitchell's diagram that is in the temporal range of the gap model is reasonably uniform with respect to major large periodic components in the climate. This range of Mitchell's diagram is designed to be consistent with Quaternary climate history published in Figure A.2 of the U.S. Committee for the Global Atmospheric Research Program (1975). The range of gap model application is slightly longer than the major astronomically dictated climate variations (1 year and 1 day with harmonically related peaks at 6 and 3 months and at 12 and 6 hours). It does overlap with the longer 2500-year periodicities that Mitchell associated with the "Neo-glacial (Little Ice-Age) cycle of the Holocene." There also is some overlap with the "Quaternary ice-volume shifts" of periods near 2,000 years. The behavior of forests in response to these longer periodicities is the subject of the attempts to reconstruct pollen chronologies. Even though the periodicities in climate are not as pronounced over the range of the gap models, the periodic content of the models is quite rich in this range (Fig. 10.2b). The models display a strong 500-year periodicity that is associated with changes in the composition of the canopy trees. There is the 200-year canopy replacement cycle that is evident in Figure 5.2 and that is the subject of the discussions of the regeneration cycle (e.g., Watt 1947). There also is a

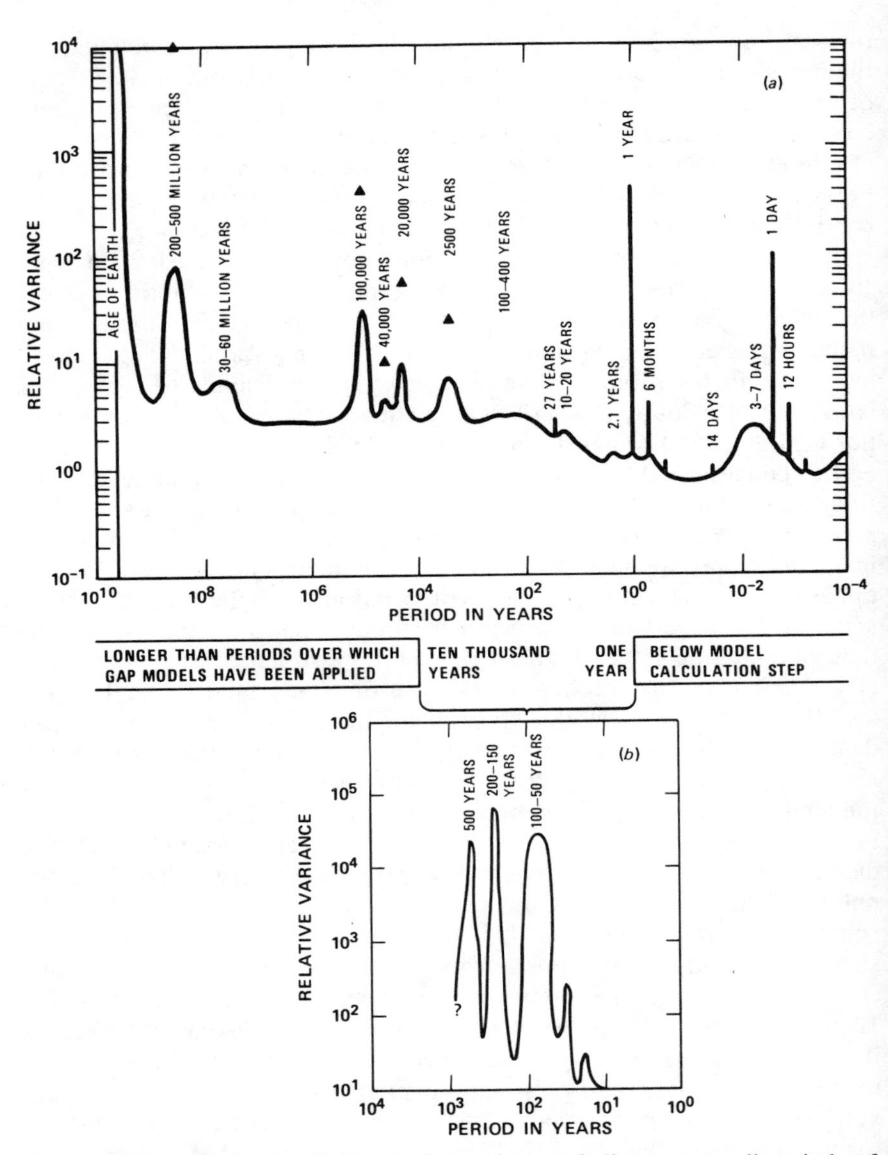

Figure 10.2. (a) Estimate of the relative variance of climate over all periods of variation, from those comparable to the age of the earth to about 1 hour. Strictly periodic components of variation are represented by spikes of arbitrary width. Solid triangles indicate the scaling relationship between the spikes and the amplitude of other features of the spectrum if these latter (somewhat less strictly periodic) features were represented as spikes. (From Mitchell [1976].) (b) Equivalent graph for periodic features of the FORET model. Graph transformed from spectral analysis of FORET model in Emanuel, et al (1978b).

100- to 50-year periodicity that appears to be related to mortality events in subcanopy trees. To date, gap models have not been inspected for periodic components longer than ~500 years (*see* Emanuel, et al [1978b]).

In general, the range of application of gap models is that of a macroscale model (Delcourt, et al [1982]) that has been successfully applied in some cases over shorter time scales and at smaller space scales. The periodic response of one of the more important driving variables in the models (climate) is reasonably uniform over the range to which gap models have been applied (1 year to about 10^4 years); however, the models demonstrate a fairly rich periodic spectra in this range. These important periodicities in the model performance are associated with canopy replacement processes and already have been discussed in some detail. Given a notion of the range of application of theories that are derived from gap models and a general inspection of the dynamics of both the models and one of the important driving variables in this range, it is appropriate to inspect some of the theoretical implications of these models.

Some Consequences of Gap Models

Many of the consequences of gap models as theoretical constructs for predicting forest dynamics are derived from the view that forests can be considered to be a dynamic mosaic. Some of the consequences become apparent when the model can be used to explore the dynamic response that is expected of a forest over longer time periods. In my opinion, the more important theoretical implications are:

(1) A wide range of forest dynamic processes can be described by models that incorporate dynamics of each tree's growth in a set of populations of trees and that incorporate the birth and death phenomena in these populations: The functions that were discussed in Chapter 3 and that are the major parts of the gap models are simple. They were developed for maximal ease in parameter estimation, particularly for tree species that were not studied in detail. They also were designed to approximate general functional shapes of important relationships among trees in forests. In the original design of the gap models ranging from the Botkin, et al (1972) model to the present set of models, the computer codes that implemented the models always have been designed in a modular fashion that would allow any of the functions in Chapter 3 to be replaced with more species-specific or more mechanistic functions. The lack of obvious model sensitivity to such changes has provided little actual incentive to incorporate this increased detail. As a whole, the applications and predictive abilities of gap models (even in extrapolating independent validation data sets) have been quite good.

(2) Relatively simple categorizations of the adaptive traits of tree species ("roles") can produce complex webs of interactions among species as they compete for the occupancy of space: The interplay among the four roles of trees that was discussed in Chapter 5 and based on a matrix of two regeneration traits by two mortality traits includes a wide array of ecological interactions. These interactions include both within- and between-role competition, mutualism, and asymmetric negative/positive interactions that resemble parasitism or predation in a strictly mathematical sense. These interactions also are understood to take place on a background of stochastic variation, particularly with regard to the tree regeneration process. The implication of these results is that the community theory, which is based largely on the analysis of animal population dynamics using the Lotka-Volterra equations, has analogous application in exploring the stability of tree communities, as well as more abstract properties such as species diversity (Huston, 1979). Furthermore, the complexity of the pattern of interactions among trees (and, presumably, plants in general) is as great as that reported for animals. This complexity of tree interactions is masked in the usual trophic structure representation of food webs.

(3) The biomass dynamics of landscapes can be developed from an understanding of the pattern of biomass changes on small patches and from an understanding of the synchrony of these patches across the landscape: This idea has ample antecedents in ecology and is best attributed to the classic work of Watt (1947). The pattern of change in the biomass over time for a landscape that comprised of a number of patches that are all disturbed at the same time would (in the case of a monospecies forest) feature a rise to an overshoot and a settling to a quasi-equilibrium average value. The same idealized response for a more diverse forest would not be expected to demonstrate the overshoot response. The dynamic responses of landscapes in which contagion is an important aspect of the individual patch dynamics are considerably more difficult to anticipate, and additional field work and modelling are both needed.

(4) A consequence of the view of the mosaic landscape as the summation of the dynamics of the patches that comprise the landscape is a categorization of three broad classes of natural landscapes based on their behavior: These categories are the intrinsically nonequilibrium landscape, the effectively nonequilibrium landscape, and the quasi-equilibrium landscape. These classes are discussed in some detail in Chapter 7. Strategies of management for the three types of landscapes can be quite different. Many policies that are used to manage natural landscapes assume that they will behave as quasi-equilibrium landscapes. The size of the units managed under these policies often is small enough so that one would expect them to behave as effectively nonequilibrium systems.

(5) For animal species that select particular habitat patches of a mosaic landscape, the resultant animal communities on effectively nonequili-

brium landscapes have many of the statistical attributes associated with the theory of island biogeography: The relationship to the slope parameters of the species/area curve for islands is an important relationship that may tie the dynamic-mosaic concept of landscapes to island biogeography theory.

(6) Several current problems require an understanding of the dynamics of forests under altered conditions and provide incentive to further develop forest dynamics models: Three of these problems (the effects of air pollutants, climate change, and carbon dioxide fertilization on forests) were discussed in Chapter 9. There will be aspects of any of these problems that are effectively out of the range of time and space scales, in which gap models probably are best applied. Almost regardless of the scale of the models and data sets that eventually may be used to attack these problems, experimentation with even such simple models as a gap model indicates that there will be a need for an integrated research approach. It is with a renewed interest in the synthesis of data collection, experimentation, and ecosystem modelling that the best hope in solving such difficult problems as understanding the consequences of climate change or of regional-scale pollutant effects will lie.

Epilogue

> We must face the fact that we are on the brink of times when man may be able to magnify his intellectual and inventive capability, just as in the nineteenth century he used machines to magnify his physical capacity.*

The gap models that have been discussed in this volume are examples of a potentially larger class of models that explicitly include multiple levels of organization in their formulation. In particular, gap models simulate: 1) The growth of individual trees; 2) The demography of a multispecies assemblage of individuals; and (3) The changes in the abiotic environment that are induced by, or influential on, the dynamics of the two biotic scales. Some of the tree models discussed in Chapter 2 also are multiple-level models as are models for other systems (e.g., elephant populations—Wu and Botkin [1978]; Peruvian Indians—Weinstein, et al [1983]). I expect that there will be more such models in the near future.

These models are products of the general availability of high-speed digital computers over the past decades. Computers increase the ranges of scales and complexities that can be simultaneously considered in an ecological model. In a gap model, the computer rapidly performs the millions of routine calculations that are needed to interact and grow every tree on a simulated plot. These calculations mimic the descriptions of how forests (or other ecological systems) function that have been used in ecology since its inception. The models are a technologically augmented extension of the traditional interest in interactions that is ecology's historical focus.

---

* From Alexander, C.: *Notes on The Synthesis of Form*. Cambridge, Mass. Harvard University Press, 1964, 216 pages.

The quantification of biological and physical phenomena at the scales of resolution in which the phenomena are typically studied and understood make these models' parameters and structures appear to be realistic, with respect to the ecological features of the system. In gap models, the model-formulation step has involved consideration and simplification of ecological concepts. The model-testing step has been oriented toward data and data structures. This prioritization of the use of concepts and data may arise because data sets with real power to test complex hypotheses are somewhat rare in forest systems.

Much needs to be done in extending the use of such complex computer models in ecological systems. The models have been used in systems that are dominated by long-lived organisms. The numerical stability of the models, when applied to fast subsystems, could be a considerable problem. Hopefully, some of these problems will attract the interest of mathematicians and numerical analysts.

While the models represent an increase in the range of scales that usually are explicitly considered, there certainly are phenomena with scales that are so disparate that they must necessarily be considered separately. The magnitude of the scale difference at which this is the case is not known. It is important to unravel the consequences of both scales and the interactions of scales in ecosystems. The confusion regarding ecological succession seems to have its origin in a misunderstanding or a miscommunication of scales. This problem also may lay at the root of other debates in the field as well.

For the ecologist, I think the present value of these models lies in enlivening our understanding of the consequences of interactions in natural systems and in providing a tool for projecting the longer-term consequences of some of our ideas about the ways in which ecosystems function.

Appendix

Below is a listing of the FORET model written in the FORTRAN IV computer language.

```
      COMMON/FOREST/NTREES(100),DBH(700),IAGE(700),KSPRT(100),
     > NEWTR(100),SUMLA(700),NEW(100),SWTCH(5)
      COMMON/PARAM/AAA(100,6),DMAX(100),DMIN(100),B3(100),B2(100),
     > ITOL(100),AGEMX(100),CURVE(100),G(100),SPRTND(100),SPRTMN(100),
     > SPRTMX(100),SWITCH(100,5),KTIME(100)
      COMMON/CONST/NSPEC,SOILO,DEGD
      COMMON/RAN/YFL
      COMMON/DEAD/NOGRO(700),NTEMP(700)
      COMMON/COUNT/NTOT,NYEAR
      COMMON/TEMP/DTEMP(700),ITEMP(700)
      COMMON/PASSX/A(1101,36)
      COMMON/SEED/USEED(11)
      INTEGER USEED
      DIMENSION Z(2)
      LOGICAL SWITCH,SWTCH
      DATA NCT/0/
C.....REFERENCE FOR RANDOM NUMBER GENERATOR - DESIGN AND USE OF COMPUTER
C.....SIMULATION MODELS - EMSHOFF AND SISSON - THE MACMILLAN COMPANY
C.....
C.....USEED(1) - KILL  TEST MAXIMUM AGE FOR EACH TREE
C.....USEED(2) - KILL- TEST GT .368 FOR TREE TO DIE
C.....USEED(3) - SPROUT- SELECT SPECIES TO SPROUT
C.....USEED(4) - SPROUT- SELECT NUMBER OF TREES TO SPROUT
C.....USEED(5) - SPROUT- USED TO CALCULATE DBH FOR SPROUTS
C.....USEED(6) - BIRTH- SELECT A SPECIES FROM 1 TO 3 TIMES
C.....USEED(7) - BIRTH- USED IN SMALL MAMMAL SWITCH
C.....USEED(8) - BIRTH- USED IN SEEDING RATE SWITCH
C.....USEED(9) - BIRTH- DETERMINES THE NUMBER OF SEEDLINGS TO PLANT
C.....USEED(10) - BIRTH- SELECT SEEDLINGS (PARTICULAR SPECIES)
C.....USEED(11) - BIRTH- USED TO CALCULATE DBH FOR SEEDLINGS
C.....
C.....
C.....INITIALIZE SEED FOR RANDOM NUMBER GENERATOR
C.....
      USEED(1) = 73313
      USEED(2) = 85325
      USEED(3) = 27254
      USEED(4) = 91124
      USEED(5) = 57597
      USEED(6) = 17911
      USEED(7) = 79393
      USEED(8) = 35717
      USEED(9) = 71519
      USEED(10) = 17741
      USEED(11) = 91297
      NSEED1 = 31933
      NSEED2 = 36973
      CALL INPUT
      IPLOT = 0
      KLAST = 1
      NYEAR = 600
      NYR1 = NYEAR+1
      KTIMES = 1
      N3 = NSPEC+3
C.....
C.....AM - MEAN VALUE FOR GROWING DEGREE DAYS
```

```
C.....S - STANDARD DEVIATION FOR GROWING DEGREE DAYS
C.....
      AM = 5858.
      S = 655.98
      DO 10 IVA = 1,NYR1
         DO 10 IVS = 1,N3
            A(IVA,IVS) = 0.
   10       CONTINUE
C.....
C.....SOILQ - THE MAXIMUM BIOMASS RECORDED FOR FORESTS IN THE AREA
C.....
C.....
C.....SOILQ IS IN KG PER 1/12 HA
C.....
      SOILQ = 50833.33333
C.....
C.....DEGD - THE GROWING DEGREE-DAYS FOR A SITE USING A 42 DEG F BASE
C.....
      DEGD = AM
C.....
C.....NSPEC - NUMBER OF SPECIES FOR SIMULATION
C.....
   20 CONTINUE
      CALL PLOTIN(IPLOT)
      WRITE(6,9000) IPLOT
      KYR = 0
      CALL OUTPUT(KYR,IPLOT)
      DO 60 JJ=1,KTIMES
         CALL INIT
         DO 50 I=1,NYEAR
            KYR = I
            NCT = NCT+1
            IF (NCT.EQ.2) GO TO 30
            CALL GGNORD(NSEED1,NSEED2,Z)
            GO TO 40
   30       Z(1) = Z(2)
            NCT = 0
   40       DEGD = AM+S*Z(1)
            CALL KILL
            CALL BIRTH(KYR)
            CALL GROW
            CALL OUTPUT(KYR,IPLOT)
   50       CONTINUE
   60    CONTINUE
      IF (IPLOT.NE.KLAST) GO TO 20
C.....
C.....OUTPUT AVERAGED SPECIES BIOMASS, TOTAL BIOMASS, STEMS, LEAF AREA
C.....
      ZNYR = KLAST
      DO 80 IV1 = 1,NYR1
         TIMEX = IV1-1
         WRITE(6,9002) TIMEX
C        WRITE(4,9001) TIMEX
         DO 70 IV2 = 1,N3
            A(IV1,IV2) = A(IV1,IV2)/ZNYR
C           WRITE(4,9001)A(IV1,IV2)
   70       CONTINUE
   80    CONTINUE
      CALL PLT
      STOP
 9000 FORMAT(/,'  PLOT NUMBER  ',I4)
 9001 FORMAT(2E10.3)
 9002 FORMAT(' ',2E10.3)
      END

      SUBROUTINE INPUT
      COMMON/PARAM/AAA(100,6),DMAX(100),DMIN(100),B3(100),B2(100),
     > ITOL(100),AGEMX(100),CURVE(100),G(100),SPRTND(100),SPRTMN(100),
     > SPRTMX(100),SWITCH(100,5),KTIME(100)
      COMMON/CONST/NSPEC,SOILQ,DEGD
      INTEGER SENSIT
      DIMENSION NSELCT(100)
C.....
C.....THE VARIABLE REDUCE AND THE SENSIT ARRAY HAVE BEEN USED IN
C.....     CONNECTION WITH SO2 STUDIES. THE GROWTH RATE OF INDIVIDUAL
C.....     SPECIES CAN BE REDUCED ACCORDING TO SO2 SENSITIVITY.
C.....
C.....NSPEC - NUMBER OF SPECIES
C.....     NSELCT    CONTAINS THE SPECIES NUMBER FOR THOSE SPECIES
C.....     THAT WILL BE USED IN THE SIMULATION
C.....
      READ(5,9000) NSPEC,(NSELCT(I),I=1,NSPEC)
C.....
C.....INPUT INDIVIDUAL SPECIES INFORMATION
C.....AAA - SPECIES NAME
C.....DMAX - MAXIMUM GROWING DEGREE DAYS
C.....DMIN - MINIMUM GROWING DEGREE DAYS
C.....B3 - INDIVIDUAL SPECIES CONSTANT USED IN GROW
C.....B2 - INDIVIDUAL SPECIES CONSTANT USED IN GROW
C.....ITOL - LIGHT TOLERANCE CLASS
C.....AGEMX - MAXIMUM AGE OF SPECIES
C.....CURVE - DENOTES TYPE OF CURVE FOR CALCULATING BIOMASS
C.....     (NOT CURRENTLY USED)
C.....G - SCALES THE GROWTH RATE OF EACH SPECIES
C.....SPRTND - TENDENCY TO STUMP SPROUT
```

```
C.....SPRTMN - MINIMUM SIZE TREE THAT WILL SPROUT
C.....SPRTMX - MAXIMUM SIZE TREE THAT WILL SPROUT
C.....SWITCH - REPRODUCTION SWITCHES USED IN BIRTH
C.....KTIME - DESIGNATES SEED SOURCE LIMITATION FOR CERTAIN SPECIES
C.....NUM - INDIVIDUAL SPECIES NUMBER
C.....SENSIT - DENOTES SENSITIVITY TO SULFUR DIOXIDE
C.....
      J = 1
      REDUCE = 0.
      DO 10 K=1,100
         READ(5,9001) (AAA(J,I),I=1,6),DMAX(J),DMIN(J),B3(J),B2(J),
     >     ITOL(J),AGEMX(J),CURVE(J),G(J),SPRTND(J),SPRTMN(J),SPRTMX(J),
     >     (SWITCH(J,I),I=1,5),KTIME(J),NUM,SENSIT
         IF (NSELCT(J).NE.NUM) GO TO 10
         IF (REDUCE.NE.0.) G(J) = G(J)*(1.00-(SENSIT*REDUCE))
         WRITE(6,9002) (AAA(J,I),I=1,6),DMAX(J),DMIN(J),B3(J),B2(J),
     >     ITOL(J),AGEMX(J),CURVE(J),G(J),SPRTND(J),SPRTMN(J),SPRTMX(J),
     >     (SWITCH(J,I),I=1,5),SENSIT,NUM
         IF (NUM.EQ.NSELCT(NSPEC)) GO TO 20
         J = J+1
   10    CONTINUE
   20 CONTINUE
      RETURN
 9000 FORMAT(40I2)
 9001 FORMAT(6A4,F6.0,F5.0,F4.4,F5.2,I1,F4.0,F2.0,F5.1,F2.0,F4.1,F4.0,
     > 5L1,I4,I5/I1)
 9002 FORMAT(' ',6A4,F6.0,F5.0,F5.4,F5.2,I1,F4.0,F2.0,F5.1,F2.0,F4.1,
     > F4.0,5L1,I4,2I5)
      END

      SUBROUTINE PLOTIN(IPLOT)
      COMMON/FOREST/NTREES(100),DBH(700),IAGE(700),KSPRT(100),
     > NEWTR(100),SUMLA(700),NEW(100),SWTCH(5)
      COMMON/CONST/NSPEC,SOILQ,DEGD
C.....
C.....INITIALIZE VARIABLES TO START SIMULATION ON BARE PLOT
C.....NTREES CONTAINS NUMBER OF TREES FOR EACH SPECIES
C.....DBH CONTAINS DIAMETER AT BREAST HEIGHT FOR EACH TREE
C.....
      IPLOT = IPLOT+1
      DO 10 I=1,NSPEC
         NTREES(I) = 0
         DBH(I) = 0.
   10    CONTINUE
      NSPE1 = NSPEC+1
      DO 20 I=NSPE1,700
         DBH(I) = 0.
   20    CONTINUE
      RETURN
      END

      SUBROUTINE INIT
      COMMON/FOREST/NTREES(100),DBH(700),IAGE(700),KSPRT(100),
     > NEWTR(100),SUMLA(700),NEW(100),SWTCH(5)
      COMMON/CONST/NSPEC,SOILQ,DEGD
      COMMON/DEAD/NOGRO(700),NTEMP(700)
C.....
C.....INITIALIZE VARIABLES TO ZERO FOR REPLICATE PLOTS
C.....
C.....KSPRT IS USED TO FLAG THE TREES THAT CAN SPROUT
C.....NOGRO IS USED TO FLAG THE TREES THAT DON'T GROW
C.....IAGE CONTAINS THE AGE FOR EACH TREE
C.....
C.....
      DO 10 I=1,NSPEC
         NOGRO(I) = 0
         KSPRT(I) = 1
         IAGE(I) = 0
   10    CONTINUE
      NSPE1 = NSPEC+1
      DO 20 I=NSPE1,700
         NOGRO(I) = 0
         IAGE(I) = 0
   20    CONTINUE
      RETURN
      END

      SUBROUTINE RANDOM(NSEED)
      COMMON/RAN/YFL
C.....
C.....SUBROUTINE RANDOM CALLS THE UNIFORM RANDOM NUMBER GENERATOR AND
C.....     RETURNS THE RANDOM NUMBER IN YFL
C.....
      YFL = RANDU(NSEED)
      RETURN
      END

      SUBROUTINE KILL
      COMMON/FOREST/NTREES(100),DBH(700),IAGE(700),KSPRT(100),
     > NEWTR(100),SUMLA(700),NEW(100),SWTCH(5)
      COMMON/PARAM/AAA(100,6),DMAX(100),DMIN(100),B3(100),B2(100),
     > ITOL(100),AGEMX(100),CURVE(100),G(100),SPRTND(100),SPRTMN(100),
     > SPRTMX(100),SWITCH(100,5),KTIME(100)
      COMMON/CONST/NSPEC,SOILQ,DEGD
      COMMON/RAN/YFL
      COMMON/DEAD/NOGRO(700),NTEMP(700)
```

```
      COMMON/COUNT/NTOT,NYEAR
      COMMON/TEMP/DTEMP(700),ITEMP(700)
      COMMON/SEED/USEED(11)
      INTEGER USEED
      LOGICAL SWITCH,SWTCH
      KNT = 0
      DO 30 I=1,NSPEC
         IF (NTREES(I).EQ.0) GO TO 30
         NL = KNT+1
         NU = NTREES(I)+KNT
         DO 20 K=NL,NU
            CALL RANDOM(USEED(1))
C.....
C.....KILL TREES BASED ON PROBABILITY THAT ONLY 1% REACH MAXIMUM AGE
C.....
            IF (YFL.LE.(4.605/AGEMX(I))) GO TO 10
C.....
C.....CHECK TO SEE IF THERE WAS ANY GROWTH FOR TREE
C.....
            IF (NOGRO(K).EQ.0) GO TO 20
            CALL RANDOM(USEED(2))
            IF (YFL.GT.0.368) GO TO 20
   10       CONTINUE
            NTREES(I) = NTREES(I)-1
C.....
C.....CHECK TO SEE IF DEAD TREE CAN STUMP SPROUT. SET KSPRT = -1
C.....IF TREE CAN SPROUT
C.....
            IF (DBH(K).GT.SPRTMN(I).AND.DBH(K).LT.SPRTMX(I)) KSPRT(I) =
     >       -1
            DBH(K) = -1.0
   20       CONTINUE
         KNT = NU
   30    CONTINUE
C.....
C.....REWRITE DIAMETERS AND AGES TO ELIMINATE DEAD TREES
C.....
      K = 0
      DO 40 I=1,700
         IF (DBH(I).EQ.0.) GO TO 50
         IF (DBH(I).LT.0.) GO TO 40
         K = K+1
         DBH(K) = DBH(I)
         IAGE(K) = IAGE(I)
         NOGRO(K) = NOGRO(I)
   40    CONTINUE
   50 NTOT = K
      IF (NTOT.EQ.0) RETURN
      NTOT1 = K+1
      DO 60 I=NTOT1,NU
         DBH(I) = 0.
         IAGE(I) = 0
         NOGRO(I) = 0
   60    CONTINUE
      RETURN
      END

      SUBROUTINE SPROUT
      COMMON/FOREST/NTREES(100),DBH(700),IAGE(700),KSPRT(100),
     > NEWTR(100),SUMLA(700),NEW(100),SWTCH(5)
      COMMON/PARAM/AAA(100,6),DMAX(100),DMIN(100),B3(100),B2(100),
     > ITOL(100),AGEMX(100),CURVE(100),G(100),SPRTND(100),SPRTMN(100),
     > SPRTMX(100),SWITCH(100,5),KTIME(100)
      COMMON/CONST/NSPEC,SOILQ,DEGD
      COMMON/RAN/YFL
      COMMON/DEAD/NOGRO(700),NTEMP(700)
      COMMON/COUNT/NTOT,NYEAR
      COMMON/TEMP/DTEMP(700),ITEMP(700)
      COMMON/SEED/USEED(11)
      INTEGER USEED
      LOGICAL SWITCH,SWTCH
C.....
C.....SMALLEST AVERAGE STUMP SPROUT IS .1 CM
C.....
      SIZE = .1
C.....
C.....SUM TOTAL NUMBER OF TREES
C.....
      NTOT = 0
      DO 10 I=1,NSPEC
         IF (NTREES(I).EQ.0) GO TO 10
         NTOT = NTOT+NTREES(I)
   10    CONTINUE
C.....
C.....DETERMINE WHICH SPECIES CAN SPROUT
C.....
      NW = 0
      DO 20 I=1,NSPEC
         IF (SPRTND(I).LE.0.) GO TO 20
         IF (KSPRT(I).GE.0) GO TO 20
         NW = NW+1
         NEW(NW) = I
   20    CONTINUE
C.....
C.....CHECK FOR SPROUTS
```

```
C.....
      IF (NW.EQ.0) GO TO 90
      DO 30 J=1,NTOT
         ITEMP(J) = IAGE(J)
         DTEMP(J) = DBH(J)
         NTEMP(J) = NOGRO(J)
   30    CONTINUE
C.....
C.....CHOOSE RANDOM NUMBER OF SPROUTS
C.....
      CALL RANDOM(USEED(3))
      NW = NW*YFL+1.0
C.....
C.....SELECT SPECIES TO SPROUT
C.....
      NSPC = NEW(NW)
      NSUM = 0
      DO 40 I=1,NSPC
   40    NSUM = NSUM+NTREES(I)
C.....
C.....SPRTND IS THE TENDENCY FOR THE ITH SPECIES TO STUMP OR
C.....ROOT SPROUT. THE VALUE OF SPRTND IS THE AVERAGE NUMBER
C.....OF SPROUTS THAT MIGHT OCCUR WITH A TREE DEATH
C.....
      CALL RANDOM(USEED(4))
C.....
C.....RANDOMLY SELECT NUMBER OF TREES TO SPROUT
C.....
      NSPRT = YFL*SPRTND(NSPC)+1
      NL = NSUM+1
      NUP = NTOT
      DO 50 I=1,NSPRT
         NSUM = NSUM+1
         NTREES(NSPC) = NTREES(NSPC)+1
         NTOT = NTOT+1
         IF (NTOT.GT.700) CALL ERR
         ITEMP(NSUM) = 0
         CALL RANDOM(USEED(5))
         DTEMP(NSUM) = SIZE+.1*(1.0-YFL)**3
         NTEMP(NSUM) = 0
C.....
C.....STORE DIAMETERS AND AGES FOR NEW SPROUTS
C.....
   50    CONTINUE
      IF (NL.GT.NUP) GO TO 70
      N1 = NSUM+1
      DO 60 J=NL,NUP
         DTEMP(N1) = DBH(J)
         ITEMP(N1) = IAGE(J)
         NTEMP(N1) = NOGRO(J)
         N1 = N1+1
   60    CONTINUE
C.....
C.....REINITIALIZE ORIGINAL DIAMETERS AND AGES
C.....
   70 DO 80 L=1,NTOT
         IAGE(L) = ITEMP(L)
         NOGRO(L) = NTEMP(L)
   80    DBH(L) = DTEMP(L)
   90 CONTINUE
C.....
C.....REINITIALIZE SPROUT SWITCH FOR EACH SPECIES
C.....
      DO 100 I=1,NSPEC
  100    KSPRT(I) = 1
      RETURN
      END

      SUBROUTINE GROW
      COMMON/FOREST/NTREES(100).DBH(700).IAGE(700).KSPRT(100).
     > NEWTR(100),SUMLA(700),NEW(100),SWTCH(5)
      COMMON/PARAM/AAA(100,6),DMAX(100),DMIN(100),B3(100),B2(100),
     > ITOL(100),AGEMX(100),CURVE(100),G(100),SPRTND(100),SPRTMN(100),
     > SPRTMX(100),SWITCH(100,5),KTIME(100)
      COMMON/CONST/NSPEC,SOILQ,DEGD
      COMMON/RAN/YFL
      COMMON/DEAD/NOGRO(700),NTEMP(700)
      COMMON/COUNT/NTOT,NYEAR
      COMMON/TEMP/DTEMP(700),ITEMP(700)
      LOGICAL SWITCH,SWTCH
C.....
C.....DBH IS IN CENTIMETERS
C.....EACH TREE IS REQUIRED TO ADD A 1.0 MM GROWTH RING EACH YEAR
C.....
      TINC = .1
      PHI = 1.
C.....
C.....CALCULATE TOTAL NUMBER OF TREES
C.....
      NTOT = 0
      DO 10 I=1,NSPEC
   10    NTOT = NTOT+NTREES(I)
      IF (NTOT.EQ.0) RETURN
C.....
C.....SUM LEAF AREA OF ALL TREES THAT ARE OF APPROXIMATELY
```

```
C.....THE SAME HEIGHT
C.....
      DO 20 I=1,700
         NOGRO(I) = 0
   20    SUMLA(I) = 0.
      NL = 1
      SBIO = 0.
      DO 40 J=1,NSPEC
         IF (NTREES(J).EQ.0) GO TO 40
         NU = NL+NTREES(J) -1
         DO 30 K=NL,NU
C.....
C.....CALCULATE STAND BIOMASS
C.....
            SBIO = SBIO+.1193*DBH(K)**2.393
C.....
C.....HEIGHT PROFILE IS CALCULATED IN .1 METER UNITS
C.....
            IHT = (B2(J)*DBH(K)-B3(J)*DBH(K)**2)/10.+1.
            IF (IHT.GT.700) GO TO 90
            SUMLA(IHT) = SUMLA(IHT)+1.9283295E-4*DBH(K)**2.129
   30       CONTINUE
         NL = NL+NTREES(J)
   40    CONTINUE
      DO 50 J=1,699
         J1 = 700-J
         SUMLA(J1) = SUMLA(J1)+SUMLA(J1+1)
   50    CONTINUE
C.....
C.....CALCULATE AMOUNT OF GROWTH FOR EACH TREE
C.....
      NL = 1
      DO 80 I=1,NSPEC
         IF (NTREES(I).EQ.0) GO TO 80
         NU = NL+NTREES(I)-1
         DO 70 J=NL,NU
            HT = B2(I)*DBH(J)-B3(I)*DBH(J)**2
            IHT = HT/10.+2.
            SLAR = SUMLA(IHT)
            AL = PHI*EXP(-SLAR*.25)
            GR = (137.+.25*B2(I)**2/B3(I))*(0.5*B2(I)/B3(I))
            DNC = G(I)*DBH(J)*(1.0-(137.*DBH(J)+B2(I)*DBH(J)**2-B3(I) *
     >       DBH(J)**3)/GR)/(274.+3.0*B2(I)*DBH(J)-4.0*B3(I)*DBH(J)**2)
     >       *(1.0-SBIO/SOILQ)*4.0*(DEGD-DMIN(I))*(DMAX(I)-DEGD)/
     >       (DMAX(I)-DMIN(I))**2
            DINC = 2.24*(1.-EXP(-1.136*(AL-.08)))*DNC
C.....
C.....CHECK FOR SHADE-TOLERANT
C.....
            IF (ITOL(I).LT.2) DINC=(1.0-EXP(-4.64*(AL-.05)))*DNC
C.....
C.....CHECK INCREMENT LESS THAN 1.0 MM REQUIRED GROWTH
C.....
            IF (DINC.LT.TINC) DINC = 0.0
            IF (DINC.NE.0.) GO TO 60
            NOGRO(J) = -1
   60       DBH(J) = DBH(J)+DINC
   70       CONTINUE
         NL = NL+NTREES(I)
   80    CONTINUE
      RETURN
   90 WRITE(6,9000)
      STOP
 9000 FORMAT('1 IHT EXCEEDED 700')
      END

      SUBROUTINE BIRTH(KYR)
      COMMON/FOREST/NTREES(100),DBH(700),IAGE(700),KSPRT(100),
     > NEWTR(100),SUMLA(700),NEW(100),SWTCH(5)
      COMMON/PARAM/AAA(100,6),DMAX(100),DMIN(100),B3(100),B2(100),
     > ITOL(100),AGEMX(100),CURVE(100),G(100),SPRTND(100),SPRTMN(100),
     > SPRTMX(100),SWITCH(100,5),KTIME(100)
      COMMON/CONST/NSPEC,SOILQ,DEGD
      COMMON/RAN/YFL
      COMMON/DEAD/NOGRO(700),NTEMP(700)
      COMMON/COUNT/NTOT,NYEAR
      COMMON/TEMP/DTEMP(700),ITEMP(700)
      COMMON/SEED/USEED(11)
      INTEGER USEED
      LOGICAL SWITCH,SWTCH
C.....
C.....SAPLINGS ENTER THE PLOT AT AVERAGE SIZE OF 1.27 CM DBH
C.....
      SIZE = 1.27
      CALL RANDOM(USEED(6))
C.....
C.....SELECT A SPECIES FROM 1 TO 3 TIMES
C.....
      NPLANT = 3.*YFL+1.
      CALL RANDOM(USEED(7))
C....
C.... RAT  RANDOM NUMBER IS FIXED FOR YEAR - USED IN SWITCH 4
C....
      RAT = YFL
      DO 140 JK=1,NPLANT
```

```
   10     AREA = 0.
          NL = 1
C.....
C.....CALCULATE LEAF AREA FOR EACH SPECIES
          DO 30 J=1,NSPEC
              IF (NTREES(J).EQ.0) GO TO 30
              NU = NL+NTREES(J)-1
              DO 20 K=NL,NU
                  AREA = AREA+1.9283295E-4*DBH(K)**2.129
   20             CONTINUE
              NL = NL+NTREES(J)
   30         CONTINUE
C.....
C.....TOTAL NUMBER OF TREES IN STAND
C.....
          NTOT = NL-1
C.....
C.....SWITCH 1 IS TRUE IF THE SPECIES REQUIRES LEAF LITTER FOR
C.....SUCCESSFUL RECRUITMENT
C.....SWITCH 2 IS TRUE IF THE SPECIES REQUIRES MINERAL SOIL
C.....SWITCH 3 IS TRUE IF THE SPECIES RECRUITMENT IS REDUCED BY HOT YEA
C.....SWITCH 4 IS TRUE IF THE SPECIES IS A PREFERRED FOOD OF DEER
C.....OR SMALL MAMMALS
C.....SWITCH 5 REDUCES SEEDING RATE OF DESIRABLE MAST
          DO 40 J=1,5
              SWTCH(J) = .TRUE.
   40         CONTINUE
C.....SET SWITCHES BASED ON VALUE OF BIOMASS, DEGD, AND RANDOM NUMBER
C.....
          IF (AREA.GE..1) SWTCH(1) = .FALSE.
          IF (AREA.LE..2) SWTCH(2) = .FALSE.
          IF (DEGD.LE.5858.) SWTCH(3) = .FALSE.
          IF (RAT.GT..5) SWTCH(4) = .FALSE.
          CALL RANDOM(USEED(8))
          IF (YFL.GE..5) SWTCH(5) = .FALSE.
C.....
C.....CHECK SWITCHES TO DETERMINE IF SEEDING IS ALLOWED
C.....
          NW = 0
          DO 60 J=1,NSPEC
              DO 50 K=1,5
                  IF (SWITCH(J,K).AND.SWTCH(K)) GO TO 60
   50             CONTINUE
              IF (KYR.GT.KTIME(J).AND.NTREES(J).LE.0) GO TO 60
              NW = NW+1
              NEWTR(NW) = J
   60         CONTINUE
C.....
C.....CHECK TO SEE IF THERE ARE ANY NEW TREES
C.....
          IF (NW.EQ.0) GO TO 130
C.....
C.....CALCULATE AGE AND DIAMETER FOR NEW TREES
C.....
          DO 70 I=1,NTOT
              ITEMP(I) = IAGE(I)
              DTEMP(I) = DBH(I)
              NTEMP(I) = NOGRO(I)
   70         CONTINUE
C.....
C.....DETERMINE THE NUMBER OF SEEDLINGS TO PLANT  0 TO 8
C.....
          CALL RANDOM(USEED(9))
          MPLANT = 8.*YFL
C.....
C.....SELECT SEEDLINGS (PARTICULAR SPECIES)
C.....
          CALL RANDOM(USEED(10))
          NW = NW*YFL+1.0
          NSP = NEWTR(NW)
          NSUM = 0
          DO 80 I=1,NSP
   80         NSUM = NSUM+NTREES(I)
C.....
C.....PLANT RANDOM NUMBER OF SEEDLINGS
C.....
          NL = NSUM+1
          NUP = NTOT
          DO 90 J=1,MPLANT
              NTOT = NTOT+1
              IF (NTOT.GT.700) CALL ERR
              NSUM = NSUM+1
              NTREES(NSP) = NTREES(NSP)+1
              ITEMP(NSUM) = 0
C.....
              CALL RANDOM(USEED(11))
              DTEMP(NSUM) = SIZE+.3*(1.0-YFL)**3
              NTEMP(NSUM) = 0
   90         CONTINUE
          IF (NL.GT.NUP) GO TO 110
          N1 = NSUM+1
          DO 100 L=NL,NUP
              DTEMP(N1) = DBH(L)
              ITEMP(N1) = IAGE(L)
              NTEMP(N1) = NOGRO(L)
```

```
              N1 = N1+1
  100         CONTINUE
C.....
C.....REINITIALIZE ORIGINAL DBH AND AGE ARRAYS - INCLUDING NEW TREES
C.....
C.....
  110     DO 120 I=1,NTOT
              IAGE(I) = ITEMP(I)
              DBH(I) = DTEMP(I)
              NOGRO(I) = NTEMP(I)
  120         CONTINUE
  130     CONTINUE
          CALL SPROUT
        IF(NW .NE. 0  .AND.AREA .LT. 0.2) GO TO 10
  140     CONTINUE
C.....
C.....INCREMENT AGES
C.....
        DO 150 I=1,NTOT
          IAGE(I) = IAGE(I)+1
  150     CONTINUE
        RETURN
        END

        SUBROUTINE OUTPUT(KYR,IPLOT)
        COMMON/FOREST/NTREES(100),DBH(700),IAGE(700),KSPRT(100),
       > NEWTR(100),SUMLA(700),NEW(100),SWTCH(5)
        COMMON/PASSX/A(1101,36)
        COMMON/COUNT/NTOT,NYEAR
        COMMON/CONST/NSPEC,SOILQ,DEGD
        DIMENSION BAR(100)
C       DIMENSION TYR(1101)
C.....
C.....SUBROUTINE OUTPUT CALCULATES AND STORES SPECIES BIOMASS, TOTAL
C.....     BIOMASS,TOTAL NUMBER OF STEMS, AND LEAF AREA
C.....BIOMASS IS ORIGINALLY CALCULATED IN KG PER 1/12 HA AND THEN
C.....      CONVERTED TO METRIC TONS/HA (.012).
C.....
        KYR1 = KYR+1
        AREA=0.0
        TBAR=0.0
        NTOT = 0
        NL = 1
        DO 20 I =1,NSPEC
          BAR(I)=0.0
          IF (NTREES(I).EQ.0) GO TO 20
          NU = NL+NTREES(I)-1
          DO 10 J=NL,NU
            BAR(I) = BAR(I)+.1193*DBH(J)**2.393
            AREA=AREA+1.9283295E-4*DBH(J)**2.129
   10       CONTINUE
          TBAR=TBAR+BAR(I)
          NL = NU+1
          NTOT = NTOT +NTREES(I)
   20     CONTINUE
C.....J.C.L.CHANGES MUST BE MADE TO USE THE FOLLOWING WRITE STATEMENT
        TBAR=TBAR*.012
        DO 30 IV1 = 1,NSPEC
          BAR(IV1) = BAR(IV1)*.012
          A(KYR1,IV1) = A(KYR1,IV1) +BAR(IV1)
   30     CONTINUE
        ATOT=NTOT
        N1 = NSPEC+1
        N2 = N1+1
        N3 = N2+1
        A(KYR1,N1)=A(KYR1,N1)+TBAR
C       TYR(KYR1) = TBAR
        A(KYR1,N2)=A(KYR1,N2)+ATOT
        A(KYR1,N3)=A(KYR1,N3)+AREA
C       IF(KYR1 .EQ. 1101) WRITE(9,2000) (TYR(I),I=1,1101)
C2000   FORMAT(8E10.3)
C       WRITE(9) (BAR(IV1),IV1=1,NSPEC),(NTREES(IV1),IV1=1,NSPEC),
C      1TBAR,ATOT,AREA
        RETURN
        END

        SUBROUTINE ERR
        WRITE(6,9000)
        STOP
 9000   FORMAT('1 THE NUMBER OF TREES HAS EXCEEDED 700')
        END

        FUNCTION RANDU(SEED)
        REAL*4 RNEW
        INTEGER PLUS/Z1000000/,IBIT/Z40000000/,HEX/Z10003/
        INTEGER OLD,NEW,SEED
        LOGICAL*1 NORMAL/.FALSE./
        EQUIVALENCE(NEW,RNEW)
CCCCCCCC
C.....
C.....  THIS SUBROUTINE CALCULATES UNIFORM RANDOM NUMBERS
C.....
CCCCCCCC
C   RANDON NUMBER GENERATOR
```

```
      OLD=HEX*SEED
      SEED = IABS(OLD)
      NEW=OLD/256
      IF (NEW.LE.0) NEW=NEW+PLUS
      NEW=NEW+IBIT
      RANDU = RNEW+0.0
      RETURN
      END

      SUBROUTINE GGNORD(NSEED1,NSEED2,Z)
      DIMENSION Z(1)
      DATA PI2/0.62831853E01/
C.....
C.....CALCULATES NORMALLY DISTRIBUTED RANDOM NUMBERS
C.....
      K = 0
      A1 = RANDU(NSEED1)
      A2 = RANDU(NSEED2)
      K = K+1
      Z(K) = SQRT(-.2E01*ALOG(A1))*SIN(PI2*A2)
      K = K+1
      Z(K) = SQRT(-0.2E01*ALOG(A1))*COS(PI2*A2)
      RETURN
      END

      SUBROUTINE PLT
      COMMON/CONST/NSPEC,SOILQ,DEGD
      COMMON/PASSX/A(1101,36)
      COMMON/COUNT/NTOT,NYEAR
      DIMENSION X(1101),Y(1101),IXVAR(37),IYVAR(37)
C.....
C.....SUBROUTINE PLT PLOTS THE PERCENTAGE BIOMASS FOR SELECTED SPECIES
C.....
      YR = NYEAR
      NYEAR = NYEAR+1
      NU = NYEAR-4
      NPTS = NYEAR-8
      READ(5,9000) NN
      DO 10 I=1,NN
         READ(5,9000) IXVAR(I),IYVAR(I)
   10    CONTINUE
      N1 = NSPEC+1
      DO 20 IQ = 2,NYEAR
         A(IQ,IYVAR(1)) = 100.*(1.-A(IQ,IYVAR(1))/A(IQ,N1))
   20    CONTINUE
      DO 30 IQ = 2,NYEAR
         DO 30 IR = 2,NN
            IB = IR-1
            A(IQ,IYVAR(IR)) = A(IQ,IYVAR(IB))-100.*A(IQ,IYVAR(IR))/A(IQ,
     >         N1)
   30       CONTINUE
      CALL ESDPLT(7,'THARP',1,1)
      CALL BGNPL(-1)
      CALL PAGE(14.,11.)
      CALL TITLE(' ',1,'YEARS',5,'PERCENT OF TOTAL BIOMASS$',100,9.,10.)
      CALL FRAME
      CALL GRAF(0.,'SCALE',YR,0.,'SCALE',100.)
      DO 60 II = 1,NN
         DO 50 IJK = 5,NU
            IJK1 = IJK-4
            IJK9 = IJK+4
            X(IJK1) = IJK-1
            SUM = 0.
            DO 40 MMM = IJK1,IJK9
               SUM = SUM+A(MMM,IYVAR(II))
   40          CONTINUE
            Y(IJK1) = SUM/9.
   50       CONTINUE
         Y(1) = Y(2)
         CALL CURVE(X,Y,NPTS,0)
   60    CONTINUE
      CALL ENDPL(1)
      CALL DONEPL
      STOP
 9000 FORMAT(2I5)
      END
```

References

Aber, J. D., D. B. Botkin, and J. M. Melillo. 1978. Predicting the effects of differing harvesting regimes on forest floor dynamics in northern hardwoods. Can. J. For. Res. 8:306–315.

Aber, J. D., D. B. Botkin, and J. M. Melillo. 1979. Predicting the effects of different harvesting regimes on productivity and yield in northern hardwoods. Can. J. For. Res. 9:10–14.

Aber, J. D., and J. M. Mellilo. 1980. Litter decomposition: Measuring relative contributions of organic matter and nitrogen to forest soils. Can. J. Bot. 58:416–421.

Aber, J. D., and J. M. Melillo. 1982a. FORTNITE: A computer model of organic matter and nitrogen dynamics in forest ecosystems. University of Wisconsin Research Bulletin. R3130.

Aber, J. D., and J. M. Melillo. 1982b. Nitrogen immobilization in decaying hardwood leaf litter as a function of initial nitrogen and lignin content. Can. J. Bot. 60:2263–2269.

Acevedo, M. F. 1978. On a Markovian model of forest succession: Its application to tropical forests. Department of Electrical Engineering and Computer Science, University of California, Berkeley. M.E. thesis.

Adee, K. T., and H. B. Wood. Regeneration and succession following canopy dieback in a *Metrosideros* rain forest on the island of Hawaii. (Unpublished manuscript.)

Adem, J. 1981. Numerical simulation of the annual cycle of climate during the ice ages. J. Geophys. Res. 86:12015–12034.

Albrektson, A. 1980. Total tree production as compared to conventional forestry production. IN T. Persson (ed.), Structure and Function of Northern Coniferous Forests—An Ecosystem Study. Ecol. Bull (Stockholm) 32:315–327.

Alexander, C. 1964. Notes on the Synthesis of Form. Harvard University Press, Cambridge. 216 pp.

Allen, T. F. H., S. M. Bartell, and J. F. Koonce. 1977. Multiple stable configurations in ordination of phytoplankton community change rates. Ecology 58:1076–1084.

Allen, T. F. H., and T. B. Starr. 1982. Hierarchy: Perspectives for Ecological Complexity. University of Chicago Press, Chicago. 310 pp.

Anderson, H. E. 1968. Sundance fire: An analysis of fire phenomena. U.S.D.A. Forest Service Intermountain Forest and Range Experimental Station, Ogden, Utah. 43 pp.

Anderson, M. C. 1971. Radiation and crop structure. pp. 412–466. IN Z. Sestak, J. Catsky, and P. G. Jarvis (eds.), Plant Photosynthetic Production. Dr. W. Junk, The Hague.

Anderson, S. T. 1970. The relative pollen productivity and pollen representation of north European trees, and correction factors for tree pollen spectra. Dan. Geol. Unders. Afh. Raekke 2 (96):99p.

Anderson, S. T. 1973. The differential pollen productivity of trees and its significance for the interpretation of a pollen diagram from a forested region. pp. 109–115. IN H. J. B. Birks and R. G. West (eds.), Quaternary Plant Ecology. Blackwell, Oxford.

Andrewartha, H. G., and L. C. Birch. 1954. The Distribution and Abundance of Animals. University of Chicago Press, Chicago. 782 pp.

Anonymous 1973. A resource and management survey of the Cotter River catchment. Report prepared for the Forest Branch, Department of Capital Territory, by the Consultancy Group, Department of Forestry, Australian National University, Canberra.

Arney, J. D. 1971. Computer simulation of Douglas-fir tree and stand growth. School of Forestry, Oregon State University, Corvallis. Ph.D. thesis. 79 pp.

Arney, J. D. 1974. An individual tree model for stand simulation in Douglas-fir. p. 38–46. IN J. Fries (ed.), Growth Models for Tree and Stand Simulation. Department of Forest Yield Research, Royal College of Forestry, Stockholm. Res. Notes 30.

Ash, J. B. 1981. The *Nothofagus* Blume (Fagaceae) on New Guinea. pp. 355–397. IN J. L. Gressitt (ed.), Ecology and Biogeography of New Guinea. Dr. W. Junk, The Hague.

Ashe, W. W. 1897. Forests of North Carolina. North Carolina Geol. Surv. Bull. 6:139–224.

Ashe, W. W. 1911. Chestnut in Tennessee. Tennessee Geol. Surv. Ser. 10B. 35 pp.

Ashton, D. H. 1970. Fire and vegetation. pp. 1–6. IN The Second Fire Ecology Symposium. Monash University Pub. No. 28. Forest Commission, Victoria, Australia.

Ashton, P. S. 1969. Speciation among tropical forest trees: Some deductions in the light of recent evidence. Biol. J. Linn. Soc. 1:155–196.

Atkinson, I. A. E. 1970. Successional trends in the coastal and lowland forest of Mauna Loa and Kilauea volcanoes, Hawaii. Pacific Sci. 24:387–400.

Attiwill, P. M. 1979. Nutrient cycling in a *Eucalyptus obliqua* (L'Herit.) forest. III. Growth, biomass, and net primary production. Aust. J. Bot. 27:439–458.

Aubreville, A. 1933. La foret de la Cote d'Ivoire. Bull. Comm. Afr. Occid. Franc. 15:205–261.

Aubreville, A. 1938. La foret colonial: Les forets de l'Afrique occidentale francaise. Ann. Acad. Sci. Colon., Paris, 9:1–245.

Auclair, A. N., and G. Cottam. 1971. Dynamics of black cherry (*Prunus serotina* Erhr.) in Southern Wisconsin oak forests. Ecol. Monogr. 41:153–177.

Bacastow, R. B., and C. D. Keeling. 1973. Atmospheric carbon dioxide and radiocarbon in the natural carbon cycle: II. Changes from A.D. 1700 to 2070 as deduced from a geochemical model. pp. 86–135. IN G. M. Woodwell and E. V. Pecan (eds.), Carbon and the Biosphere. CONF-720510. National Technical Information Service, Springfield, Virginia.

Bacastow, R. B., and C. D. Keeling. 1981. Atmospheric carbon dioxide concentration and the observed airborne fraction. pp. 103–112. IN B. Bolin (ed.), Carbon Cycle Modeling—SCOPE 16. John Wiley, New York.

Baker, F. S. 1949. A revised tolerance table. J. For. 47:179–181.

Baker, H. G. 1970. Evolution in the tropics. Biotropica 2:101–111.

Ballard, L. A. T. 1963. Physical barriers to germination. Seed Sci. Technol. 1:285–303.

Barden, L. S., and F. W. Woods. 1973. Characteristics of lightning fires in southern Appalachian forests. Proc. Annu. Tall Timber Fire Ecol. Conf. 13:345–361.

Barney, R. J. 1969. Interior Alaska wildfires, 1956–1965. Pacific Northwest Forest and Range Experimental Station, U.S.D.A. Forest Service Institute of Northern Forestry, Juneau, Alaska.

Bassett, J. R. 1964. Tree growth as affected by soil moisture availability. Soil Sci. Proc. 28:436–438.

Bazzaz, F. A. 1979. The physiological ecology of plant succession. Annu. Rev. Ecol. Syst. 10:351–372.

Bazzaz, F. A., and S. T. A. Pickett. 1980. Physiological ecology of tropical succession: a comparative review. Annu. Rev. Ecol. Syst. 11:287–310.

Beard, J. S. 1944. Climax vegetation in tropical America. Ecology 25:127–158.

Beard, J. S. 1955. The classification of tropical American vegetation types. Ecology 36:89–100.

Bedinger, M. S. 1971. Forest species as indicators of flooding in the lower White River Valley, Arkansas. U.S. Geol. Surv. Prof. Pap. 750-C:C248–253.

Bega, V. 1974. *Phytophthora cinnamomi*: Its distribution and possible role in 'ohi'a decline on the island of Hawaii. Plant Dis. Reporter 58:1069–1073.

Bell, C. R. 1970. Seed distribution and germination experiment. pp. D-177—182. IN H. T. Odum and R. F. Pigeon (eds.), A Tropical Rain Forest, Book 2. TID-24270. National Technical Information Service, Springfield, Virginia.

Bella, I. E. 1970. Simulation of growth, yield and management of aspen. Faculty of Forestry, University of British Columbia, Vancouver. Ph.D. thesis. 190 pp.

Bernabo, J. C., and T. Webb III. 1977. Changing patterns in the Holocene pollen record of northeastern North America: A mapped summary. Quat. Res. 8:64–96.

Bierzychudek, P. 1982. Life histories and demography of shade-tolerant temperate forest herbs: A review. New Phytol. 90:757–776.

Birks, H. J. B. 1976. Late-Wisconsin vegetational history at Wolf Creek, Central Minnesota. Ecol. Monogr. 46:395–429.

Biswell, H. H. 1967. Forest fire in perspective. Proc. Annu. Tall Timbers Fire Ecol. Conf. 7:43–64.

Biswell, H. H. 1974. Effects of fire on chaparral. pp. 321–365. IN T. T. Kozlowski and C. E. Ahlgren (eds.), Fire and Ecosystems. Academic Press, New York.

Bledsoe, L. J. 1976. Linear and nonlinear approaches for ecosystem dynamic modeling. pp. 283–298. IN B. C. Patten (ed.), Systems Analysis and Simulation in Ecology, Vol. 4. Academic Press, New York.

Bolin, B. (ed.). 1981. Carbon Cycling Modeling—SCOPE 16. John Wiley, New York.

Bormann, F. H., and G. E. Likens. 1979a. Pattern and Process in a Forested Ecosystem. Springer-Verlag, New York. 253 pp.

Bormann, F. H., and G. E. Likens. 1979b. Catastrophic disturbance and the steady state in northern hardwood forests. Am. Sci. 67:660–669.

Bosch, C. A. 1971. Redwoods: A population model. Science 162:345–349.

Botkin, D. B., J. F. Janak, and J. R. Wallis. 1972a. Some ecological consequences of a computer model of forest growth. J. Ecol. 60:849–873.

Botkin, D. B., J. F. Janak, and J. R. Wallis. 1972b. Rationale, limitations and assumptions of a northeastern forest growth simulator. IBM J. Res. Develop. 16:101–116.

Box, E. O. 1981. Macroclimate and Plant Forms: An Introduction to Predictive Modeling in Phytogeography. Dr. W. Junk, The Hague. 258 pp.

Bradley, R. 1725. Dictionarie oeconomique: or, the family dictionary,etc. Done into English from the second edition by M. Chomel: With considerable alterations and improvements; in two volumes, Volume II. From H to Z. Printed for D. Midwinter, at the Three Crowns in St. Paul's Church-Yard, London.

Bradshaw, R. H. W., and T. Webb III. 1983. Pollen/tree relationships in Wisconsin and Michigan, U.S.A. (unpublished manuscript).

Braun, E. L. 1950. Deciduous Forests of Eastern North America. Blakiston Co., Philadelphia. 596 pp.

Bray, J. R. 1956. Gap-phase replacement in a maple-basswood forest. Ecology 37:598–600.

Brian, P. W. 1949. The production of antibiotics by microorganisms in relation to biological equilibria in the soil. Symp. Soc. Exp. Biol. 3:357–372.

Brown, M. J., and F. D. Podger. 1982. Short note on the apparent anomaly between observed and predicted vegetation types in southwest Tasmania. Aust. J. Ecol. 7:203–205.

Brown, R. T., and J. R. Roti. 1963. The "Solidago factor" in jack pine seed germination. Bull. Ecol. Soc. 44:113.

Brunig, E. F. 1973. Some further evidence on the amount of damage attributed to lightning and wind-throw in *Shorea albida* forest in Sarawak. Commonwealth Forestry Review 52:260–265.

Bryson, R. A., and J. A. Dutton. 1961. Some aspects of the variance spectra of tree rings and varves. Ann. N.Y. Acad. Sci. 95:580–604.

Bryson, R. A., and W. M. Wendland. 1967. Tentative climatic patterns from late-glacial and post-glacial episodes in central North America. pp. 271–298. IN W. J. Mayer-Oakes (ed.), Life, Land, and Water. University of Manitoba Press, Winnipeg.

Bryson, R. A., and J. E. Kutzbach. 1974. On the analysis of pollen-climate canonical transfer functions. Quat. Res. 4:162–174.

Burbidge, N. T., and M. Gray. 1976. Flora of the Australian Capital Territory. Australian National University Press, Canberra.

Burger, H. 1953. Holtz, Blattmenge and Zwachs. Fichten im gleichalterigen Hochwald. Mitt. Schweiz. Anst. Forstl. Versuchsw 29:38–130.

Burgess, I. P., A. Floyd, J. Kikkawa, and V. Pattemore. 1975. Recent developments in the silviculture and management of sub-tropical rain forest in New South Wales. Proc. Ecol. Soc. Aust. 9:74–84.

Burgess, R. L. 1981. United States. pp. 67–104. IN E. J. Kormondy, and J. F. McCormick (eds.), Handbook of Contemporary Developments in World Ecology. Greenwood Press, Westport, Connecticut.

Burton, P. J. 1980. Light regimes and *Metrosideros* regeneration in a Hawaiian montane rain forest. Botany Department, University of Hawaii, Honolulu, M.Sc. thesis. 378 pp.

Cairns, J., M. Dahlberg, M. Dickson, K. L. Smith, and N. T. Waller. 1969. The relationship of freshwater protozoan communities to the MacArthur-Wilson equilibrium model. Am. Nat. 103:430–454.

Capen, D. E. 1981. The use of multivariate statistics in studies of wildlife habitat. General Technical Report RM-87. Rocky Mountain Forest and Range Experiment Station, U.S.D.A. Forest Service, Fort Collins, Colorado. 249 pp.

Cattelino, P. J., I. R. Noble, R. O. Slatyer, and S. R. Kessell. 1979. Predicting the multiple pathways of plant succession. Environ. Manag. 3:41–50.

Chase, C. D., R. E. Pfeifer, and J. S. Spence, Jr. 1970. The growing timber resource of Michigan. U.S.D.A. Forest Service Res. Bull. NC-9. 62 pp.

Chew, R. M. 1974. Consumers as regulators of ecosystems: An alternative to energetics. Ohio J. Sci. 74:359–370.

Chittenden, A. K. 1905. Forest conditions of Northern New Hampshire. Bull. Bur. For. U.S. Dept. Agric. 55. 100 pp.

Clarke, F. L. 1875. Decadence of Hawaiian forest. All About Hawaii 1:19–20.

Clements, F. E. 1916. Plant succession: An analysis of the development of vegetation. Carnegie Inst. Pub. 242. Washington, D.C. 512 pp.

Clements, F. E. 1928. Plant Succession and Indicators. Wilson, New York. 953 pp.

Clements, F. E. 1936. Nature and structure of the climax. J. Ecol. 24:252–284.

Clutter, J. L. 1963. Compatible growth and yield models for loblolly pine. For. Sci. 9:354–371.

Codwell, R. K., and D. J. Futuyma. 1971. On the measurement of niche breadth and overlap. Ecology 52:567–576.

Cody, M. J. 1968. On methods of resource division in grassland bird communities. Am. Nat. 102:107–147.

Cogbill, C. V., and G. E. Likens. 1974. Acid precipitation in the Northeastern United States. Water Resour. Res. 10:1133–1137.

Cole, H. S. 1969. Objective reconstruction of the paleoclimatic record through application of eigenvectors of the present-day pollen spectra and climate to the late-Quaternary pollen stratigraphy. University of Wisconsin, Madison. Ph.D. dissertation.

Cole, L. C. 1960. Competitive exclusion. Science 312:348–349.

Conklin, H. C. 1954. An ethnoecological approach to shifting agriculture. N.Y. Acad. Sci. 17:133–142.

Connell, J. H., and R. O. Slatyer. 1977. Mechanisms of succession in natural communities and their role in community stability and organization. Am. Nat. 111:1119–1144.

Connell, J. H. 1978. Diversity in tropical rain forests and coral reefs. Science 199:1302–1310.

Connor, E. F., and E. D. McCoy. 1979. The statistics and biology of the species-area relationship. Am. Nat. 113:791–833.

Cooper, A. W. 1981. Aboveground biomass accumulation and net primary production during the first 70 years of succession in *Populus grandidentata* stands on poor sites in northern lower Michigan. pp. 339–360. IN D. C. West, H. H. Shugart, and D. B. Botkin (eds.), Forest Succession: Concepts and Application. Springer-Verlag, New York.

Cooper, W. S. 1913. The climax forest of Isle Royale, Lake Superior, and its development. I. Bot. Gaz. 55:1–44.

Costin, A. B. 1954. A study of the Monaro Region of New South Wales. Government Printer, Sydney. 658 pp.

Cox, D. L. 1980. A note on the queer history of "niche." Bull. Ecol. Soc. Am. 61:201–202.

Craig, A. J. 1972. Pollen influx to laminated sediments: A pollen diagram from northeastern Minnesota. Ecology 53:46–57.

Craig, R. B., D. L. DeAngelis, and K. R. Dixon. 1979. Long- and short-term dynamic optimization models with application to the feeding strategy of the loggerhead shrike. Am. Nat. 113:31–51.

Cremer, K. W. 1972. Morphology and development of primary and accessory buds of *Eucalyptus regnans*. Aust. J. Bot. 20:175–196.

Crossley, D. A. 1976. The roles of terrestrial saprophagous arthropods in forest soils: Current status of concepts. pp. 49–56. IN W. J. Mattson (ed.), The Role of Arthropods in Forest Ecosystems. Springer-Verlag, New York.

Crow, T. R., and P. L. Weaver. 1977. Tree growth in a moist tropical forest of Puerto Rico. U.S.D.A. Forest Service Res. Pap. ITF-22.

Crow, T. R. 1980. A rain forest chronicle: A 30-year record of change in structure and composition at El Verde, Puerto Rico. Biotropica 12:42–55.

Crow, T. R., and D. F. Grigal. 1980. A numerical analysis of arborescent communities in the rain forest of the Luquillo Mountains, Puerto Rico. Vegetatio 40:135–146.

Curtis, J. T. 1956. The modification of mid-latitude grasslands and forests by man. pp. 721–736. IN W. L. Thomas (ed.), Man's Role in Changing the Face of the Earth. University of Chicago Press, Chicago.

Curtis, J. T. 1959. The Vegetation of Wisconsin. University of Wisconsin Press, Madison. 657 pp.

Curtis, R. O. 1967. A method of estimation of gross yield of Douglas-fir. For. Sci. Monogr. 13:1–24.

Darlington, P. J. 1957. Zoogeography. John Wiley, New York.

Daubenmire, R. 1968. Plant Communities: A Textbook of Plant Synecology. Harper and Row, New York. 300 pp.

Davis, B. N. K. 1975. The colonization of isolated patches of nettles (*Urtica dioica* L.) by insects. J. Appl. Ecol. 12:1–14.

Davis, M. B., and J. C. Goodlett. 1960. Comparison of the present vegetation with pollen spectra in surface samples from Brownington Pond, Vermont. Ecology 41:346–357.

Davis, M. B. 1963. On the theory of pollen analysis. Am. J. Sci. 261:897–912.

Davis, M. B. 1968. Pollen grains in lake sediments: Redeposition increased by seasonal water circulation. Science 162:796–799.

Davis, M. B. 1969. Palynology and environmental history during the Quaternary. Am. Sci. 57:317–332.

Davis, M. B. 1973. Redeposition of pollen grains in lake sediments. Limnol. Oceanogr. 18:44–52.

Davis, M. B. 1981. Quaternary history and the stability of forest communities. pp. 132–153. IN D. C. West, H. H. Shugart, and D. B. Botkin (eds.), Forest Succession: Concepts and Application. Springer-Verlag, New York.

Davis, R. B. 1967. Pollen studies of near-surface sediment in Maine lakes. pp. 143–173. IN E. J. Cushing and H. E. Wright, Jr. (eds.), Quaternary Paleoecology. Yale University Press, New Haven.

Davis, R. B. 1974. Stratigraphic effects of tubificids in profundal lake sediments. Limnol. Oceanogr. 19:466–488.

Davis, R. B., and T. Webb III. 1975. The contemporary distribution of pollen in eastern North America. Quat. Res. 5:395–434.

DeAngelis, D. L. 1975. Stability and connectance in food web models. Ecology 56:238–243.

DeAngelis, D. L., R. H. Gardner, and H. H. Shugart. 1981. Productivity of forest ecosystems studied during IBP: The woodlands data set. pp. 567–672. IN D. E. Reichle (ed.), Dynamic Properties of Forest Ecosystems. Cambridge University Press, Cambridge.

DeCosta, J. H., D. D. Wade, and J. E. Deeming. 1968. The Gaston fire. U.S.D.A Forest Service Res. Pap. SE-43. 36 pp.

Delcourt, H. R. 1979. Late Quaternary vegetation history of the eastern Highland Rim and adjacent Cumberland Plateau of Tennessee. Ecol. Monogr. 49:218–237.

Delcourt, P. A., H. R. Delcourt, R. C. Brister, and L. E. Lackey. 1980. Quaternary vegetation history of the Mississippi Embayment. Quat. Res. 7:218–237.

Delcourt, P. A., and H. R. Delcourt. 1981. Vegetation maps for Eastern North America: 40,000 Yr B.P. to the present. pp. 123–165. IN R. C. Romans (ed.), Geobotany II. Plenum Pub. Co., New York.

Delcourt, H. R., P. A. Delcourt, and T. Webb III. 1982. Dynamic plant ecology: The spectrum of vegetational change in space and time. Quat. Sci. Rev. 1:153–176.

Delvaux, J. 1971. Des tables de production aux bilans energetiques. pp. 177–184. IN P. Duvigneaud (ed.), Productivity of Forest Ecosystems. Unesco, Paris.

DeMichele, D. W., and P. J. H. Sharpe. 1973. An analysis of the mechanics of guard cell motion. J. Theor. Biol. 41:77–96.

Denslow, J. S. 1980. Gap partitioning among tropical rain forest trees. Biotropica 12(Suppl.):47–55.

Diamond, J. M. 1976. Island biogeography and conservation: Strategy and limitations. Science 193:1027–1029.

Doyle, T. W. 1981. The role of disturbance in the gap dynamics of a montane rain forest: An application of a tropical forest succession model. pp. 56–73. IN D. C. West, H. H. Shugart, and D. B. Botkin (eds.), Forest Succession: Concepts and Application. Springer-Verlag. New York.

Dress, P. E. 1970. A system for the stochastic simulation of even-aged forest stands of pure species composition. Purdue University, West Lafayette, Indiana. Ph.D. thesis. 267 pp.

Drury, W. H., and I. C. T. Nisbet. 1973. Succession. J. Arnold Arbor. 54:331–368.

Dueser, R. D., and H. H. Shugart. 1979. Niche pattern in a forest-floor small-mammal fauna. Ecology 59:89–98.

Edson, M. M., T. C. Foin, and C. M. Knap. 1981. "Emergent properties" and ecological research. Am. Nat. 18:593–596.

Egler, F. E. 1939. Vegetation zones of Oahu, Hawaii. Empire For. J. 18:1–14.

Egler, F. E. 1954. Vegetation science concepts. I. Initial floristic composition—a factor in old-field vegetation development. Vegetatio 4:412–417.

Ek, A. R., and R. A. Monserud. 1974a. FOREST: A computer model for the growth and reproduction of mixed species forest stands. Research Report A2635. College of Agricultural and Life Sciences, University of Wisconsin, Madison. 90 pp.

Ek, A. R., and R. A. Monserud. 1974b. Trials with program FOREST: Growth and reproduction simulation for mixed species even- or uneven-aged forest stands. pp. 56–73. IN J. Fries (ed.), Growth Models and Forest Stand Simulation. Department of Forest Yield Research, Royal College of Forestry, Stockholm. Res. Notes 30.

Elton, C. S. 1927. Animal Ecology. Sidgwick and Jackson, London.

Emanuel, W. R., and R. J. Mulholland. 1976. Energy-based model for Lago Pond, Georgia. IEEE Trans. Automatic Control 20:98–101.

Emanuel, W. R., H. H. Shugart, and D. C. West. 1978b. Spectral analysis and forest dynamics: Long-term effects of environmental perturbations. pp. 195–210. IN H. H. Shugart (ed.), Time Series and Ecological Processes. Society of Industrial and Applied Mathematics, Philadelphia.

Emanuel, W. R., D. C. West, and H. H. Shugart. 1978a. Spectral analysis of forest model time series. Ecol. Model. 4:313–326.

Eshelman, S., and J. L. Stanford. 1977. Tornados, funnel clouds and thunderstorm damage in Iowa during 1974. Iowa State J. Res. 51:327–361.

Ewart, A. J. 1925. Handbook of Forest Trees for Victorian Foresters. Government Printer, Melbourne. 523 pp.

Faegri, K., and J. Iverson. 1975. Textbook of Pollen Analysis. Hafner, New York.

Farquhar, G. D., and S. von Caemmerer. 1982. Modeling of photosynthetic response to environmental conditions. pp. 549–587. IN O. L. Lange, P. S. Nobel, C. B. Osmond, and H. Ziegler (eds.), Physiological Plant Ecology II. Springer-Verlag, Berlin.

Farrell, T. P., and D. H. Ashton. 1973. Ecological studies on the Bennison High Plains. Victorian Nat. 90:286–298.

Forbes, S. A. 1880. On some interactions of organisms. Bull. Ill. State Lab. Nat. Hist. 1:1–18.

Forcella. F., and T. Weaver. 1977. Biomass and productivity of the subalpine *Pinus albicaulis—Vaccinium scoparium* association in Montana. Vegetatio 35:95–105.

Forcier, L. K. 1975. Reproductive strategies and the co-occurrence of climax tree species. Science 189:808–810.

Forman, R. T. T., and M. Godron. 1981. Patches and structural components for a landscape ecology. BioScience 31:733–740.

Foster, H. D., and W. W. Ashe. 1980. Chestnut oak in the southern Appalachians. U.S.D.A. Forest Service Circ. 105. 27 pp.

Foster, R. B. 1977. *Tachigalia versicolor* is a suicidal neotropical tree. Nature 268:624–626.

Fowells, H. A. 1965. Silvics of Forest Trees of the United States. U.S.D.A. Forest Service Handbook No. 271. Government Printing Office, Washington, D.C. 762 pp.

Fox, J. F. 1977. Alternation and coexistence of tree species. Am. Nat. 11:69–89.

Francis, W. D. 1954. Australian Rain-Forest Trees. Forest and Timber Bureau, Halstead Press, Sydney.

Francis, W. D. 1970. Australian Rain-Forest Trees (2nd ed.). Australian Government Printer, Sydney. 506 pp.

Franklin, J. F., F. C. Hall, C. T. Dyrness, and C. Maser. 1972. Federal research natural areas in Oregon and Washington. A guidebook for scientists and educators. Pacific N.W. Forest and Range Experiment Station, Portland, Oregon. 54 pp.

Franklin, J. F. 1979. Ecosystem studies in the Hoh River drainage, Olympic National Park. p. 1–8. IN E. E. Starkey, J. F. Franklin, and J. W. Matthews (eds.), Ecological Research in National Parks of the Pacific Northwest. Oregon State University Forest Research Laboratory Publication, Corvallis, Oregon.

Franklin, J. F., and M. A. Hemstrom. 1981. Aspect of succession in the coniferous forests of the Pacific Northwest. pp. 212–229. IN D. C. West, H. H. Shugart, and D. B. Botkin (eds.), Forest Succession: Concepts and Application. Springer-Verlag, New York.

Fritts, H. C. 1976. Tree Rings and Climate. Academic Press, New York.567 pp.

Fujii, K. 1969. Numerical taxonomy of ecological characteristics and the niche concept. Syst. Zool. 18:151–153.

Fujimori, T., S. Kawanabe, H. Saito, C. C. Grier, and T. Shidei. 1976. Biomass and primary production in forests of three vegetation zones of the northwestern United States. J. Jpn. For. Soc. 58:360–373.

Gaffney, P. M. 1975. The roots of the niche concept. Am. Nat. 109:490.

Gams, H. 1918. Prinzipienfragen der Vegetations forschung. Vjschr. naturf. Ges. Zurich. 63:293–493.

Gardner, C. A. 1957. The fire factor in relation to the vegetation of western Australia. W. Aust. Nat. 5:166–173.

Gardner, M. R., and W. R. Ashby. 1970. Connectance of large dynamic (cybernetic) systems: Critical values for stability. Nature 228:784.

Gause, E. F. 1934. The Struggle for Existence. Williams and Wilkins, Baltimore. (Reprinted 1964 by Hafner, New York.)

Gerrish, G., and D. Mueller-Dombois. 1980. Behavior of native and non-native plants in two tropical rain forests on Oahu, Hawaiian Islands. Phytocoenologia 8:237–295.

Gilbert, F. S. 1980. The equilibrium theory of island biogeography: Fact or fiction? J. Biogeogr. 7:209–235.

Gilbert, J. M. 1958. Forest succession in the Florentine Valley, Tasmania. Pap. and Proc. Roy. Soc. Tas. 93:129–151.

Gill, A. M. 1975. Fire and the Australian flora: A review. Aust. For. 38:4–25.

Gill, A. M. 1981. Adaptive responses of Australian vascular plant species to fires. pp. 243–272. IN A. M. Gill, R. H. Groves, and I. R. Noble (eds.), Fire and the Australian Biota. Australian Academy of Science, Canberra.

Gilpin, M. E. 1975. Stability of feasible predator-prey systems. Nature 254:173.

Gimingham, G. H. 1978. *Calluna* and its associated species: Some aspects of coexistence in communities. pp. 35–42. IN E. van der Maarel and M. J. A. Werger (eds.), Plant Species and Plant Communities. Dr. W. Junk, The Hague.

Givnish, T. J. 1978. On the adaptive significance of compound leaves, with particular reference to tropical trees. pp. 351–380. IN P. B. Tomlinson and N. H. Zimmerman (eds.), Tropical Trees as Living Systems. Cambridge University Press, Cambridge.

Gleason, H. A. 1917. The structure and development of the plant association. Bull. Torrey Bot. Club 43:463–481.

Gleason, H. A. 1927. Further views on the succession concept. Ecology 8:299–326.

Gleason, H. A. 1939. The individualistic concept of the plant association. Am. Midl. Nat. 21:92–110.

Goh, B. S. 1979. Stability in models of mutualism. Am. Nat. 113:261–275.

Gomez-Pompa, A., C. Vasquez-Yanes, and S. Guevara. 1972. The tropical rain forest: A nonrenewable resource. Science 177:762–765.

Goodall, D. W. 1972. Building and testing ecosystem models. pp. 173–214. IN J. N. R. Jeffers (ed.), Mathematical Models in Ecology. Blackwell Scientific Publications, Oxford.

Goodlett, J. C. 1954. Vegetation adjacent to the border of the Wisconsin drift in Potter County, Pennsylvania. Proc. Annu. Tall Timber Fire Ecol. Conf. 4:111–125.

Gorham, E., P. M. Vitousek, and W. A. Reiners. 1979. The regulation of chemical budgets over the course of terrestrial ecosystem succession. Annu. Rev. Ecol. Syst. 10:53–84.

Gorman, M. 1978. Island Ecology. Champman and Hall, London. 79 pp.

Greeley, W. B., and W. W. Ashe. 1907. White oak in the southern Appalachians. U.S.D.A. Forest Service Circ. 105. 27 pp.

Green, D. G. 1976. Nova Scotia forest history: Evidence from statistical analysis of pollen data. Dalhousie University, Halifax, Nova Scotia. Ph.D. thesis.

Green, D. G. 1981. Time series and postglacial forest ecology. Quat. Res. 15:265–277.

Green, R. H. 1971. A multivariate approach to the Hutchinsonian niche: Bivalve mollusks of central Canada. Ecology 55:734–783.

Gregory, P. H. 1973. The Microbiology of the Atmosphere. Leonard Hill, Aylesbury, England.

Greig-Smith, P. 1982. A. S. Watt, F.R.S. A biographical note. pp. 9–10. IN E. I. Newman (ed.), The Plant Community as a Working Mechanism. Blackwell Scientific Publications, Oxford.

Grigal, D. F., and R. A. Goldstein. 1971. An integrated ordination-classification analysis of an intensively sampled oak-hickory forest. J. Ecol. 59:481–592.

Griggs, R. F. 1946. The timberlines of northern America and their interpretation. Ecology 27:275–289.

Grime, J. P. 1974. Vegetation classification by reference to strategies. Nature 250:26–31.

Grime, J. P. 1979a. Plant Strategies and Vegetation Processes. John Wiley, Chichester, England. 222 pp.

Grime, J. P. 1979b. Competition and the struggle for existence. pp. 123–140. IN R. M. Anderson, B. D. Turner, and L. R. Taylor (eds.), Population Dynamics. Blackwell Scientific Publications, Oxford.

Grinnell, J. 1917. The niche-relationships of the California thrasher. Auk 34:427–433.

Grinnell, J. 1928. Presence and absence of animals. University of California Chron., Oct. 1928. pp. 429–450.

Grisebach, A. 1838. Ueber den Einfluss des climas auf die Begrnzung der natralichen Floren. Linnaea 12:159–200.

Grubb, P. J. 1977. The maintenance of species-richness in plant communities: The importance of the regeneration niche. Biol. Rev. 52:107–145.

Gutierrez, L. T., and W. R. Fey. 1980. Ecosystem Succession. A General Hypothesis and a Test Model of a Grassland. M.I.T. Press, Cambridge. 231 pp.

Hack, J. T., and J. C. Goodlett. 1960. Geomorphology and forest ecology of a mountain region in the central Appalachians. U.S. Geol. Surv. Prof. Pap. 347. 64 pp.

Haefner, J. W. 1978. Ecosystem assembly grammars: Generative capacity and empirical adequacy. J. Theor. Biol. 73:293–318.

Haines, D. A., V. J. Johnson, and W. A. Main. 1975. Wildfire atlas of the north central states. U.S.D.A. Forest Service Gen. Tech. Pap. NC-16.

Hall, N., R. D. Johnston, and G. M. Chippendale. 1970. Forest Trees of Australia. Australia Government Printing Service, Canberra. 334 pp.

Hall, W. L. 1907. The waning hardwood supply and the Appalachian forest. U.S.D.A. Forest Service Circ. 116. 16 pp.

Hardin, G. 1960. The competitive exclusion principle. Science 131:1292–1297.

Hare, F. K., and M. K. Thomas. 1979. Climate Canada. John Wiley, Toronto.

Harley, J. L. 1952. Association between microorganisms and higher plants (mycorrhiza). Annu. Rev. Microbiol. 6:367–386.

Harlow, W. M., and E. S. Harrar. 1941. Textbook on Dendrology. McGraw Hill, New York.

Harlow, W. M., and E. S. Harrar. 1969. Textbook of Dendrology. McGraw-Hill, New York.

Harper, J. L. 1967. A Darwinian approach to plant ecology. J. Ecol. 55:247–270.

Harper, J. L. 1977. The Population Biology of Plants. Academic Press, London. 892 pp.

Harper, J. L. 1982. After description. pp. 11–26. IN E. I. Newman (ed.), The Plant Community as a Working Mechanism. Blackwell Scientific Publications, Oxford.

Harris, V. T. 1952. An experimental study of habitat selection by prairie and forest races of the deermouse, *Peromyscus maniculatus*.Contrib. Lab. Vertebr. Biol. 56. University of Michigan Press, Ann Arbor.

Harris, W. F., R. S. Kinerson, and N. T. Edwards. 1973. Comparison of belowground biomass in natural deciduous forests and loblolly pine plantations. pp. 29–38. IN J. K. Marshall (ed.), The Belowground Ecosystem. Range Science Department Science Series No. 26. Colorado State University, Fort Collins.

Hatch, C. R. 1971. Simulation of an even-aged red-pine stand in northern Minnesota. University of Minnesota, St. Paul. Ph.D. thesis.

Hegyi, F. 1972. Dry matter distribution in jack pine stands in northern Ontario. For. Chron. 488:193–197.
Hegyi, F. 1974. A simulation model for managing jack-pine stands. pp. 74–87. IN J. Fries (ed.), Growth Models for Tree and Stand Simulation. Department of Forest Yield Research, Royal College of Forestry, Stockholm. Res. Notes. 30.
Heide, K. M., and R. H. W. Bradshaw. 1982. The pollen-tree relationship within forests of Wisconsin and Upper Michigan, U.S.A. Rev. Paleobot. Palynol. 36:1–23.
Heinselman, M. L. 1973. Fire in the virgin forests of the Boundary Water Canoe Area, Minnesota. Quat. Res. 3:329–382.
Heinselman, M. L. 1981. Fire and succession in the conifer forests of northern North America. pp. 374–405. IN D. C. West, H. H. Shugart, and D. B. Botkin (eds.), Forest Succession: Concepts and Application. Springer-Verlag, New York.
Henderson, R. E. 1981. An overview of the atmospheric processes and modeling aspects of acid rain. Final report prepared for U.S. Department of Energy by the Mitre Corporation, McLean, Virginia.
Henderson, W., and C. W. Wilkins. 1975. The interaction of bush fires and vegetation. Search 6:130–133.
Henry, J. D., and J. M. A. Swan. 1974. Reconstructing forest history from live and dead plant material: An approach to the study of forest succession in southwest New Hampshire. Ecology 55:772–783.
Hepting, G. H. 1968. Diseases of forest and tree crops caused by air pollutants. Phytopathology 58:1098–1101.
Hespenheide, H. A. 1971. Flycatcher habitat selection in the eastern deciduous forest. Auk 88:61–74.
Hillis, W. E., and A. G. Brown (eds.). 1978. Eucalypts for Wood Production. Griffin Press, Adelaide, South Australia, 434 pp.
Holdridge, L. R. 1967. Life Zone Ecology (rev. ed.). Tropical Science Center, San Jose, Costa Rica. 206 pp.
Holliday, I., and G. Watton. 1975. A Field Guide to the Banksias. Rigby Ltd. Adelaide, South Australia. 166 pp.
Hool, J. N. 1966. A dynamic programing—Markov chain approach to forest production control. For. Sci. Monogr. 12:1–26.
Hopkins, M. S., J. Kikkawa, A. W. Graham, J. G. Tracey, and L. J. Webb. 1977. An ecological basis for the management of rain forests. pp. 57–66. IN R. Monroe and N. C. Stevens (eds.), The Border-Ranges: A Land-use Conflict in Regional Perspective. Royal Soc. Queensland, St. Lucia, Queensland, Australia.
Horn, H. S. 1966. Measurement of overlap in comparative ecological studies. Am. Nat. 100:419–424.
Horn, H. S. 1971. The Adaptive Geometry of Trees. Princeton University Press, Princeton. 144 pp.
Horn, H. S. 1975a. Forest succession. Sci. Am. 232:90–98.
Horn, H. S. 1975b. Markovian properties of forest succession. pp. 196–211. IN M. L. Cody and J. M. Diamond (eds.), Ecology and Evolution of Communities. Harvard University Press, Cambridge.
Horn, H. S. 1976. Succession. pp. 187–204. IN R. M. May (ed.), Theoretical Ecology. Blackwell Scientific Publications, Oxford.

Horn, H. S. 1981. Some causes of variety in patterns of secondary succession. pp. 24–35. IN D. C. West, H. H. Shugart, and D. B. Botkin (eds.), Forest Succession: Concepts and Application. Springer-Verlag, New York.

Hosaka, E. Y. 1939. Ecological and floristic studies in Kipapa Gulch, Oahu. Bishop Mus. Occ. Pap. 13:175–232.

Humboldt, A. von. 1807. Ideen Zu einer Geographie der Pflanzen nebst einem Naturgemlde der Tropenlnder. Cotta, Tbingen. 182 pp.

Hurlbert, S. H. 1971. The nonconcept of species diversity: A critique and alternative parameters. Ecology 52:577–587.

Huston, M. A. 1979. A general hypothesis of species diversity. Am. Nat. 113:81–101.

Hutchinson, G. E. 1942. Addendum to R. L. Lindeman. The trophic-dynamic aspect of ecology. Ecology 23:417–418.

Hutchinson, G. E. 1944. Limnological studies in Connecticut. VII. A critical examination of the supposed relationship between phytoplankton periodicity and chemical changes in lake waters. Ecology 25:3–26 (see footnote 5, p. 20).

Hutchinson, G. E. 1948. Circular causal systems in ecology. Ann. N.Y. Acad. Sci. 50:221–246.

Hutchinson, G. E. 1957. Concluding remarks. Cold Spring Harbor Symposium. Quant. Biol. 22:415–427.

Hutchinson, G. E. 1965. The ecological theater and the evolutionary play. Yale University Press, New Haven. 134 pp.

Ilvessalo, Y. 1920. Entragstafeln fr Kiefern-, Fichten- und Birkebestnde in der Sdhalfe von Finland. Acta. For. Fenn., No. 15. 96 pp.

Ilvessalo, Y. 1937. Pera-Pohjolan luonnon normaalien metsikoiden kasva ja kehitys. Metsatieteellisen Tutkimuslaitoken Julkaisuja 24:1–168.

Iverson, J. 1941. Land occupation in Denmark's stone age: A pollen analytical study of the influence of farmer culture on the vegetation development. Dan. Geol. Unders. Afh. Raekke 2 66:1–68.

Iverson, J. 1949. The influence of prehistoric man on vegetation. Dan. Geol. Unders. Afh. Raekke 4 6:5–22.

Iverson, J. 1964. Reprogressive vegetational succession in the post-glacial. J. Ecol. 52(Suppl.):59–70.

Iwaki, H., and T. Totsuka. 1959. Ecological and physiological studies on the vegetation of Mt. Shiwagare. II. On the crescent-shaped 'dead-trees-strips' in the Yatsugetke and Chichibu mountains. Bot. Mag. Tokyo 72:255–260.

Jackson, W. D. 1968. Fire, air, water and earth—an elemental ecology of Tasmania. Proc. Ecol. Soc. Aust. 3:9–16.

Jacobi, J. D. 'ohi'a dieback in Hawaii: A comparison of adjacent dieback and non-dieback rain forest stands. (unpublished manuscript).

Jacobs, M. R. 1955. Growth habits of the Eucalypts. Forestry and Timber Bureau, Canberra. 262 pp.

James, F. C. 1971. Ordinations of habitat relationships among breeding birds. Wilson Bull. 83:215–236.

Janssen, C. R. 1967. A comparison between the recent regional pollen rain and the sub-recent vegetation in four major vegetation types in Minnesota (USA). Rev. Paleobot. Palynol. 2:331–342.

Janzen, D. H. 1969. Seed-eaters versus seed size, number, toxicity and dispersal. Evolution 23:1–27.

Janzen, D. H. 1971. Seed predation by animals. Annu. Rev. Ecol. Syst. 2:465–492.

Janzen, D. H. 1979. How to be a fig. Annu. Rev. Ecol. Syst. 10:13–52.

Johnson, E. A., and J. S. Rowe. 1975. Fire in the subarctic wintering ground of the Beverly caribou herd. Am. Midl. Nat. 94:1–14.

Johnson, R. H. 1910. Determinate evolution in the color pattern of lady-beetles. Carnegie Inst. Washington Publ. 122. 104 pp.

Johnson, W. C., and D. M. Sharpe. 1976. Forest dynamics in the northern Georgia Piedmont. For. Sci. 22:307–322.

Johnston, D. W., and E. P. Odum. 1956. Breeding bird populations in relation to plant succession on the Piedmont of Georgia. Ecology 37:50–62.

Jones, E. W. 1945. The structure and reproduction of the virgin forest of the north temperate zone. New Phytol. 45:130–148.

Jones, E. W. 1950. Some aspects of natural regeneration in the Benin rain forest. Empire For. Rev. 29:108–124.

Kasanaga, H., and M. Monsi. 1954. On the light-transmission of leaves and its meaning for the production of dry matter in plant communities. Jpn. J. Bot. 14:302–324.

Kauffman, S. A., R. M. Shymko, and K. Trabert. 1978. Control of sequential compartment formation in *Drysophila*. Science 199:259–269.

Kazimirov, N. I., and R. M. Morozova. 1973. Biological Cycling of Matter in Spruce Forests of Karelia. Academy of Sciences, Navka, Leningrad-Branch. 168 pp.

Keeling, C. D., R. B. Bacastow, A. E. Bainbridge, C. A. Ekdahl, P. R. Guenther, L. S. Waterman, and J. F. S. Chin. 1976. Atmospheric carbon dioxide variations at Mauna Loa Observatory, Hawaii. Tellus 28:538–551.

Ker, J. W., and J. H. G. Smith. 1955. Advantages of the parabolic expression of height-diameter relationships. For. Chron. 31:235–246.

Kercher, J. R., M. C. Axelrod, and G. E. Bingham. 1980. Forecasting effects of sulfur dioxide pollution on growth and succession. pp. 200–202. IN P. R. Miller (ed.), Proceedings of the Symposium on Effects of Air Pollutants on Mediterranean and Temperate Forest Ecosystems. U.S. Forest Service Gen. Tech. Rep. PSW-43.

Kessell, S. R. 1976. Gradient modeling: A new approach to fire modeling and wilderness resource management. Environ. Manag. 1:39–48.

Kessell, S. R. 1979a. Gradient Modeling: Resource and Fire Management. Springer-Verlag, New York. 432 pp.

Kessell, S. R. 1979b. Phytosociological inference and resource management. Environ. Manag. 3:29–40.

Kessell, S. R., and M. W. Potter. 1980. A quantitative succession model for nine Montana forest communities. Environ. Manag. 4:227–240.

Kessell, S. R. 1981a. Application of gradient analysis concepts to resource management modeling. Proc. Ecol. Soc. Aust. 11:163–173.

Kessell, S. R. 1981b. The challange of modeling post-disturbance plant succession. Environ. Manage. 5:5–13.

Kessell, S. R., R. B. Good, and M. W. Potter. 1982. Computer modeling in natural area management. Special Pub. No. 9. Australian National Parks and Wildlife Service, Canberra.

King, J. E. 1981. Late Quaternary vegetation history of Illinois. Ecol. Monogr. 51:43–62.

Kira, T., and T. Shidei. 1967. Primary production and turnover of organic matter in different forest ecosystems of the western Pacific. Jpn. J. Ecol. 17:70–87.

Kitchell, J. F., R. V. O'Neill, D. Webb, G. W. Gallepp. S. M. Bartell, J. F. Koonce, and B. S. Ausmus. 1979. Consumer regulation of nutrient cycling. BioScience 29:28–34.

Kliejunas, J. T., and W. H. Ko. 1973. Root rot of 'ohi'a (*Metrosideros collina* subsp. *polymorpha*) caused by *Phytophythora cinnamomi*. Plant Dis. Rep. 57:383–384.

Kliejunas, J. T., and W. H. Ko. 1974. Deficiency of inorganic nutrients as contributing factor to 'ohi'a decline. Phytopathology 64:891–896.

Klopfer, P. 1965. Behavioral aspects of habitat selection. A preliminary report on stereotype in foliage preferences in birds. Wilson Bull. 75:15–22.

Koestler, A. 1967. The Ghost in the Machine. Macmillan, New York.

Kohler, W. 1947. Gestalt psychology. Liveright Pub. Co., New York. 482 pp.

Koppen, W. 1931. Grundriss der Klimakunde. W. deGruyter, Berlin.

Kozlowski, T. T. 1971a. Growth and Development of Trees, Vol. I. Seed Germination, Ontogeny and Shoot Growth. Academic Press, New York. 443 pp.

Kozlowski, T. T. 1971b. Growth and Development of Trees, Vol. II. Cambial Growth, Root Growth and Reproductive Growth. Academic Press, New York. 514 pp.

Kozlowski, T. T., and C. E. Ahlgren (eds.). 1974. Fire and Ecosystems. Academic Press, New York. 542 pp.

Kozlowski, T. T. 1981. Impacts of air pollution of forest ecosystems. BioScience 30:88–93.

Kramer, P. J., and T. T. Kozlowski. 1960. The Physiology of Trees. McGraw-Hill, New York.

Kramer, P. J. 1981. Carbon dioxide concentration, photosynthesis, and dry matter production. BioScience 31:29–33.

Kurcheva, G. F. 1960. The role of invertebrates in the decomposition of oak litter. Pocroredenie 4:16–23.

Lacey, C. J. 1974. Rhizomes in tropical eucalypts and their role in recovery from fire damage. Aust. J. Bot. 22:29–38.

Laemmlen, F. F., and R. V. Bega. 1972. Decline of 'ohi'a and koa forests in Hawaii. Phytopathology 62:770.

Lawlor, L. R., and J. Maynard Smith. 1976. Coevolution and the stability of competing species. Am. Nat. 110:79–99.

Lawlor, L. R. 1979. Direct and indirect effects of n-species competition. Oecologia 43:355–364.

Lawlor, L. R. 1980. Structure and stability in natural and randomly constructed competitive communities. Am. Nat. 116:394–400.

Lawton, R. O. (personal communication). Letter dated January 28, 1982.Department of Biology, The University of Alabama in Huntsville, Huntsville.

Leak, W. B. 1969a. Sapling stand development: A compound exponential process. For. Sci. 16:177–180.

Leak, W. B. 1969b. Stocking of northern hardwood forest based on exponential dropout rate. For. Chron. 45:1–4.

Leak, W. B. 1970. Successional change in northern hardwoods predicted by birth and death simulation. Ecology 5:794–801.

Leak, W. B., and R. E. Graber. 1974. Forest vegetation related to elevation in the White Mountains in New Hampshire. U.S.D.A. Forest Service Res. Pap. NE-299.

Leak, W. B., and R. E. Graber. 1976. Seedling input, death and growth in uneven-aged northern hardwoods. Can. J. For. Res. 6:368–374.

Lee, J. J., and D. L. Inman. 1975. The ecological role of consumers—An aggregated systems view. Ecology 56:1455–1458.

Lee, Y. 1967. Stand models for lodgepole pine and limits to their application Faculty of Forestry, University of British Columbia, Vancouver. 333 pp Ph.D. thesis.

Leigh, E. G. 1975. Structure and climate in tropical rain forests. Annu. Rev. Ecol Syst. 6:667–686.

Levin, S. A. 1976. Population dynamic models in heterogeneous environments Annu. Rev. Ecol. Syst. 7:287–310.

Levine, S. H. 1976. Competitive interactions in ecosystems. Am. Nat. 110:903-910.

Levins, R. 1968. Evolution in Changing Environments. Princeton Universit Press, Princeton. 120 pp.

Lewton-Brain, L. 1909. The Maui forest trouble. Hawaiian Planters Record 1:92-95.

Liebig, J. 1840. Chemistry and its Application to Agriculture and Physiology Taylor and Walton, London.

Likens, G. E., and T. J. Butler. 1981. Recent acidification of precipitation i North America. Atmos. Environ. 15:1103–1109.

Lin, J. Y. 1970. Growing space index and stand simulation of young wester hemlock in Oregon. School of Forestry, Duke University, Durham, Nort Carolina. 182 pp. Ph.D. thesis.

Lin, J. Y. 1974. Stand growth simulation models for Douglas-fir and wester hemlock in the northwestern United States. pp. 102–118. IN J. Fries (ed. Growth Models for Tree and Stand Simulation. Department of Forest Yiel Research, Royal College of Forestry, Stockholm. Res. Notes 30.

Lindeman, R. L. 1942. The trophic-dynamic aspect of ecology. Ecology 23:399-418.

Lindsay, A. A. 1972. Tornado tracks in the presettlement forests of Indiana. Proc Indiana Acad. Sci. 82:181.

Little, E. L. 1954. Checklist of native and naturalized trees of the United State (including Alaska). U.S.D.A. Forest Service Handbook No. 41. 472 pp.

Little, E. L. 1971. Atlas of United States trees, Vol. I: Conifers and Importa Hardwoods. Government Printing Office, Washington, D.C.

Little, E. L., Jr., and F. H. Wadsworth. 1964. Common trees of Puerto Rico an the Virgin Islands. U.S.D.A. Forest Service Agricultural Handbook No. 24 548 pp.

Lohm, U., and T. Persson (eds.). 1976. Soil Organisms as Components of Ecosy tems. Proc., 6th Int., Colloq. Soil Zool., Ecol. Bull. 25. 614 pp.

Loomis, R. S., and W. A. Williams. 1969. Productivity and the morphology of crop stands: Patterns with leaves. pp. 24–47. IN Physiological Aspects of Crop Yield. American Society of Agronomy, Crop Science Society of America, Madison, Wisconsin.

Loucks, O. L. 1970. Evolution of diversity, efficiency and community stability. Am. Zool. 10:17–25.

Loucks, O. L., A. R. Ek., W. C. Johnson, and R. A. Monserud. 1981. Growth, aging and succession. pp. 37–86. IN D. E. Reichle (ed.), Dynamic Properties of Forest Ecosystems. Cambridge University Press, Cambridge.

Lugo, A. 1970. Photosynthetic studies of four species of rain forest seedlings. pp. I-81–102. IN H. T. Odum and R. F. Pigeon (eds.), A Tropical Rain Forest. Book 3. Office of Information Service, U.S.A.E.C. (Available as TID-24270 from the National Technical Information Service, Springfield, Virginia.)

Luke, R. H., and A. G. McArthur. 1978. Bushfires in Australia. Australian Government Publishing Service, Canberra. 359 pp.

Lutz, H. J., and F. S. Griswold. 1939. The influence of tree roots on soil morphology. Am. J. Soil Sci. 237:389–400.

Lutz, H. J., and A. C. Cline. 1947. Results of the first thirty years of experimentation in the Harvard forest, 1908–1938. Part 1. The conversion of stands of old field origin by various methods of cutting and subsequent treatments. Harvard Forest Bull. 23. 182 pp.

Lyford, W. H., and D. W. MacLean. 1966. Mound and pit microrelief in relation to soil disturbance and tree distribution in New Brunswick, Canada. Harv. For. Pap. 15.

Lyon, H. L. 1909. The forest disease on Maui. Hawaiian Planter's Record 1:151–159.

Lyon, H. L. 1918. The forests of Hawaii. Hawaiian Planters Record 20:267–281.

Lyon, H. L. 1919. Some observations on the forest problems of Hawaii. Hawaiian Planters Record 21:289–300.

MacArthur, R. H., and E. O. Wilson. 1963. An equilibrium theory of insular zoogeography. Evolution 17:373–387.

MacArthur R. H., and J. H. Connell. 1966. The Biology of Populations. John Wiley, New York. 200 pp.

MacArthur, R. H., and R. Levins. 1967. The limiting similarity, convergence and divergence of coexisting species. Am. Nat. 101:377–385.

MacArthur, R. H., and E. O. Wilson. 1967. The Theory of Island Biogeography. Princeton University Press, Princeton. 203 pp.

MacArthur, R. M. 1958. Population ecology of some warblers of northeastern coniferous forests. Ecology 39:599–619.

MacArthur, R. M., and E. Pianka. 1966. On optimal use of a patch environment. Am. Nat. 100:603–609.

MacLean, D. A., and R. W. Wein. 1976. Biomass of jack pine and mixed hardwood stands in northeastern New Brunswick. Can. J. For. Res. 6:441–447.

MacMahon, J. A. 1981. Successional processes: Comparisons among biomes with special reference to probable roles of and influences on animals. pp. 277–304. IN D. C. West, H. H. Shugart, and D. B. Botkin (eds.), Forest Succession: Concepts and Application. Springer-Verlag, New York.

Maguire, B., Jr. 1967. A partial analysis of the niche. Am. Nat. 101:515–523.

Maher, L. J. 1977. Palynological studies in the western arm of Lake Superior. Quat. Res. 7:14–44.

Major, J. 1951. A functional, factorial approach to plant ecology. Ecology 32:392–412.

Manabe, S., and R. J. Stouffer. 1980. Sensitivity of a global climate model to an increase in CO_2 in the atmosphere. J. Geophys. Res. 85:5529–5554.

Mankin, J. B., R. V. O'Neill, H. H. Shugart, and B. W. Rust. 1977. The importance of validation in ecosystem analysis. pp. 63–71. IN G. S. Innis (ed.), New Directions in the Analysis of Ecological Systems. Part I. Simulation Councils of America, LaJolla, California.

Margalef, R. 1958. Information theory in ecology. Gen. Syst. 3:36–71.

Margalef, R. 1963. On certain unifying principles in ecology. Am. Nat. 97:357–374.

Margalef, R. 1968. Perspectives in Ecological Theory. University of Chicago Press, Chicago. 111 pp.

Marie-Victorin, F. 1929. Le dynamisme dans la flore du Quebec. Contrib. Inst. Bot. Univ. Montreal No. 13.

Marks, P. 1971. The role of *Prunus pensylvanica* L. in the rapid regeneration of disturbed sites. Yale University, New Haven, Connecticut. Ph.D. thesis.

Marks, P. L. 1974. The role of pin cherry (*Prunus pensylvanica* L.) in the maintenance of stability in northern hardwood ecosystems. Ecol. Monogr. 44:73–88.

Marquis, D. A. 1967. Clear-cutting in northern hardwoods. U.S. Forest Service Res. Pap. NE-85. 13 pp.

Martin, A. C., H. S. Zim, and A. L. Nelson. 1951. American Wildlife and Plants, A Guide to Wildlife Food Habits: The Use of Trees, Shrubs, Weeds, and Herbs by Birds and Mammals of the United States. Dover Publishing Company, Inc., New York.

Martinka, C. J. 1976. Fire and elk in Glacier National Park. Proc. Tall Timber Fire Ecol. Conf. 14:377–389.

Martinka, R. R. 1972. Structural characteristics of Blue Grouse territories in southwestern Montana. J. Wildlife Manag. 36:498–510.

Matte, V. 1971. *Pinus radiata* plantations in Chile: Present situation and future possibilities. pp. 217–223. IN P. Duvigneaud (ed.), Productivity of Forest Ecosystems. Unesco, Paris.

Mattson, W. J., and N. D. Addy. 1975. Phytophagous insects as regulators of forest primary production. Science 190:515–522.

Mattson, W. J. 1977. The Role of Arthropods in Forest Ecosystems. Springer-Verlag, New York. 104 pp.

May, R. M. 1973. Stability and Complexity in Model Ecosystems. Princeton University Press, Princeton.

May, R. M. 1975. Some notes on estimating the competition matrix. Ecology 56:737–741.

May, R. M. 1975. Stability and Complexity in Model Ecosystems. (2nd ed.), Princeton University Press, Princeton.

May, R. M. 1976. Simple mathematical models with very complex dynamics. Nature 261:459–467.

May, R. M. 1981. Models for single populations. pp. 5–29. IN R. M. May (ed.) Theoretical Ecology (2nd ed.). Blackwell Scientific Publications, Oxford.

McAndrews, J. H. 1966. Post-glacial history of prairie, savanna, and forest in northwestern Minnesota. Mem. Torrey Bot. Club. 22:1–72.

McAndrews, J. H., and D. M. Power. 1973. Palynology of the Great Lakes: The surface sediments of Lake Ontario. Can. J. Earth Sci. 10:777–792.

McArthur, A. G. 1967. Fire behavior in eucalypt forests. Comm. Aust. For. Timb. Bur. Leaflet 107.

McBrayer, J. F. 1973. Exploitation of deciduous leaf litter by *Apheloria montana* (Diplopoda: Eurydesmidae). Pedobiologia 13:90–98.

McBrayer, J. F. 1977. Contributions of Cryptozoa to forest nutrient cycles. pp. 70–77. IN W. J. Mattson (ed.), The Role of Arthropods in Forest Ecosystems. Springer-Verlag, New York.

McCormick, J. 1968. Succession. Via 1:1–16.

McIntosh, R. P. 1980. The relationship between succession and the recovery process in ecosystems. pp. 11–62 IN J. Cairns (ed.), The Recovery Process in Damaged Ecosystems. Ann Arbor Science, Ann Arbor.

McIntosh, R. P. 1981. Succession and ecological theory. pp. 10–23. IN D. C. West, H. H. Shugart, and D. B. Botkin (eds.), Forest Succession: Concepts and Application. Springer-Verlag, New York.

McLaughlin, S. B., D. C. West, H. H. Shugart, and D. S. Shriner. 1978. Air pollution effects on forest succession: Application of a mathematical model. 71st Annual Meeting of the Air Pollution Control Assoc. 78–24.5. 16 pp.

McNaughton, S. J., and L. L. Wolf. 1970. Dominance and the niche in ecological systems. Science 167:131–139.

Medway, L. 1972. Phenology of a tropical rain forest in Malaya. Biol. J. Linn. Soc. 4:117–146.

Meetenmeyer, V. 1978. Macroclimate and lignin control of decomposition rates. Ecology 59:465–472.

Meeuwig, R. O. 1979. Growth characteristics of pinyon-juniper stands in the western Great Basin. U.S.D.A. Forest Service Res. Pap. INT-238. 31 pp.

Meyer, W. H. 1937. Yield of even-aged stands of sitka spruce and western hemlock. U.S.D.A. Agric. Tech. Bull. 544. 68 pp.

Mielke, D. L., H. H. Shugart, and D. C. West. 1977. User's manual for FORAR, a stand model for upland forests of southern Arkansas. ORNL/TM-5767. Oak Ridge National Laboratory, Oak Ridge, Tennessee.

Mielke, D. L., H. H. Shugart, and D. C. West. 1978. A stand model for upland forests of southern Arkansas. ORNL/TM-6225. Oak Ridge National Laboratory, Oak Ridge, Tennessee.

Miller, P. R., and J. R. McBride. 1975. Effects of air pollutants of forests. pp. 196–236. IN J. B. Mudd and T. T. Kozlowski (eds.), Responses of Plants to Air Pollution. Academic Press, New York.

Mitchell, H. L., and R. F. Chandler. 1939. The nitrogen nutrition and growth of certain deciduous trees of the northern United States. Black Rock For. Bull. 11.

Mitchell, J. M., Jr. 1976. An overview of climatic variability and its causal mechanisms. Quat. Res. 6:481–493.

Mitchell, K. J. 1969. Simulation of the growth of even-aged stands white spruce. School of Forestry, Yale University Bull. 75. 48 pp.

Mitchell, K. J. 1975a. Dynamics and simulated yield of Douglas-fir. For. Sci. Monogr. 17. (Suppl. to For. Sci. 21). 39 pp.

Mitchell, K. J. 1975b. Stand description and growth simulation from low-level stereophotos of tree crowns. J. For. 73:12–16, 45.

Monsi, M., and Y. Oshima. 1955. A theoretical analysis of the succession process of plant community, based on the production of matter. Jpn. J. Bot. 15:60–82.

Monsi, M., and T. Saeki. 1953. Ber den Lichtfaktor in den Pflanzengesellschaften und seine Bedeutung fur die Stoffproduktion. Jpn. J. Bot. 14:22–52.

Moore, D. M. 1960. Trees of Arkansas. Arkansas Forestry Commission, Little Rock. 128 pp.

Moran, J. M. 1972. An analysis of periglacial climatic indicators of late-glacial time in North America. University of Wisconsin, Madison. Ph.D. dissertation.

Morrison, D. F. 1967. Multivariate Statistical Methods. McGraw-Hill, New York. 415 pp.

Morrow, P. A. 1976. The significance of phytophagous insects in the *Eucalyptus* forests of Australia. pp. 19–29. IN W. J. Mattson (ed.), The Role of Arthropods in Forest Ecosystems. Springer-Verlag, New York.

Morrow, P. A. 1977. Host specificity of insects in a community of three codominant *Eucalyptus* species. Aust. J. Ecol. 2:89–106.

Moser, J. W., and O. F. Hall. 1969. Deriving growth and yield functions for uneven-aged forest stands. For. Sci. 15:183–188.

Mueller-Dombois, D., and H. Ellenberg. 1974. Aims and Methods of Vegetation Ecology. John Wiley, New York. 547 pp.

Mueller-Dombois, D. 1980. The 'ohi'a dieback phenomenon in the Hawaiian rain forest. pp. 153–161. IN J. Cairns (ed.), The Recovery Process in Damaged Ecosystems. Ann Arbor Science, Ann Arbor.

Mueller-Dombois, D. 1982. Canopy dieback in indigenous forests of Pacific Islands: Hawaii, Papua New Guinea and New Zealand. Hawaiian Bot. Soc Newletter (21:2–6).

Mueller-Dombois, D., J. D. Jacobi, R. C. Cooray, and N. Balakrishnan. 1981 'Ohi'a rain forest study, final report. CPSU (University of Hawaii, Botany Department), Tech. Rep. No. 20. 117 pp. *f* 3 map sheets.

Munro, D. D. 1974. Forest growth models. A prognosis. pp. 7–21. IN J. Fries (ed.), Growth Models for Tree and Stand Simulation. Department of Forest Yield Research, Royal College of Forestry, Stockholm. Res. Notes 30.

Namkoong, G., and J. H. Roberds. 1974. Extinction probabilities and the changing age structure of redwood forests. Am. Nat. 108:355–368.

Neiland, B. J. 1958. Forest and adjacent burn in the Tillamook burn area of northwestern Oregon. Ecology 39:660–671.

Neill, W. E. 1974. The community matrix and the interdependence of the competition coefficient. Am. Nat. 108:399–408.

Newnham, R. M. 1964. The development of a stand model for Douglas-fir. Faculty of Forestry, University of British Columbia, Vancouver. 201 pp. Ph.D thesis.

Nicolis, G., and J. F. G. Auchmuty. 1974. Dissipative structures, catastrophes and pattern formation: A bifurcation analysis. Proc. Nat. Acad. Sci. USA 71:2748–2751.

Niering, W. A., and F. E. Egler. 1955. A shrub community of *Viburnum lentago* stable for twenty-five years. Ecology 36:356–360.

Noble, I. R., and R. O. Slatyer. 1977. Post fire succession in Mediterranean ecosystems. pp. 27–36. IN H. A. Mooney and A. C. E. Conrad (eds.), Pro ceedings, Symposium on the Environmental Consequences of Fire and Fuel Management in Mediterranean Ecosystems. U.S.D.A. Forest Service Gen. Tech. Rept. W0–3.

Noble, I. R., and R. O. Slatyer. 1978. The effect of disturbances on plant succession. Proc. Ecol. Soc. Aust. 10:135–145.

Noble, I. R., and R. O. Slatyer. 1980. The use of vital attributes to predict successional changes in plant communities subject to recurrent disturbances. Vegetatio 43:5–21.

Noble, I. R., G. A. V. Bary, and A. M. Gill. 1980. McArthur's fire danger meters expressed as equations. Aust. J. Ecol. 5:201–203.

Noble, I. R. In press. The role of fire in the evolution of species interactions, competition and succession. IN H. A. Mooney (ed.), The Role of Past and Present Fire Frequency on Ecosystem Development and Management (book in preparation).

Northeastern Forest Experiment Station. 1971. Oak Symposium Proceedings. U.S.D.A. Northeastern Forest Experiment Station, Upper Darby, Pennsylvania. 161 pp.

Noy-Meir, I. 1982. Stability of plant-herbivore models and possible applications to savanna. pp. 591–609. IN B. J. Huntley and B. H. Walker (eds.), Ecology of Tropical Savannas. Springer-Verlag, Heidelberg.

Odum, E. P. 1959. Fundamentals of Ecology (2nd Ed.). W. B. Saunders & Co., Philadelphia. 546 pp.

Odum, E. P. 1960. Organic production and turnover in old field succession. Ecology 41:34–49.

Odum, E. P. 1962. Relationships between structure and function in the ecosystem. Jpn. J. Ecol. 12:108–118.

Odum, E. P. 1968. Energy flow in ecosystems: A historical review. Am. Zool. 8:11–18.

Odum, E. P. 1969. The strategy of ecosystem development. Science 164:262–270.

Odum, E. P. 1971. Fundamentals of Ecology. W. B. Saunders, Philadelphia. 574 pp.

Odum, E. P., J. T. Finn, and E. H. Franz. 1979. Perturbation theory and the subsidy-stress gradient. BioScience 29:349–352.

Odum, H. T., and R. C. Pinkerton. 1955. Time's speed regulator, the optimum efficiency for maximum output in physical and biological systems. Am. Sci. 43:331–343.

Odum, H. T. 1956. Efficiencies, size of organisms and community structure. Ecology 37:592–597.

Odum, H. T., and R. F. Pigeon (eds.). 1970. A Tropical Rain Forest. Technical Information Center, Oak Ridge, Tennessee. 1,650 pp.

Odum, H. T. 1971. Environment, Power, and Society. Wiley-Interscience, New York. 331 pp.

Ogden, J. G. III. 1966. Forest history of Ohio: I. Radiocarbon dates and pollen stratigraphy of Silver Lake, Logan County, Ohio. Ohio J. Sci. 66:387–400.

Ogden, J. G. III. 1969. Correlation of contemporary and late Pleistocene pollen records in the reconstruction of post-glacial environments in northeastern North America. Mitt. Int. Vere. Limnol. 17:64–77.

Ogden, J. G. III. 1977. Limiting factors in paleoenvironmental reconstruction. pp. 29–42. IN R. C. Romans (ed.), Geobotany. Plenum Press, New York.

Old, K. M., G. A. Kile, and C. P. Ohmart (eds.). 1981. Eucalypt Dieback in Forests and Woodlands. C.S.I.R.O., Melbourne. 285 pp.

Oldeman, R. A. A. 1978. Architecture and energy exchange of dicotyledonous trees in the forest. pp. 525–560. IN P. B. Tomlinson and M. H. Zimmerman (eds.), Tropical Trees as Living Systems. Cambridge University Press, Cambridge.

Oliver, C. D., and E. P. Stephens. 1977. Reconstruction of a mixed-species forest in central New England. Ecology 58:562–572.

Oliver, C. D. 1981. Forest development in North America following major disturbances. For. Ecol. Manag. 3:153–168.

Olson, J. S. 1958. Rates of succession and soil changes on southern Lake Michigan sand dunes. Bot. Gaz. 119:125–170.

Olson, J. S. 1963. Energy storage and the balance of producers and decomposers in ecological systems. Ecology 44:322–331.

Olson, J. S., and G. Christofolini. 1966. Model simulation of Oak Ridge vegetation succession. ORNL-4007. Oak Ridge National Laboratory, Oak Ridge, Tennessee. pp. 106–107.

O'Neill, R. V. 1976. Ecosystem persistence and heterotrophic regulation. Ecology 57:1244–1253.

Oosting, H. J. 1942. An ecological analysis of the plant communities of Piedmont, North Carolina. Am. Midl. Nat. 28:1–126.

Oosting, H. J. 1956. The Study of Plant Communities (2nd Ed.). W. H. Freeman and Co., San Francisco. 440 pp.

Opler, P. A., H. G. Baker, and G. W. Frankie. 1980. Plant reproductive characteristics during secondary succession in neotropical lowland forest ecosytems. Biotropica 12(Suppl.):40–46.

Oshima, Y., M. Kimura, H. Iwake, and S. Kuroiwa. 1958. Ecological and physiological studies on the vegetation of Mt. Shimagare. I. Preliminary survey of the vegetation of Mt. Shimagare. Bot. Mag. Tokyo 71:289–300.

Ovington, J. D. 1957. Dry matter production by *Pinus sylvestris* L. Ann. Bot., Lond. n.s. 21:287–314.

Ovington, J. D., and H. A. I. Madgwick. 1959. The growth and composition of natural stands of birch. I. Dry matter production. Plant Soil 10:271–283.

Owen, D. D. 1860. Second Report of a Geological Reconnaissance of the Middle and Southern Counties of Arkansas. Sherman, Philadelphia. 207 pp.

Paijmans, K. (ed.). 1976. New Guinea Vegetation. Australian National University Press, Canberra. 213 pp.

Paine, R. T. and S. A. Levin. 1981. Intertidal landscapes: Disturbance and the dynamics of pattern. Ecol. Monogr. 51:145–178.

Papp, R. P., J. T. Kliejunas, R. S. Smith, Jr., and R. F. Scharpf. 1979. Association of *Plagithmysus bilineatus* (Coleoptera: Cerambycidae) and *Phytophthora cinnamomi* with the decline of 'ohi'a—lehua forests on the island of Hawaii. For. Sci. 25:187–196.

Pardo, R. 1973. AFA's social register of big trees. Am. For. 79:21–47.

Parkhurst, D. F., and O. L. Loucks. 1972. Optimal leaf size in relation to environment. J. Ecol. 60:505–537.

Pastor, J., and J. G. Blackheim. 1984. Distribution and cycling of nutrients in an aspen (*Populus tremuloides* Michx.)—mixed hardwood—spodosol ecosystem in northern Wisconsin. Ecology 65:(in press).

Pastor, J., J. D. Aber, C. A. McClaugherty, and J. M. Melillo. 1984. Aboveground production and N and P cycling along a nitrogen mineralization gradient on Blackhawk Island, Wisconsin. Ecology 65:(in press).

Patrick, R. 1967. The effect of invasion rate, species pool and the size of area on the structure of the diatom community. Proc. Nat. Acad. Sci. U.S.A. 58:1335–1342.

Patten, B. C. 1975. Ecosystem linearization: An evolutionary design problem. Am. Nat. 109:529–539.

Patten, D. T. 1963. Vegetational pattern in relation to environments in the Madison Range, Montana. Ecol. Monogr. 33:375–406.

Peattie, D. C. 1950. A Natural History of Trees of Eastern and Central North America. Houghton Mifflin, Boston.

Peden, L. M., J. S. Williams, and W. E. Frayer. 1973. A Markov model for stand projection. For. Sci. 19:303–314.

Peet, R. K. 1981. Changes in biomass and production during secondary succession. pp. 325–338. IN D. C. West, H. H. Shugart, and D. B. Botkin (eds.). Forest Succession: Concepts and Application. Springer-Verlag, New York.

Phipps, R. L. 1967. Annual growth of suppressed chestnut oak and red maple, a basis for hydrological inference. U.S. Geol. Surv. Prof. Pap. 485-C. 27 pp.

Phipps, R. L. 1979. Simulation of wetlands forest vegetation dynamics Ecol. Model. 7:257–288.

Pianka, E. R. 1981. Competition and niche theory. pp. 167–198. IN R. M. May (ed.), Theoretical Ecology: Principles and Applications (2nd Ed.). Blackwell Scientific Publications, Oxford.

Pickett, S. T. A. 1976. Succession: An evolutionary interpretation. Am. Nat. 110:108–119.

Pickett, S. T. A., and J. N. Thompson. 1978. Patch dynamics and the design of nature reserves. Biol. Conserv. 13:27–37.

Pielou, E. C. 1969. An Introduction to Mathematical Ecology. Wiley-Interscience, New York. 286 pp.

Pielou, E. C. 1972. Niche width and overlap: A method for measuring them. Ecology 53:687–692.

Pielou, E. C. 1981. The usefulness of ecological models: A stock-taking. Q. Rev. Biol. 56:17–31.

Pinchot, G. 1906. The lumber cut of the United States: 1906. U.S.D.A. Forest Service Circ. 122. 42 pp.

Pinchot, G. 1907. Consumption of tanbark and tanning extract in 1906. U.S.D.A. Forest Service Circ. 119. 9 pp.

Platt, T., and K. L. Denman. 1975. Spectral analysis in ecology. Annu. Rev. Ecol. Syst. 6:189–210.

Plonski, W. L. 1960. Normal yield tables for black spruce, jack pine, aspen, white birch, tolerant hardwoods, white pine and red pine for Ontario. Ont. Dept. Lands Forests Silvic. Ser. Bull. 2. 29 pp.

Podger, F. D. 1972. *Phytophthora cinnamomi,* a cause of lethal disease in indigenous plant communities in Western Australia. Phytopathology 62:972–981.

Polunin, N. 1937. The birch forests of Greenland. Nature 140:939–940.

Poole, A. L. 1937. A brief ecological survey of the Pukekura State Forest, South Westland. N. Z. J. For. 4:30–57.

Poore, M. E. D. 1964. Integration in the plant community. J. Ecol. 52(Suppl.):213–226.

Poore, M. E. D. 1968. Studies in Malaysian rain forest. I. The forest on Triassic sediments in Vengka Forest Preserve. J. Ecol. 56:143–196.

Potter, M. W., S. R. Kessell, and P. J. Cattelino. 1979. FORPLAN: A *FOR*est *P*lanning *LAN*guage and simulator. Environ. Manag. 3:59–72.

Preston, F. W. 1962. The canonical distribution of commonness and rarity: Part I. Ecology 43:185–215.

Preussner, K. 1976. New tables of diameter distribution for pure stands of *Picea abies* on the basis of a stochastic model. Bertrge fr die Forstwirtschaft. 10:93–97.

Prigogine, I., and G. Nicolis. 1971. Biological order, structure and instabilities. Q. Rev. Biophys. 4:107–148.

Pryor, L. D., and R. M. Moore. 1954. Plant communities. pp. 162–177. IN H. L. White (ed.), Canberra, A Nation's Capital. Angus and Robertson, Sydney.

Pryor, L. D. 1968. Trees in Canberra. Department of the Interior, Canberra. 19[illegible] pp.

Pryor, L. D., and L. A. S. Johnson. 1971. A classification of the Eucalypts. Australia National University Press, Canberra.

Pryor, L. D. 1976. Biology of *Eucalyptus*. Edward Arnold Pub. Ltd., London. 8[illegible] pp.

Purdie, R. W., and R. O. Slatyer. 1976. Vegetative succession after fire in sclerophyll woodland communities in south-eastern Australia. Aust. J. Ecol 1:223–236.

Raffa, K. F., and A. A. Berryman. 1983. The role of host plant resistance in the colonization behavior and ecology of bark beetles (Coleoptera: Scolytidae) Ecol. Monogr. 53:27–49.

Raunkiaer, C. 1934. The Life Forms of Plants and Statistical Plant Geography Clarendon, Oxford. 632 pp.

Raup, H. M., and R. E. Carlson. 1941. The history of land use in the Harvard Forest. Harvard For. Bull. 20. 632 pp.

Raup, H. M. 1957. Vegetational adjustment to the instability of site. Proc. Pap Union Conserv. Nature Nat. Resour. 36–48.

Reed, K. L. 1980. An ecological approach to modeling growth of forest trees. For Sci. 26:35–50.

Reichle, D. E., and D. A. Crossley, Jr. 1967. Investigation of heterotrophic productivity in forest insect communities. pp. 563–587. IN K. Petrusewicz (ed.) Secondary Productivity of Terrestrial Ecosystems. Panstowowe Wydawnictws Naukowe, Warsaw.

Reichle, D. E., B. E. Dinger, N. T. Edwards, W. F. Harris, and P. Sollins. 1973 Carbon flow and storage in a forest ecosystem. pp. 345–365. IN G. M Woodwell and E. V. Pecan (eds.), Carbon and the Biosphere. Proc., 24th Brookhaven Symp. Biol. Technical Information Center, Oak Ridge, Tennessee.

Reifsnyder, W. E., L. P. Herrington, and K. W. Spalt. 1967. Thermophysical properties of bark of shortleaf, longleaf and red pine. Yale University School For. Bull. No. 70.

Rhoades, D. F. 1979. Evolution of plant chemical defenses against herbivores. pp

3–54. IN G. A. Rosenthal and D. H. Janzen (eds.), Herbivores, Their Interaction with Secondary Plant Metabolites. Academic Press, New York.

Rice, E. L. 1964. Inhibition of nitrogen-fixing and nitrifying bacteria by seed plants (1). Ecology 45:825–837.

Richards, P., and G. B. Williamson. 1975. Treefalls and patterns of understory species in a wet lowland tropical forest. Ecology 56:1226–1229.

Richards, P. W. 1952. The Tropical Rain Forest: An Ecological Study. Cambridge University Press, London. 450 pp.

Richards, P. W. 1969. Speciation in the tropical rain forest and the concept of the niche. Biol. J. Linn. Soc. 1:149–153.

Robbins, R. G., and R. Pullen. 1965. Vegetation of the Wabag-Tari area. C.S.I.R.O. Aust. Land Res. Ser. 15:100–115.

Roberts, A. 1974. The stability of feasible ecosystems. Nature 251:607.

Rodin, L. E., and N. I. Bazilevich. 1967. Production and mineral cycling in terrestrial vegetation. Oliver and Boyd, Edinburgh. 288 pp.

Rondenx, J. 1977. Construction et utilisation de tarifs de cubage "peuplements" pour l'epicea (*Picea abies* Karst.) en Ardenne meridionale. Bull. Rech. Agron. Gembloux 12:330–348.

Roughgarden, T. 1979. Theory of Population Genetics and Evolutionary Ecology: An Introduction. Macmillan, New York. 634 pp.

Rowe, J. S. 1977. Forest Regions of Canada. Canadian Forest Service Publ. 1300. Ottawa.

Runkle, J. R. 1981. Gap regeneration in some old-growth forests of the eastern United States. Ecology 62:1041–1051.

Salisbury, E. J. 1926. The geographical distribution of plants in relation to climatic factors. Geogr. J. 67:312–335.

Salt, G. W. 1979. A comment on the use of the term "emergent properties." Am. Nat. 113:145–148.

Sando, R. W., and D. A. Haines. 1972. Fire weather and the behavior of the Little Sioux fire. U.S.D.A. Forest Service Res. Pap. NC-76. 6 pp.

Sargent, C. S. 1884. Report on the forests of North America. Misc. Doc. 42. Part 9. Government Printing Office, Washington, D.C. 612 pp.

Satoo, T., R. Kunagi, and Z. Kumekawa. 1956. Materials for the study of growth in stands. 3. Amount of leaves and the production of wood in an aspen (*Populus davidiana*) second growth in Hakkaido. Bull. Tokyo Univ. For. 52:33–58.

Satoo, T., and M. Senda. 1958. Materials for the studies of growth in stands. 4. Amount of leaves and production of wood in a young plantation of *Chamaecyparis obtusa*. Bull. Tokyo Univ. For. 54:71–100.

Satoo, T., K. Negisi, and M. Senda. 1959. Materials for the studies of growth in stands. 5. Amount of leaves and growth in plantations of *Zelkowa serata* applied with crown thinning. Bull. Tokyo Univ. For. 55:101–123.

Saxton, W. T. 1925. Phases of vegetation under monsoon conditions. J. Ecol. 17:197–222.

Schmidt, R. L. 1970. A history of pre-settlement fires on Vancouver Island as determined from Douglas-fir ages. IN Tree Ring Analysis with Special Reference to Northwest America. University of British Columbia, Faculty of Forestry Bulletin 7:107–108.

Schoener, T. W. 1974. Resource partitioning in ecological communities. Science 185:27–39.

Schowalter, T. D. 1981. Insect herbivore relationship to the state of the hos plant: Biotic regulation of ecosystem nutrient cycling through ecological suc cession. Oikos 37:126–130.

Schulze, E.-D. 1982. Plant life forms and their carbon, water and nutrient rela tions. pp. 615–676. IN O. L. Lange, P. S. Nobel, C. B. Osmond, and H Ziegler (eds.), Physiological Plant Ecology II. Springer-Verlag, Berlin.

Scurfield, G. 1960. Air pollution and tree growth. For. Abstr. 21:339–349.

Shaler, N. S. 1891. The origin and nature of soils. 12th Annual Report, U.S Geological Survey 1890–1891. Part I. Geology. pp. 213–345.

Shanks, R. E., and E. E. C. Clebsch. 1962. Computer program for the estimatio of forest stand weight and mineral pool. Ecology 43:339–341.

Shugart, H. H., and B. C. Patten. 1972. Niche quantification and the concept o niche pattern. pp. 283–327. IN B. C. Patten (ed.), Systems Analysis and Simu lation in Ecology, Vol. II. Academic Press, New York.

Shugart, H. H., T. R. Crow, and J. M. Hett. 1973. Forest succession models: A rationale and methodology for modeling forest succession over large regions For. Sci. 19:203–212.

Shugart, H. H., and D. C. West. 1977. Development of an Appalachian deciduou forest succession model and its application to assessment of the impact of th chestnut blight. J. Environ. Manag. 5:161–179.

Shugart, H. H. (ed.). 1978. Time series and ecological processes. SIAM press Philadelphia. 303 p.

Shugart, H. H., and R. V. O'Neill. 1979. Systems Ecology. Dowden, Hutchinso and Ross, Stroudsburg, Pennsylvania. 368 pp.

Shugart, H. H., and D. C. West. 1979. Size and pattern of simulated forest stands For. Sci. 25:120–122.

Shugart, H. H., and D. C. West. 1980. Forest succession models. BioScienc 30:308–313.

Shugart, H. H., A. T. Mortlock, M. S. Hopkins, and I. P. Burgess. 1980a. A computer simulation model of ecological succession in Australian sub-tropica rain forest. ORNL/TM-7029. Oak Ridge National Laboratory, Oak Ridge, Tennessee. 48 pp.

Shugart, H. H., W. R. Emanuel, D. C. West, and D. L. DeAngelis. 1980b Environmental gradients in a simulation model of a beech-yellow-poplar stand Math. Biosci. 50:163–170.

Shugart, H. H., and I. R. Noble. 1981. A computer model of succession and fire response of the high-altitude *Eucalyptus* forest of the Brindabella Range, Australian Capital Territory. Aust. J. Ecol. 6:149–164.

Shugart, H. H., and D. C. West. 1981. Long-term dynamics of forest ecosystems. Am. Sci. 69:647–652.

Shugart, H. H., D. C. West, and W. R. Emanuel. 1981a. Patterns and dynamics of forests: An application of simulation models. pp. 74–94. IN D.C. West, H. H. Shugart, and D. B. Botkin (eds.), Forest Succession: Concepts and Application. Springer-Verlag, New York.

Shugart, H. H., M. S. Hopkins, I. P. Burgess, and A. T. Mortlock. 1981b. The development of a succession model for subtropical rain forest and its application to assess the effects of timber harvest at Wiangarree State Forest, New South Wales. J. Environ. Manag. 11:243–265.

Shugart, H. H., and S. B. McLaughlin. 1984. Modeling SO_2 effects on forest growth and community dynamics. IN W. E. Winner, H. A. Mooney, and R. A. Goldstein (eds.), Sulfur Dioxide and Vegetation. Stanford University Press, Stanford (in press).

Shugart, H. H., and S. W. Seagle. 1984. Modeling forest landscapes and the role of disturbance in ecosystems and communities. IN S. T. A. Pickett and P. S. White (eds.), Natural Disturbance: An Ecological and Evolutionary Perspective. Academic Press, New York (in press).

Siegel, S. 1956. Nonparametric statistics for the Behavioral Sciences. McGraw-Hill, New York. 312 pp.

Silvertown, J. W. 1980a. The evolutionary ecology of mast seeding in trees. Biol. J. Linn. Soc. 14:235–250.

Silvertown, J. W. 1980b. Leaf-canopy induced seed dormancy in a grassland flora. New Phytol. 85:109–118.

Silvertown, J. W. 1982. Introduction to Plant Population Ecology. Longman, London. 209 pp.

Simberloff, D. S. 1974. Equilibrium theory of island biogeography and ecology. Annu. Rev. Ecol. Syst. 5:161–182.

Simberloff, D. S., and E. O. Wilson. 1970. Experimental zoogeography of islands: A two-year record of colonization. Ecology 51:934–937.

Simberloff, D. S., and L. G. Abele. 1976a. Island biogeography theory and conservation practice. Science 191:285–286.

Simberloff, D. S., and L. G. Abele. 1976b. Island biogeography and conservation: Strategy and limitations. Science 193:1032.

Simberloff, D. S., and W. Boecklen. 1981. Santa Rosalia reconsidered:
Size ratios and competition. Evolution 35:1206–1228.

Simpfendorfer, K. J. 1975. An Introduction to Trees for South Eastern Australia. Inkata Press, Melbourne. 377 pp.

Siren, G. 1955. The development of spruce forest on raw humus sites in northern Finland and its ecology. Acta. For. Fennica 62:1–363.

Skeen, J. N. 1976. Regeneration and survival of woody species in a naturally created forest opening. Bull. Torrey Bot. Club 103:259–265.

Smalley, G. W., and R. L. Bailey. 1974. Yield tables and stand structure for loblolly pine plantations in Tennessee, Alabama, and Georgia highlands. U.S.D.A. Forest Service Res. Pap. 50–96. 81 pp.

Smith, R. F. 1970. The vegetation structure of a Puerto Rican rain forest before and after short-term gamma-irradiation. pp. D103-D40. IN H. T. Odum and R. F. Pigeon (eds.), A Tropical Rain Forest: A Study of Irradiation and Ecology at El Verde, Puerto Rico. U.S. Atomic Energy Commission, Oak Ridge, Tennessee.

Smith, O. L. 1980. The influence of environmental gradients on ecosystem stability. Am. Nat. 116:1–24.

Smith, T. M., H. H. Shugart, and D. C. West. 1981a. Use of forest simulation models to integrate timber harvest and nongame bird management. pp. 501–510. IN Transactions of the 46th North American Wildlife and Natural Resources Conference. Wildlife Management Institute, Washington, D.C.

Smith, T. M., H. H. Shugart, and D. C. West. 1981b. FORHAB: A forest simulation model to predict habitat structure. pp. 104–113. IN D. E. Capen (ed.), The Use of Multivariate Statistics in Studies of Wildlife Habitat. General Technical

Report RM-87. Rocky Mountain Forest and Range Experiment Station, U.S.D.A. Forest Service, Fort Collins, Colorado.
Smith, W. H. 1981. Air Pollution and Forests. Springer-Verlag, New York. 379 pp.
Sollins, P., W. F. Harris, and N. T. Edwards. 1976. Simulating the physiology of a temperate deciduous forest. pp. 174–220. IN B. C. Patten (ed.), Systems Analysis and Simulation in Ecology, Vol. 4. Academic Press, New York.
Solomon, D. S. 1974. Simulation of the development of natural and silviculturally treated stands of even-aged northern hardwoods. pp. 327–352. IN J. Fries (ed.), Growth Models for Trees and Stand Simulation. Department of Forest Yield Research, Royal College of Forestry, Stockholm. Res. Notes. 30.
Solomon, D. S. 1977. A growth model of natural and silviculturally treated stands of even-aged northern hardwoods. U.S.D.A. Forest Service Tech. Report. NE-36. 30 pp.
Solomon, A. M., and J. B. Harrington, Jr. 1979. Palynology models. pp. 338–361. IN R. L. Edmonds (ed.), Aerobiology: The Ecological Systems Approach. Dowden, Hutchinson and Ross, Stroudsburg, Pennsylvania.
Solomon, A. M., H. R. Delcourt, D. C. West, and T. J. Blasing. 1980. Testing a simulation model for reconstruction of prehistoric forest stand dynamics. Quat. Res. 14:275–293.
Solomon, A. M., D. C. West, and J. A. Solomon. 1981. Simulating the role of climate change and species immigration in forest succession. pp. 154–177. IN D. C. West, H. H. Shugart, and D. B. Botkin (eds.), Forest Succession: Concepts and Application. Springer-Verlag, New York.
Solomon, A. M., and H. H. Shugart. 1984. Integrating forest-stand simulations with paleoecological records to examine long-term forest dynamics. Proc., European Sci. Fund. Workshop on Forest Dynamics, Uppsala, Sweden, March 1983. Elsevier, Amsterdam, Holland (in press).
Sousa, W. P. 1980. The response of a community to disturbance: The importance of successional age and species' life histories. Oecologia 45:72–81.
Springett, B. P. 1978. On the ecological role of insects in Australian eucalypt forests. Aust. J. Ecol. 3:129–139.
Sprugel, D. G. 1976. Dynamic structure of wave-regenerated *Abies balsamea* forest in the northeastern United States. J. Ecol. 64:889–912.
Sprugel, D. G., and F. H. Bormann. 1981. Natural disturbance and the steady state in high-altitude balsam fir forests. Science 211:390–393.
Spurr, S. H. 1956. Natural restocking of forests following the 1938 hurricane in central New England. Ecology 37:443–451.
Spurr, S. H., and B. V. Barnes. 1973. Forest Ecology. The Ronald Press Co., New York. 571 pp.
Stearns, F. S. 1949. Ninety years of forest change in a northern hardwood forest in Wisconsin. Ecology 30:350–358.
Stephens, E. P. 1956. The uprooting of trees: A forest process. Soil Sci. Soc. Am. Proc. 20:113–116.
Sugihara, G. 1981. $S = CA^z$, $z = 1/4$: A reply to Connor and McCoy. Am. Nat. 117:790–793.
Sukachev, V. 1964. Chapter VII. Dynamics of forest biogeocoenoses. pp. 535–571. IN V. Sukachev and N. Dylis (eds.), Fundamentals of Forest Biogeo-

coenology. Oliver and Body, Edinburgh. (A translation of: Osnovy lesnoi biogeotsenologii. Nauka, Moscow.)

Sullivan, A. D., and J. L. Clutter. 1972. A simultaneous growth and yield model for loblolly pine. For. Sci. 18:76–86.

Suzuki, T., and T. Umemura. 1967a. Forest transition as a stochastic process. III. J. Jpn. For. Soc. 49:208–210.

Suzuki, T., and T. Umemura. 1967b. Forest transition as a stochastic process. IV. J. Jpn. For. Soc. 49:402–404.

Suzuki, T., and T. Umemura. 1974a. Forest transition as a stochastic process. II. pp. 327–352. IN J. Fries (ed.), Growth Models for Trees and Stand Simulation. Department of Forest Yield Research, Royal College of Forestry, Stockholm. Res. Notes 30.

Suzuki, T., and T. Umemura. 1974b. Forest transition as a stochastic process. V. J. Jpn. For. Soc. 56:195–204.

Swain, A. M. 1973. A history of fire and vegetation in northeastern Minnesota as recorded by lake sediments. Quat. Res. 3:383–397.

Swain, A. M. 1978. Environmental changes during the last 2,000 years in north-central Wisconsin: Analysis of pollen, charcoal, and seeds from varved lake sediments. Quat. Res. 10:55–68.

Swaine, M. D., and P. Greig-Smith. 1980. An application of principal components analysis to vegetation change in permanent plots. J. Ecol. 68:33–41.

Swank, W. T., J. B. Waide, D. A. Crossley, Jr., and R. L. Todd. 1981. Insect defoliation and nitrate export from forest ecosystems. Oecologia 51:297–299.

Swift, M. J., O. W. Heal, and J. M. Anderson. 1979. Decomposition in Terrestrial Ecosystems. University of California Press, Berkeley.

Switzer, G. I., I. E. Nelson, and W. J. Smith. 1966. The characterization of dry matter and nitrogen accumulation by loblolly pine (*Pinus taeda* L.). Proc. Am. Soc. Soil Sci. 30:114–119.

Tanner, J. T. 1942. Present status of the ivory-billed woodpecker. Wilson Bull. 54:57–58.

Tansley, A. G. 1929. Succession: The concept and its values. pp. 677–686. Proc. Int. Congress Plant Science, 1926.

Tansley, A. G. 1935. The use and abuse of vegetational concepts and terms. Ecology 16:284–307.

Tauber, H. 1965. Differential pollen dispersal and the interpretation of pollen diagrams. Dan. Geol. Unders. Afh. Raekke 2 89:1–69.

Tauber, H. 1977. Investigations of aerial pollen transport in a forested area. Dan. Bot. Ark. 32:1–121.

Teal, J. 1962. Energy flow in the salt marsh ecosystem of Georgia. Ecology 43:614–624.

Terborgh, J. 1976. Island biogeography and conservation: Strategy and limitations. Science 193:1029–1030.

Tharp, M. L. 1978. Modeling major perturbations on a forest ecosystem. University of Tennessee, Knoxville. M.S. Thesis.

Thornthwaite, C. W. 1948. An approach towards a rational classification of climate. Geogr. Rev. 38:55–94.

Thornthwaite, C. W., and J. R. Mather. 1957. Publications in Climatology. Laboratory of Climatology 10:185–311.

Tilly, L. J. 1968. The structure and dynamics of Cone Spring. Ecol. Monogr. 38:169–197.

Tinbergen, N. 1951. The Study of Instinct. Oxford University Press, London. 228 pp.

Tiren, L. 1927. On barrytans storlek hos tallbestand. Swedish Institute of Experimental Forestry Report No. 23. pp. 295–336.

Travis, C. C., and W. M. Post. 1979. Dynamics and comparative statistics of mutualistic communities. J. Theor. Biol. 78:553–571.

Tregonning, K., and A. Roberts. 1979. Complex systems which evolve towards homeostasis. Nature 281:563–564.

Trimble, G. R., and E. H. Tryon. 1966. Crown encroachment into openings cut in Appalachian hardwood stands. J. For. 64:104–108.

Trimble, J. L., and C. R. Shriner. 1981. Inventory of United States growth models. ORNL/Sub-80/13819/1. National Technical Information Service, Springfield, Virginia. 133 pp.

Troll, C. 1948. Der asymmetrische Aufbau der Vegetationszonen und Begetationsstufen auf der Nord- und Sdhalbkugel. Ber. Geobot. Forsch. Inst. Rbel 1947:46–83.

U.S. Department of Commerce. 1968. Climatic Atlas of the United States. Government Printing Office, Washington, D.C.

U.S. Geological Survey. 1965. Monthly average temperatures for January and July. The National Atlas of the United States of America (1970). U.S. Government Printing Office, Washington, D.C.

Usher, M. B. 1979. Markovian approaches to ecological succession. J. Anim. Ecol. 48:413–426.

Van Cleve, K. 1973. Short-term growth response to fertilization in young quaking aspen. J. For. 71:12.

Van Cleve, K., and L. A. Viereck. 1981. Forest succession in relation to nutrient cycling in the boreal forest of Alaska. pp. 185–211. IN D. C. West, H. H. Shugart, and D. B. Botkin (eds.), Forest Succession: Concepts and Application. Springer-Verlag, New York.

van der Drift, J. 1958. The role of soil fauna in the breakdown of forest litter. Proc. 15th Intern. Cong. Zool. pp. 357–360.

van der Pijl, L. 1972. Principles of Dispersal in Higher Plants, (2nd ed.) Springer-Verlag, Berlin. 162 pp.

Van Dyne, G. M. 1966. Ecosystems, systems ecology, and systems ecologists. ORNL-3957. Oak Ridge National Laboratory, Oak Ridge, Tennessee. 29 pp.

Van Hise, C. R. 1904. A treatise on metamorphism. U.S. Geol. Surv. Monogr. 47.

Van Steenis, C. G. G. J. 1958. Rejuvenation as a factor for judging the status of vegetation types: The biological nomad theory. pp. 212–215. IN Study of Tropical Vegetation. Proc., Kandy Symposium. Unesco, Paris.

Van Voris, P., R. V. O'Neill, W. R. Emanuel, and H. H. Shugart. 1980. Functional complexity and ecosystem stability. Ecology 61:1352–1360.

Van Wagner, C. E. 1973. Height of crown scorch in forest fires. Can. J. For. Res. 3:373–378.

Vines, R. G. 1968. Heat transfer through bark and the resistance of trees to fires. Aust. J. Bot. 16:499–514.

von Uexkull, J. 1909. Umwelt und Innenwelt der Tiere. Springer-Verlag, Berlin.

Wade, D. D., and D. E. Ward. 1973. An analysis of the Air Force Bomb Range fire. U.S.D.A. Forest Service Res. Pap. SE-105. 38 pp.
Wadsworth, F. H. 1951. Forest management in the Luquillo Mountains. I. The setting. Carib. For. 12:93–114.
Wadsworth, F. H. 1957. Tropical rain forest: The *Dacrodynes-Sloanea* association of the West Indies. Trop. Silvic. 2:13–23.
Waggoner, P. E., and G. R. Stephens. 1970. Transition probabilities for a forest. Nature 225:1160–1161.
Walker, B. H. 1981. Is succession a viable concept in African savanna ecosystems? pp. 431–448. IN D. C. West, H. H. Shugart, and D. B. Botkin (eds.), Forest Succession: Concepts and Application. Springer-Verlag, New York.
Walker, N. E. 1957. Soil Microbiology. Butterworths, London. 262 pp.
Walter, H. 1971. Ecology of Tropical and Subtropical Vegetation. Oliver and Boyd, Edinburgh. 539 pp.
Waring, R. H., and J. F. Franklin. 1979. Evergreen coniferous forests of the Pacific Northwest. Science 204:1380–1386.
Waring, R. H., and G. B. Pitman. 1980. A simple model of host resistance to bark beetles. Oregon State University, School of Forestry, Corvallis. Research Note 65. 2 pp.
Warming, E. 1909. Oecology of Plants: An Introduction to the Study of Plant Communities. Oxford University Press, Oxford. 422 pp.
Watt, A. S. 1925. On the ecology of British beech woods with special reference to their regeneration. II. The development and structure of beech communities on the Sussex Downs. J. Ecol. 13:27–73.
Watt, A. S. 1947. Pattern and process in the plant community. J. Ecol. 35:1–22.
Watts, W. A. 1975. Vegetation record for the past 20,000 years from a small marsh on Lookout Mountain, northwestern Georgia. Geol. Soc. Am. Bull. 86:287–291.
Weatherly, A. H. 1963. Notions of niche and competition among animals with special reference to freshwater fish. Nature 197:14–17.
Weaver, H. 1974. Effects of fire on temperate forests: Western United States. pp. 279–320. IN T. T. Kozlowki and C. E. Ahlgren (eds.), Fire and Ecosystems. Academic Press, New York.
Weaver, J., and F. E. Clements. 1938. Plant Ecology. McGraw-Hill, New York. 601 pp.
Webb, L. J., J. G. Tracey, and W. T. Williams. 1972. Regeneration and pattern in the subtropical rain forest. J. Ecol. 60:675–695.
Webb, T. III, and R. A. Bryson. 1972. The late- and post-glacial sequence of climatic events in Wisconsin and east-central Minnesota: Quantitative estimates derived from fossil pollen spectra by multivariate statistical analysis. Quat. Res. 2:70–115.
Webb, T. III, and J. H. McAndrews. 1976. Corresponding patterns of contemporary pollen and vegetation in central North America. Mem. Geol. Soc. Am. 145:267–299.
Webb, T. III, R. A. Laeski, and J. C. Bernabo. 1978. Sensing vegetation with pollen data: Choosing the data. Ecology 59:1151–1163.
Webb, T. III., S. Howe, R. H. W. Bradshaw, and K. M. Heide. 1981. Estimating plant abundances from pollen percentages: The use of regression analysis. Rev. Paleobot. Palynol. 34:269–300.

Wecker, S. C. 1963. The role of early experience in habitat selection by the prairie deermouse, *Peromyscus maniculatus bairdi*. Ecol. Monogr. 33:307–325.

Weiner, J. G. 1975. Nutrient cycles, nutrient limitation and vertebrate populations. Biologist 57:104–124.

Weinstein, D. A. 1982. The long-term retention properties of forest ecosystems: A simulation investigation. University of Tennessee, Knoxville. Ph.D. dissertation. 175 pp.

Weinstein, D. A., H. H. Shugart, and D. C. West. 1982. The long-term nutrient retention properties of forest ecosystems: A simulation investigation. ORNL/TM-8472. ORNL, Oak Ridge, Tennessee. 145 pp.

Weinstein, D. A., and H. H. Shugart. 1983. Ecological modeling of landscape dynamics. pp. 29–45. IN H. Mooney and M. Godron (eds.), Disturbance and Ecosystems. Springer-Verlag, New York.

Weinstein, D. A., H. H. Shugart, and C. C. Brandt. 1983. Energy flow and the persistence of a human population: A simulation analysis. Human Ecology 11:201–225.

West, D. C., S. B. McLaughlin, and H. H. Shugart. 1980. Simulated forest response to chronic air pollution stress. J. Environ. Qual. 9:43–49.

Whitcomb, R. F., J. F. Lynch, P. A. Opler, and C. S. Robbins. 1976. Island biogeography and conservation: Strategy and limitations. Science 193:1030–1032.

White, P. S. 1979. Pattern, process, and natural disturbance in vegetation. Bot. Rev. 45:229–299.

Whitfield, D. W. A. 1972. Systems analysis. pp. 392–409. IN L. C. Bliss (ed.), Devon Island IBP Project, High Arctic Ecosystem. Department of Botany, University of Alberta, Edmonton.

Whitmore, T. C. 1975. Tropical Rain Forests of the Far East. Clarendon Press, Oxford. 282 pp.

Whittaker, R. H. 1953. A consideration of climax theory. The climax as a population and a pattern. Ecol. Monogr. 23:41–78.

Whittaker, R. H., and G. M. Woodwell. 1968. Dimension and production relations of trees and shrubs in the Brookhaven Forest, New York. J. Ecol, 56:1–25.

Whittaker, R. H., and G. M. Woodwell. 1969. Structure, production and diversity of the oak-pine forest at Brookhaven, New York. J. Ecol. 57:157–176.

Whittaker, R. H. 1970. Communities and Ecosystems. MacMillan, London and Toronto. 162 pp.

Whittaker, R. H., and P. P. Feeny. 1971. Allelochemics: Chemical interactions between species. Science 171:757–770.

Whittaker, R. H. 1974. Climax concepts and recognition. pp. 137–154.IN R. Knap (ed.), Vegetation Dynamics. Dr. W. Junk, The Hague.

Whittaker, R. H. (ed.). 1978. Classification of Plant Communities. Dr. W. Junk, The Hague. 408 pp.

Whittaker, R. H., and S. A. Levin. 1977. The role of mosaic phenomena in natural communities. Theor. Popul. Biol. 12:117–139.

Wiens, J. A. 1976. Population responses to patchy environments. Annu. Rev. Ecol. Syst. 7:81–120.

Wigginton, B. E. 1965. Trees and Shrubs for the Southeast. University of Georgia Press, Athens.

Wilkins, C. W. 1977. A stochastic analysis of the effect of fire on remote vegetation. University of Adelaide, Adelaide, South Australia. Ph.D. thesis.

Williams, W. T., G. N. Lauce, L. J. Webb, J. G. Tracey, and M. B. Dale.1969. Studies in the numerical analysis of complex rain-forest communities. III. The analysis of successional data. J. Ecol. 57:515–535.

Williams, B. K. 1981. Discriminant analysis in wildlife research: Theory and applications. pp. 59–71. IN D. E. Capen (ed.), The Use of Multivariate Statistics in Studies of Wildlife Habitat. General Technical Report RM-87. Rocky Mountain Forest and Range Experiment Station, U.S.D.A. Forest Service, Fort Collins, Colorado.

Williamson, G. B. 1975. Pattern and seral composition in an old-growth beech-maple forest. Ecology 56:727–737.

Wilson, D. S. 1976. Evolution on the level of communities. Science 192:1358–1360.

Witkamp, M., and D. A. Crossley. 1966. The role of arthropods and microflora in the breakdown of white oak litter. Pedobiologia 6:293–303.

Wood, D. L. 1982. The role of pheromones, kairomones, and allomones in the host selection and colonization behavior of bark beetles. Annu. Rev. Entomol. 27:411–446.

Woods, K. D. 1979. Reciprocal replacement and the maintenance of codominance in a beech-maple forest. Oikos 33:31–39.

Woods, K. D., and R. H. Whittaker. 1981. Canopy-understory interactionand the internal dynamics of mature hardwood and hemlock-hardwood forests. pp. 305–323. IN D. C. West, H. H. Shugart, and D. B. Botkin (eds.), Forest Succession: Concepts and Application. Springer-Verlag, New York.

Wright, H. E., Jr. 1968. The roles of pine and spruce in the forest history of Minnesota and adjacent areas. Ecology 49:937–955.

Wright, H. E., Jr. 1974. Landscape development, forest fires and wilderness management. Science 186:487–495.

Wu, L. S., and D. B. Botkin. 1978. On population processes of long-lived species. pp. 245–254. IN H. H. Shugart (ed.), Time Series and Ecological Processes. Society for Industrial and Applied Mathematics, Philadelphia.

Yamamura, N. 1976. A mathematical approach to spatial distribution and temporal succession in plant communities. Bull. Math. Biol. 38:517–526.

Yoda, K., K. Shinozaki, H. Ogawa, K. Hozumi, and T. Kira. 1965. Estimation of the total amount of respiration in woody organs of trees and forest communities. J. Biol., Osaka City Univ. 16:15–26.

Zahner, R. 1968. Water deficits and the growth of trees. pp. 191–254. IN T. T. Kozlowski (ed.), Water Deficits and Plant Growth. Academic Press, New York.

Zavitkovski, J., and M. Newton. 1968. Ecological importance of snowbrush, *Ceanothus velutinus*, in the Oregon Cascades agricultural ecology. Ecology 49:1134–1145.

Zavitkovski, J., and R. D. Stevens. 1972. Primary production of red alder ecosystems. Ecology 53:235–242.

Zoltin, R. I., and K. S. Khodashova. 1980. The Role of Animals in Biological Cycling of Forest-Steppe Ecosystems. Dowden, Hutchinson and Ross, Stroudsburg, Pennsylvania. 221 pp.

Index